T0181004

Energy, Environment, and Sustainability

Series editors

Avinash Kumar Agarwal, Department of Mechanical Engineering, Indian Institute of Technology, Kanpur, Uttar Pradesh, India
Ashok Pandey, Distinguished Scientist, CSIR-Indian Institute of Toxicology Research, Lucknow, India

This books series publishes cutting edge monographs and professional books focused on all aspects of energy and environmental sustainability, especially as it relates to energy concerns. The Series is published in partnership with the International Society for Energy, Environment, and Sustainability. The books in these series are editor or authored by top researchers and professional across the globe. The series aims at publishing state-of-the-art research and development in areas including, but not limited to:

- Renewable Energy
- Alternative Fuels
- Engines and Locomotives
- Combustion and Propulsion
- Fossil Fuels
- Carbon Capture
- Control and Automation for Energy
- Environmental Pollution
- Waste Management
- Transportation Sustainability

More information about this series at http://www.springer.com/series/15901

Nikhil Sharma · Avinash Kumar Agarwal
Peter Eastwood · Tarun Gupta
Akhilendra P. Singh
Editors

Air Pollution and Control

Springer

Editors
Nikhil Sharma
Department of Mechanical Engineering
Indian Institute of Technology Kanpur
Kanpur, Uttar Pradesh
India

Tarun Gupta
Department of Civil Engineering
Indian Institute of Technology Kanpur
Kanpur, Uttar Pradesh
India

Avinash Kumar Agarwal
Department of Mechanical Engineering
Indian Institute of Technology Kanpur
Kanpur, Uttar Pradesh
India

Akhilendra P. Singh
Department of Mechanical Engineering
Indian Institute of Technology Kanpur
Kanpur, Uttar Pradesh
India

Peter Eastwood
Dunton Technical Centre
Ford Motor Company Limited
Basildon, Essex
UK

ISSN 2522-8366 ISSN 2522-8374 (electronic)
Energy, Environment, and Sustainability
ISBN 978-981-13-5600-1 ISBN 978-981-10-7185-0 (eBook)
https://doi.org/10.1007/978-981-10-7185-0

Printed on acid-free paper

This Springer imprint is published by Springer Nature
The registered company is Springer Nature Singapore Pte Ltd.
The registered company address is: 152 Beach Road, #21-01/04 Gateway East, Singapore 189721, Singapore

Preface

Energy demand has been rising remarkably due to increasing population and urbanization. Global economy and society are significantly dependent on the energy availability because it touches every facet of human life and activities. Transportation and power generation are major examples of the energy. Without the transportation by millions of personalized and mass transport vehicles and availability of 24×7 power, human civilization would not have reached contemporary living standards.

First international conference on 'Sustainable Energy and Environmental Challenges' (SEEC-2017) was organized under the auspices of 'International Society for Energy and Environmental Sustainability' (ISEES) by the 'Center of Innovative and Applied Bioprocessing' (CIAB), Mohali, from 26–28 February 2017. The ISEES was founded at IIT Kanpur in January 2014 with an aim to spread knowledge in the fields of energy, environment, sustainability and combustion. The Society's goal is to contribute to the development of clean, affordable and secure energy resources and a sustainable environment for the society and to spread knowledge in the above-mentioned areas and spread awareness about the environmental challenges, which the world is facing today. The ISEES is involved in various activities such as conducting workshops, seminars, conferences, etc., in the domains of its interests. The Society also recognizes the outstanding works done by the young scientists and engineers for their contributions in these fields by conferring them awards under various categories.

This conference provided a platform for discussions between eminent scientists and engineers from various countries including India, USA, South Korea, Norway, Malaysia and Australia. In this conference, eminent speakers from all over the world presented their views related to different aspects of energy, combustion, emissions and alternative energy resource for sustainable development and cleaner environment. The conference started with four mini-symposiums on very topical themes, which included (i) new fuels and advanced engine combustion, (ii) sustainable energy, (iii) experimental and numerical combustion and (iv) environmental remediation and rail road transport. The conference had 14 technical sessions of topics related to energy and environmental sustainability and a panel discussion on 'Challenges, Opportunities and Directions of Technical Education & Research in the Area of Energy, Environment and Sustainability' to wrap up the three days technical extravaganza. The conference included two plenary talks, 12 keynote talks, 42 invited talks from prominent scientists, 49 contributed talks and 120 posters. Total 234 participants and speakers attended this three days conference, which hosted Dr. V. K. Saraswat, Member NITI Ayog, India, as a chief guest for the award ceremony of the ISEES. This conference laid out the roadmap for technology development, opportunities and challenges in this technology domain. The technical sessions in the conference included advances in IC engines and fuels; conversion of biomass to biofuels; combustion processes; renewable energy: prospects and technologies; waste to wealth—chemicals and fuels; energy conversion systems; numerical simulation of combustion processes; alternate fuels for IC engines; sprays and heterogeneous combustion of coal/biomass; biomass conversion to fuels and chemicals—thermochemical processes; utilization of biofuels; and environmental protection and health. All these topics are very relevant for the country and the world in present context. The society is grateful to Prof. Ashok Pandey for organizing and hosting this conference, which led to germination of this series of monographs, which included 16 books related to different aspects of energy, environment and sustainability. This is the first time that such voluminous and high quality outcome has been achieved by any society in India from one conference.

The editors would like to express their sincere gratitude to the authors for submitting their work in a timely manner and revising it appropriately at a short notice. We would like to express our special thanks to Dr. Varun Goel, Dr. R. Anand, Dr. M. Udaya Kumar, Prof. R. S. Bharj, Dr. Shijo Thomas, Dr. Niraj Kumar, Mishra, Dr. T. N. Verma, Dr. Pravesh Chandra Shukla, Dr. Chetankumar Patel, Dr. S. K. Verma, Dr. Akhilendra Pratap Singh, Dr. Joonsik Hwang, Dr. Rohit Singla, Paramvir Singh and Nikhil Sharma who reviewed various chapters of this monograph and provided their valuable suggestions to improve the manuscripts. We acknowledge the support received from various funding agencies and organizations for the successful conduct of the first ISEES conference (SEEC-2017), where these monographs germinated. These include Department of Science and Technology, Government of India (Special thanks to Dr. Sanjay Bajpai); TSI, India (Special thanks to Dr. Deepak Sharma); Tesscorn, India (Special thanks to Sh. Satyanarayana); AVL, India; Horiba, India; Springer (Special thanks to Swati Mehershi); CIAB (Special thanks to Dr. Sangwan).

This volume covers the practices and technologies that are applied to the prevention of air pollution and control. Greenhouse gas emissions, urban air quality and growing petroleum consumptions are the three challenges faced by the society anywhere in the world. This book covers all aspects related to air pollution including major sources of air pollution, measurement techniques, modelling studies and solution approach to control air pollution. This book also emphasizes on vehicles as major source of air pollution and shows the quantitative analysis of engine exhaust emissions. Focus on particulate matter as major pollutant from engines and coal-fired power plants is another important aspect of this book. Few chapters are also based on emission control techniques using different aftertreatment devices. This monograph aims to strengthen the knowledge base dealing with air pollution. This monograph is intended for air pollution practitioners, and we hope that the book would be of great interest to the professionals, postgraduate students involved in environmental studies.

Kanpur, India — Nikhil Sharma
Kanpur, India — Avinash Kumar Agarwal
Basildon, UK — Peter Eastwood
Kanpur, India — Tarun Gupta
Kanpur, India — Akhilendra P. Singh

Contents

Part I General

Introduction to Air Pollution and Its Control 3
Nikhil Sharma, Avinash Kumar Agarwal, Peter Eastwood, Tarun Gupta and Akhilendra Pratap Singh

Part II Internal Combustion Engine

Simultaneous Control of Oxides of Nitrogen and Soot in CRDI Diesel Engine Using Split Injection and Cool EGR Fueled with Waste Frying Oil Biodiesel and Its Blends 11
R. Anand

Biodiesel Soot Characteristics 45
Joonsik Hwang, Choongsik Bae, Chetankumar Patel, Tarun Gupta and Avinash Kumar Agarwal

Techniques to Control Emissions from a Diesel Engine 57
Pravesh Chandra Shukla, Tarun Gupta and Avinash Kumar Agarwal

BS-III Diesel Vehicles in Imphal, India: An Emission Perspective 73
Thokchom Subhaschandra Singh, Tikendra Nath Verma, Prerana Nashine and Chitra Shijagurumayum

Gasoline Direct Injection Engines and Particulate Emissions 87
Nikhil Sharma and Avinash Kumar Agarwal

Part III Particulate and Aerosols

Primary Organic Aerosols 109
Deepika Bhattu

Effects of VOCs on Human Health 119
Vipin Soni, Paramvir Singh, Venu Shree and Varun Goel

Thermodynamics of Carbon Nanotubes and Soot Formation 143
R. S. Bharj, Jyoti Bharj and Vishal Vasistha

Part IV Numerical/Simulations

A Study on Evolution and Modelling of Soot Formation in Diesel Jet Flames 155
M. Udayakumar and N. H. Mohamed Ibrahim

Numerical Simulation of Air Pollution Control in Hospital 185
Tikendra Nath Verma, Arvind Kumar Sahu and Shobha Lata Sinha

Part V Miscellaneous

A Review on Clean Combustion Within Porous Media 209
Niraj Kumar Mishra, P. Muthukumar and Snehasish Panigrahy

An Overview of Current Knowledge Concerning the Environmental Consequences of the Nuclear Pollution: Sources, Effects and Control 225
S. K. Verma, S. L. Sinha and D. K. Chandraker

Scope of Pyrolysis Process as a Sustainable Method to Dispose Waste Tires: A Review 247
Raghuram Kommineni, Hemanth Boddapu and Shijo Thomas

Editors and Contributors

About the Editors

Nikhil Sharma received his Masters of Technology from NIT Hamirpur, India in 2012, in Mechanical Engineering with specialization in Thermal Engineering (CFD & HT). During his Ph. D. at IIT Kanpur, he worked on "Spray, combustion, emissions and particulate investigations of a gasohol fuelled gasoline direct injection engine". He has also served at Amity University, Noida as an assistant professor in the department of Mechanical and Automation Engineering. His area of research includes alternative fuels for internal combustion engines (biodiesel, alcohols), engine emission control and particulate characterization.

Prof. Avinash Kumar Agarwal joined IIT Kanpur in 2001. His areas of interest are IC engines, combustion, alternative fuels, conventional fuels, optical diagnostics, laser ignition, HCCI, emission and particulate control, and large bore engines. He has published 230+ international journal and conference papers. Prof. Agarwal is a Fellow of SAE (2012), ASME (2013), ISEES (2015) and INAE (2015). He received several awards such as Prestigious Shanti Swarup Bhatnagar Award-2016 in Engineering Sciences, Rajib Goyal prize-2015, NASI-Reliance Industries Platinum Jubilee Award-2012; INAE Silver Jubilee Young Engineer Award-2012; SAE International's Ralph R. Teetor Educational Award-2008; INSA Young Scientist Award-2007; UICT Young Scientist Award-2007; INAE Young Engineer Award-2005.

Dr. Peter Eastwood has been associated with the technology of motor vehicle pollution control for nearly thirty years, with posts research in research and development. He was a Royal Society Post-Doctoral Research Fellow at the Institute for Physical and Theoretical Chemistry, Tuebingen, Germany, were he continued his doctoral work on exhaust sensors. He was a visiting researcher in the Physics Laboratory at Ford Motor Company's corporate research centre in Dearborn, Michigan. He is currently employed in diesel aftertreatment development at Ford Motor Company in England.

Dr. Tarun Gupta Professor and P K Kelkar Research Fellow at the Department of Civil Engineering, IIT Kanpur, Kanpur, India. He holds Doctor of Science 2004, Environmental Health, Harvard University (USA) and Master of Technology 2000, Environmental Science and Engineering, Indian Institute of Technology-Bombay (9-month research at TU-Dresden, Germany). He is teaching and carrying out research at IIT Kanpur since june, 2006. He has graduated 6 PhD and 31 M.Tech. students. He has published more than 90 articles in ISI indexed journals, 4 book chapters and filed 4 applications for Indian Patent. He is a member of INYAS (2016), PK Kelkar Research Fellowship (2015), NASI Scopus Young Scientist (2015), INSA Young Scientist (2011), INAE Young Engineer (2009), IEI Young Engineer (2008).

Dr. Akhilendra P. Singh received his Masters and PhD in Mechanical Engineering from Indian Institute of Technology Kanpur, India in 2010 and 2017 respectively. Dr. Singh has worked as a CSIR Pool Scientist at Indian Institute of Technology Kanpur from 2014 to 2017. His areas of research include advanced low temperature combustion; optical diagnostics with special reference to engine endoscopy and PIV; combustion diagnostics and engine emissions measurement. Dr. Singh has published more than 20 international journal papers and 10 international and national conference papers.

Contributors

Avinash Kumar Agarwal Engine Research Laboratory, Department of Mechanical Engineering, Indian Institute of Technology Kanpur, Kanpur, India

R. Anand Department of Mechanical Engineering, National Institute of Technology, Tiruchirappalli, Tamilnadu, India

Choongsik Bae Department of Mechanical Engineering, Korea Advanced Institute of Science and Technology, Yuseong-Gu, Daejeon, Republic of Korea

Jyoti Bharj Department of Physics, Dr B R Ambedkar National Institute of Technology, Jalandhar, Punjab, India

R. S. Bharj Department of Mechanical Engineering, Dr B R Ambedkar National Institute of Technology, Jalandhar, Punjab, India

Deepika Bhattu Paul Scherrer Institute, Villigen, Switzerland

D. K. Chandraker Reactor Design and Development Group, Bhabha Atomic Research Centre, Mumbai, Maharashtra, India

Peter Eastwood Ford Motor Company, Basildon, England, UK

Tarun Gupta Department of Civil Engineering, Indian Institute of Technology Kanpur, Kanpur, India

Hemanth Boddapu School of Nano Science and Technology, NIT Calicut, Calicut, Kerala, India

Joonsik Hwang Department of Mechanical Engineering, Korea Advanced Institute of Science and Technology, Yuseong-Gu, Daejeon, Republic of Korea

Niraj Kumar Mishra Department of Mechanical Engineering, National Institute of Technology Uttarakhand, Srinagar, India

N. H. Mohamed Ibrahim Department of Mechanical Engineering, National Institute of Technology, Tiruchirappalli, Tamil Nadu, India

P. Muthukumar Department of Mechanical Engineering, Indian Institute of Technology Guwahati, Guwahati, India

Prerana Nashine Department of Mechanical Engineering, National Institute of Technology Rourkela, Rourkela, India

Snehasish Panigrahy Department of Mechanical Engineering, Indian Institute of Technology Guwahati, Guwahati, India

Chetankumar Patel Department of Mechanical Engineering, Indian Institute of Technology Kanpur, Kanpur, India

Raghuram Kommineni School of Nano Science and Technology, NIT Calicut, Calicut, Kerala, India

Arvind Kumar Sahu Department of Mechanical Engineering, National Institute of Technology, Raipur, Chhattisgarh, India

Nikhil Sharma Engine Research Laboratory, Department of Mechanical Engineering, Indian Institute of Technology Kanpur, Kanpur, India

Chitra Shijagurumayum Department of Civil Engineering, Manipur Institute of Technology, Imphal, India

Venu Shree Department of Architecture, National Institute of Technology Hamirpur, Hamirpur, Himachal Pradesh, India

Pravesh Chandra Shukla Department of Civil Engineering, Indian Institute of Technology Kanpur, Kanpur, India

Akhilendra Pratap Singh Engine Research Laboratory, Department of Mechanical Engineering, Indian Institute of Technology Kanpur, Kanpur, India

Paramvir Singh Department of Mechanical Engineering, National Institute of Technology Hamirpur, Hamirpur, Himachal Pradesh, India

Thokchom Subhaschandra Singh Department of Mechanical Engineering, National Institute of Technology Manipur, Imphal, India

S. L. Sinha Mechanical Engineering Department, National Institute of Technology, Raipur, Chhattisgarh, India

Shobha Lata Sinha Department of Mechanical Engineering, National Institute of Technology, Raipur, Chhattisgarh, India

Vipin Soni Department of Mechanical Engineering, National Institute of Technology Hamirpur, Hamirpur, Himachal Pradesh, India

Shijo Thomas School of Nano Science and Technology, NIT Calicut, Calicut, Kerala, India

M. Udayakumar Department of Mechanical Engineering, National Institute of Technology, Tiruchirappalli, Tamil Nadu, India

Varun Goel Department of Mechanical Engineering, National Institute of Technology Hamirpur, Hamirpur, Himachal Pradesh, India

Vishal Vasistha Department of Mechanical Engineering, Dr B R Ambedkar National Institute of Technology, Jalandhar, Punjab, India

S. K. Verma Mechanical Engineering Department, National Institute of Technology, Raipur, Chhattisgarh, India

Tikendra Nath Verma Department of Mechanical Engineering, National Institute of Technology, Imphal, Manipur, India

Part I
General

Introduction to Air Pollution and Its Control

Nikhil Sharma, Avinash Kumar Agarwal, Peter Eastwood, Tarun Gupta and Akhilendra Pratap Singh

Abstract Air pollution prevention is an economic burden to a person and to a nation on a global scale. Air pollution is a threat to human and environment; therefore, it is extremely important to understand fundamental sources, causes, health effects associated with air pollution. This monograph gives an overview about air pollution and suggests the suitable preventive measures to reduce air pollution. This monograph includes air pollution from IC engines, primary organic aerosols (POAs), effect of volatile organic compounds (VOCs) on health and some advanced topics such as numerical simulation of airflow in hospital. This monograph also includes various engine technologies such as multipoint port fuel injection (MPFI), common rail direct injection (CRDI), indirect injection engine (IDI) and gasoline direct injection (GDI) techniques to reduce air pollution from road transport sector. Nuclear pollution, which is another threat for human life and environment is discussed towards end of this monograph.

Keywords Air pollution · IC engine · Particulate and aerosols
Numerical simulations

N. Sharma · A. K. Agarwal (✉) · A. P. Singh
Engine Research Laboratory, Department of Mechanical Engineering,
Indian Institute of Technology Kanpur, Kanpur 208016, India
e-mail: akag@iitk.ac.in

P. Eastwood
Ford Motor Company, Daventry, England, UK

T. Gupta
Department of Civil Engineering, Indian Institute of Technology Kanpur,
Kanpur 208016, India

N. Sharma et al. (eds.), *Air Pollution and Control*, Energy, Environment,
and Sustainability, https://doi.org/10.1007/978-981-10-7185-0_1

1 Introduction

Air pollution may be defined as the presence of undesirable material such as motor vehicle exhaust, exhaust from factories, construction in the atmosphere, which is large enough in quantity to produce adverse health effect to all living organisms and environmental. In last few decades, air pollution became a global concern due to adverse health effects and other environmental issues such as global warming, poor visibility, etc. There are primarily two sources of air pollution in the atmosphere; (1) Natural emissions; and (2) Anthropogenic emissions. Anthropogenic emissions are produced by internal combustion (IC) engines and furnaces via combustion. IC engine has made our life more convenient; however, in return for this convenience, automobiles have caused air pollution which has lead to adverse health effect. Expanding area of cities, rising traffic jams, higher energy consumption, rapid economic development are directly associated to air pollution.

Air pollution is divided into two categories—primary and secondary air pollution. Primary pollutants are those which are emitted directly from a source like exhaust of automobile, industries, burning of fossil fuels. Some of them include hydrocarbons, carbon dioxide, carbon monoxide, sulphur dioxide, nitrogen oxide, particulate matter. A secondary pollutant is a product of reaction among the primary pollutants or with water vapour and sunlight; examples are sulphuric acid, ozone, peroxy-acyl-nitrate (PAN), etc.

The **second part** of monograph is concerning air pollution from IC engines. Various technologies such as MPFI, CRDI, GDI, diesel engine and IDI have been discussed in this section. Among these technologies diesel engines are widely utilized as a power source in various applications, such as construction equipment, passenger and commercial vehicles, marine transportation and electricity generation from gensets. In spite of several advantages in diesel engines, they have high NO*x* and soot emissions, which lead to global warming, acid rain and photochemical smog. First two sections of this book deal with two techniques namely; (a) Active control and (b) Passive control techniques. Active control techniques are those which restrict the formation of the pollutants in the combustion chamber and passive control techniques refer to use of after-treatment devices for emission reduction. Active control techniques include improved combustion chamber design, use of engine control unit (ECU), exhaust gas recirculation (EGR), optimized fuel injection strategy, use of advanced combustion techniques such as homogeneous charge compression ignition (HCCI), premixed charge compression ignition (PCCI). Use of high pressure common rail direct injection (CRDI) system is another active control technique, which offers significant improvement in engine performance and emission characteristics compared to the conventional mechanical fuel injection system. Application of EGR is another important method for controlling the NOx emissions in IC engines. However, to meet the current stringent emission norms, passive control techniques are also required in addition to active control techniques. In passive control techniques, various after-treatment devices such as diesel oxidation catalyst (DOC), diesel particulate filter (DPF), NO_x absorbers, selective catalytic reduction (SCR) are employed to reduce different harmful species from engine tail pipe.

Alternatives fuels such as compressed natural gas (CNG), liquefied petroleum gas (LPG), hydrogen and renewable liquid fuels namely biodiesel, ethanol can also be used in IC engines to reduce the engine out emissions. Among these alternative fuels, biodiesel has been adopted to replace mineral diesel, which suppresses the formation of soot precursors in diffusion flames. One chapter of this monograph discussed the effect of biodiesel on soot particles in diesel engines. This chapter included the morphological characteristics of soot particles along with its chemical composition determined by the elemental analysis (EA) and thermo-gravimetric analysis (TGA).

Age of the vehicles is another issue related to air pollution because in rural areas as well as in cities of India old vehicles are still used. In one of the chapters, authors have attempted to study the emission characteristics of BS-III commercial diesel auto rickshaws operating in the city of Imphal, Manipur, India. Advanced technologies such as GDI also contribute to air pollution. Particulates from GDI engines are of different sizes such as coarse, fine and ultra-fine and they also vary in composition and origin. Particulate of different sizes is known to cause adverse health effects. In one of the chapters, fundamental aspects of both homogeneous and stratified modes of combustion of GDI engines have been discussed, in addition to wall, spray and air-guided GDI engine concepts. A section of chapter covers detailed comparison of particulate emitted by GDI and MPFI engines. Various size and concentration-based PM measurement technique and instruments available commercially are included in this section. A discussion on influence of engine load, fuel type and spray characteristics on particulate emissions are elaborated towards the end of this chapter in addition to GDI soot morphological studies.

The **third part** of the book is about particulate and POA. Atmospheric aerosols are microscopic solid particles in atmosphere and organic aerosol (OA) is a dominant component of atmospheric aerosol. OA is carbon-containing compounds and plays a critical role in affecting air quality of a region and change climate to a greater extent. OA is of two type's primary and secondary organic aerosols (SOAs). POA is freshly emitted organic mass in the atmosphere in particulate form and is an important subject of ongoing research for both air quality and climate. POA constitutes the emissions from both natural and anthropogenic aerosol particles ranging in size from a few nanometres to several tens of microns. Another class of primarily emitted volatile species, i.e. intermediate volatile organic compounds (IVOCs), present around 0.28–2.5 times of POA, potential SOA precursors, also goes unnoticed. This suggests that the policymakers and environmental regulating authorities need to take into account the secondary volatile organic compounds (SVOCs) and IVOCs causing positive and negative sampling artefacts in order to correctly account for POA source contributions. VOCs are one of them which are carcinogenic and lead to photochemical reactions. Emissions of VOCs are directly associated with large number of industrial processes, emission through transportation and various indoor and outdoor sources. A section of book focuses on the major sources of carbonyl and aromatic compounds in indoor and outdoor. Some remedial processes like photo-catalytic oxidation, plasma decomposition, chemisorption and catalytic oxidation have been described in this study through which decomposition of these contaminants can be achieved. This part of monograph contains a deep study on health effects from the carbonyl and aromatic compound.

The forth part of the book is about numerical simulation related to air pollution. Two important concepts have been touched in this section of monograph (a) modelling of soot formation in diesel engine (b) airflow in a room of the ICU. In this part of monograph, the various theories associated with the soot formation like soot inception, coagulation, agglomeration, oxidation are discussed. Also, the results of the numerical studies carried out by the authors on diesel-air flames at laboratory conditions are briefly presented. IAQ helps to maintain healthy and productive indoor environments. In this context, one chapter deals with air pollution in Healthcare place, the importance of ventilation in a hospital environment, indoor air pollutants and transmission of contaminants and airborne particle inside the infirmary. In this work, simulation of airflow in a room of the ICU has carried away to examine airflow pattern using FLUENT 15 CFD software. The study predicts room airflow information in terms of velocities, temperatures and contaminant distributions which are beneficial for infection control, building layout investigation.

In the last section of this monograph, two important aspects related to porous media and environmental issues related to nuclear power are discussed. First chapter describes the significant development to attain lower emissions and higher thermal performances. This chapter shows that changing the operating parameters and design configurations of the porous radiant burners, good emission characteristics and higher thermal efficiency can be achieved. This book chapter also summarizes the development of various porous radiant burners used in both industrial and cooking applications. The other chapter is related to environmental issues associated with nuclear power. This chapter emphasizes on radioactive wastes, which are produced from nuclear power plants. Discharges from nuclear power plant can cause substantial climatic contamination, which results in harmful health effects. In last chapter, different techniques for modelling and control of hazardous material have been presented. The modelling is in view of recreation and perception of spreading of air pollutants, estimation of the source term for atomic and compound fiascos and the hazard appraisal of unsafe substances. This chapter includes the principle of modelling the nuclear and chemical disasters, optimal control of theoretical frame with example, various modelling techniques, challenges associated with measurement of pollutants.

This monograph presents both fundamental science and applied emerging technologies for emission reduction from various sources. Particulate emitted from engines directly affect human health and environment therefore major part of this monograph is on particulate characteristics and its control techniques. Specific topics covered in the manuscript include:

- Simultaneous Control of Oxides of Nitrogen and Soot in CRDI Diesel Engine using Split Injection and Cool EGR Fuelled with Waste Frying oil Biodiesel and its Blend,
- Biodiesel Soot Characteristics,
- Techniques to Control Emissions from a Diesel Engine,
- BS-III Diesel Vehicles in Imphal, India: An emission Perspective,
- Gasoline Direct Injection Engines Technology and Particulate Emission,
- Primary Organic Aerosols,
- Effects of VOCs on Human Health,
- Thermodynamics of Carbon Nanotubes and Soot Formation,
- A Study on Evolution and Modelling of Soot Formation in Diesel Jet Fames,
- Numerical Simulation of Air Pollution Control in Hospital,
- A Review on Clean Combustion Within Porous Media,
- An Overview of Current Knowledge Concerning the Environmental Consequences of the Nuclear Pollution: Sources, Effects and Control.

The topics are organized in five different sections: (i) General, (ii) IC Engine, (iii) Particulate and Aerosols, (iv) Numerical Simulation and (v) Miscellaneous.

Part II
Internal Combustion Engine

Simultaneous Control of Oxides of Nitrogen and Soot in CRDI Diesel Engine Using Split Injection and Cool EGR Fueled with Waste Frying Oil Biodiesel and Its Blends

R. Anand

Abstract Air pollution is one of the major threats to human health and living organisms, and its control is a greater challenge due to rapid growth in population and industrialization. Diesel vehicle exhaust emissions soot and nitrogen oxides (NO_x) are the major causes of global warming, acid rain, and photochemical smog. Fuel quality improvement, low-temperature homogeneous combustions, and high turbulent combustions are the most important diesel engine emission control strategies to restrict the air pollution. Use of different gaseous fuels, CNG, LPG, hydrogen and renewable liquid fuel biodiesel, ethanol, etc., in diesel engine reduces the engine out emissions to a great extent. Common rail direct injection (CRDI) has several advantages compared to the conventional mechanical fuel injection system. The high-pressure injection enhances the air-fuel mixture to obtain the better thermal efficiency as well as lower emissions. Exhaust gas recirculation (EGR) is one of the predominant methods for controlling the NO_x emission in internal combustion engines. EGR with split injection strategy is one of the effective methods to decrease soot and NO_x emissions simultaneously without much drop in engine efficiency.

Nomenclature

ASTM	American Society for Testing and Materials
B20	20% biodiesel
B40	40% biodiesel
B60	60% biodiesel
B80	80% biodiesel
B100	Biodiesel
BMEP	Brake mean effective pressure (bar)
BSEC	Brake specific energy consumption (MJ/kWh)

R. Anand (✉)
Department of Mechanical Engineering, National Institute of Technology, Tiruchirappalli, Tamilnadu, India
e-mail: anandachu@nitt.edu

N. Sharma et al. (eds.), *Air Pollution and Control*, Energy, Environment, and Sustainability, https://doi.org/10.1007/978-981-10-7185-0_2

BSFC	Brake specific fuel consumption (kg/kWh)
aTDC	After top dead center
bTDC	Before top dead center
BTE	Brake thermal efficiency (%)
CD	Combustion duration (°CA)
CH_3OH	Methanol
CH_3ONa	Sodium methoxide
CO	Carbon monoxide (% vol.)
CO_2	Carbon dioxide (% vol.)
CRDI	Common rail direct injection
DP	Dwell period
EGR	Exhaust gas recirculation
FSN	Filter smoke number
HRR	Heat release rate (J/°CA)
ME	Methyl ester
NO	Nitric oxide (ppm)
ID	Ignition delay (°CA)
SI	Split injection
SOI	Start of injection
SOC	Start of combustion
UBHC	Unburned hydrocarbon (ppm)
WFO	Waste frying oil

1 Introduction

Unaccounted extraction and utilization of fossil fuels increase the level of carbon dioxide, and it is the main contributor for global warming. The world energy consumption had been multiplied in the middle of the years 1971–2001. It was predicted that the world energy requirement would be escalated by 53% and consumption of petroleum will ascend from 84.4 to 116 million barrels in the USA by the end of 2030 [1]. The reduction of fossil fuel resources with the constant increment in the energy consumption has spurred research interest in the search of alternative sources of energy. The major substitute energy sources of renewable are wind energy, solar energy, tidal energy, and biodiesel. Many researchers are working on biodiesel because it is renewable, nontoxic, and environmental friendly. Biodiesel, a methyl ester of fatty acid, is obtained from vegetable oils and animal fats. Biodiesel feedstock such as edible and nonedible oils is sunflower oil, palm oil; soybean oil, jatropha oil, and pongamia oil [2].

The high cost of biodiesel is the main difficulty for commercialization; however, the use of waste frying oil (WFO) rather than virgin oil to produce biodiesel is one of the cost-effective approaches. In Middle East European countries, about 0.7 – 10 million tons of rapeseed and sunflower oil based WFO have been produced per year [3, 4]. As per the Environmental Prevention Act, WFO has to be properly treated before disposing into the environment to prevent environmental pollution. But, the cost of pre-treatment processes is high. The cost of biodiesel can be reduced by using the WFO as a feedstock [5].

The various methods of transesterification reaction proposed by researchers are base-catalyzed transesterification, acid-catalyzed transesterification, enzyme-catalyzed transesterification, pyrolysis transesterification, and super critical alcohol transesterification [6]. Transesterification is the process of converting triglycerides into the ester. Freedman et al. [7] observed that the maximum yield of 93% was achievable with a reaction temperature of 60 °C, molar ratio of 6:1, and catalyst concentration of 0.5 wt%. Vicente [8] employed the response surface methodology tool for the optimization of biodiesel production from sunflower oil. The maximum yield was obtained at following condition of molar ratio of 1:6, reaction temperature of 50 °C, catalyst concentration of 1.3 wt%. Hamamre and Yamin [9] conducted the experiments on biodiesel synthesis from waste frying oil using potassium hydroxide (KOH) as a catalyst. From the results, it can be concluded that it is possible to obtain the maximum yield of 98% of biodiesel by transesterification process.

Environmental pollution and global warming are mainly due to exhaust emissions coming out from automobiles and industrialization. Diesel engine emissions can be reduced by incorporating engine modification or fuel modification. Engine modification such as injection pressure, injection timing, injection duration, fuel quantity, swirl ratio, compression ratio, piston bowl design and fuel modification such as fuels blended with alcohols or metal additives are noticed by Imtenan et al. [10], and Agarwal et al. [11] conducted the experiment on common rail direct injection (CRDI) diesel engine with different injection pressures, injection timings, and loads. It found that the particulate matter (PM) concentration increases with increase in engine load, decreased with increase in injection pressure due to smaller fuel droplet size and better evaporation. There is a drop in PM concentration while advancing of injection timing. The advanced injection timing provides more time for mixing of fuel with air. Yehliu et al. [12] investigated the experiment on CRDI diesel engine fueled with pure soybean methyl ester. The results show that the ignition characteristics of fuel could affect the start of combustion process. Soybean biodiesel was achieved the minimum PM concentration compared to other blends. This was due to the presence of oxygen in biodiesel; however, sometimes the PM concentration increased due to unburned or partially burned hydrocarbon emission.

Qi et al. [13] reported the effect of injection timing and exhaust gas recirculation (EGR) on V6 Ford Lion Engine fueled with neat biodiesel. The results unveiled that the brake specific fuel consumption (BSFC) and soot emissions were increased at higher EGR rate. However, the nitrogen oxide emission was reduced and cylinder gas peak pressure (CGPP) and heat release rate (HRR) are reduced. With retarded injection timing, the BSFC was increased and nitrogen emission was decreased. Similarly, split injection was conducted on a caterpillar scote diesel engine fueled with oxygenated fuel at no-load and full-load condition. At full load, the soot emission was reduced with no penalty of nitric oxide (NO) emission. The soot emission of split injection was less compared to single injection. The results revealed that the reduction of soot at full-load condition was due to oxygenate fuel with rich mixture combustion associated with advanced injection timing as noticed by Choi and Reitz [14]. Park et al. [15] studied the effect of multiple injection strategies on diesel engine fueled with biodiesel. The results unveiled that in a single injection, the combustion pressure and heat release rate are significantly reduced at injection timing of 30° bTDC. The emissions of carbon monoxide (CO), hydrocarbon (HC), and soot were increased due to incomplete combustion. In multiple injection strategy, the CO, HC and Soot emission reduced whereas NO emission increased due to smaller droplet size and shorter dwell period.

Exhaust gas recirculation (EGR) has suppressed the combustion temperature which reduces the NO formation and controlled the premixed combustion phase. The reduction of NO emission can be achieved through the combined effect of thermal effect (lower peak combustion temperature by circulation high specific heat gases), dilution effect (lower O_2 concentration), and chemical effect (dissociation of water (H_2O) and carbon dioxide (CO_2) combustion). Rajesh Kumar et al. [16] reported the experiment on diesel engine with different injection timings and EGR percentage up to 30%. An isobutanol-diesel blend with injection timing of 22° bTDC without EGR has found to be optimized condition for diesel engine, which gives minimum emissions with minimum BSFC compared to other blends. Yasina et al. [17] concluded the experiment on diesel engine with EGR fueled with palm oil. Brake specific fuel consumption is increased and exhaust gas temperature (EGT) is decreased with increase in EGR percentage. Increase in brake specific fuel consumption due to insufficient oxygen in cylinder leads to incomplete combustion which results in reduction of brake thermal efficiency (BTE). Tornatore et al. [18] studied the combined effect of injection timing and EGR. The NO emission and BTE were reduced when increasing the EGR rate or retardation of injection timing. Valentino et al. [19] studied the effect of injection timing and EGR for 40% butanol-diesel blends on CRDI diesel engine. The minimum NO emission and smoke was achieved at advanced injection timing and moderate injection pressure. Ozer Can et al. [20] studied the effect of EGR on diesel engine fueled with soybean biodiesel. The EGR rate varied from 5 to 15%. The results unveiled that the more BSFC and lesser BTE were occurred at 15% EGR at full-load condition. The NO and HC emissions were improved by 55 and 15%, respectively, with 15% EGR at full-load condition. There is no significant change in CO emission with higher CO_2 emission at full-load condition.

From the previous works, it could be concluded that few works had been done in the field of a transesterification reaction of waste frying oil to biodiesel production using sodium methoxide as a catalyst. Furthermore, there were fewer contributions made to the study of the split injection strategies on common rail direct injection diesel engine fueled with biodiesel and its blends. Hence, in this paper, an initiative has been done to explore the biodiesel production by using sodium methoxide as the catalyst. Moreover, split injection strategies with exhaust gas recirculation technique have been employed to study the performance, emission and combustion characteristics on CRDI diesel engine fueled with diesel and biodiesel blends.

2 Materials and Methodology

2.1 Purification of Raw Oil and Biodiesel Synthesis

In this investigation, sunflower-based waste frying oil has been considered as the feedstock. It was purchased from various restaurants in Tiruchirappalli, Tamilnadu, India. Sodium methoxide and methanol were bought from Eswarr Scientific & Co, Trichy, Tamilnadu, India. The methyl ester was produced through transesterification process at Thermal Engineering Laboratory, National Institute of Technology, Tiruchirappalli, Tamilnadu. The waste frying oil contains impurities and food particles; therefore, it was subjected to the purification process. About 100 g of oil was weighed and heated at 80 °C for 30 min, and it was kept in a separating funnel to remove the food items. The heated oil was washed with distilled water to remove the odor, followed by the sample subjected to a heating of 110 °C for removing the water content. The acid value of water-washed frying oil was measured to be 3.34 mg KOH/g by using titration method, followed by acid washing in which 1% of phosphoric acid was added to the oil and stirred for further 30 min. This resulted in reduction of acid value by 2.27 mg KOH/g [21]. The raw biodiesel was washed with distilled water for removing excess methanol and unreacted catalyst. The presence of excess methanol reduces the fuel properties of biodiesel like density, viscosity, flash point, and corrosion [22].

The biodiesel experimental setup is shown in Fig. 1. WFO was taken into three necked round bottom flask; catalyst and methanol were mixed with help of magnetic stirrer to attain homogeneity. The prepared solution was poured into the reactor, and the mixture was stirred at the following condition, temperature of 55 ° C, time of 1.5 h, molar ratio of 1:6, catalyst concentration of 1 wt%, and stirrer speed of 600 rpm. The product was allowed to settle down in a separating funnel for 24 h and the biodiesel was collected on the top side of separating funnel while glycerol settled down on the bottom portion of the funnel. It was cleaned with distilled water and heated at 110 °C to remove moisture content. Challenges and difficulties in biodiesel production lies in the collection and storage of WFO and biodiesel derived from WFO. Waste frying oil consumes more water for purification, and it is difficult to dispose glycerol and wastewater after purification process. Methanol is toxic and it should be used in a ventilated area as inhalation or ingestion of high concentrated methanol can lead to death or blindness. This process involves more chemical exposer and demands almost safety measures. The transesterification process equations are given below.

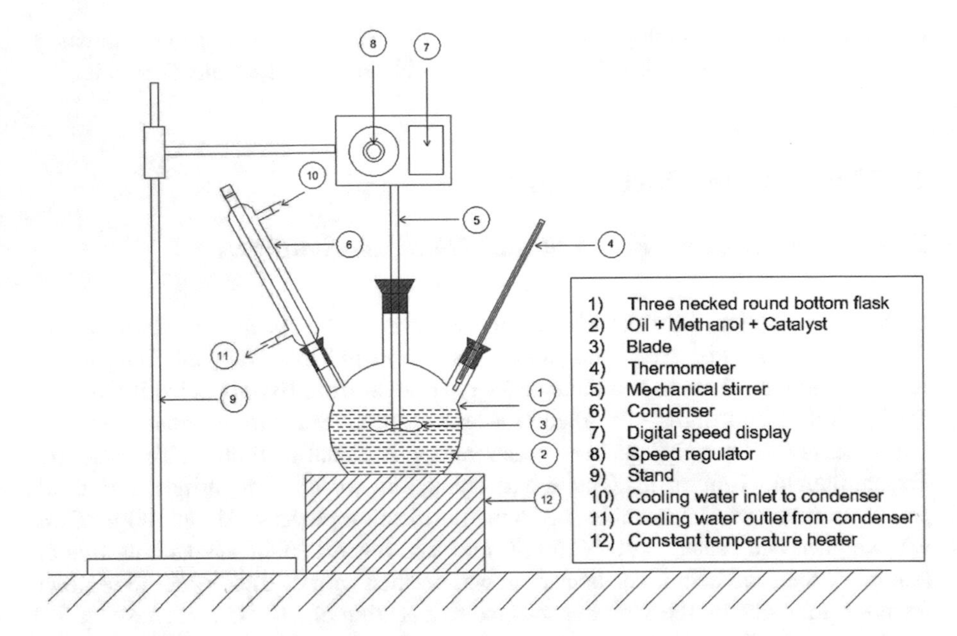

Fig. 1 Schematic view of biodiesel setup

$$TG + ROH \leftrightarrow DG + RCOOR_1 \tag{1}$$

$$DG + ROH \leftrightarrow MG + RCOOR_2 \tag{2}$$

$$MG + ROH \leftrightarrow GL + RCOOR_3 \tag{3}$$

Overall equation is given below

$$TG + 3ROH \leftrightarrow GL + 3ME \tag{4}$$

2.2 *Methyl Ester Characterizations*

Methyl ester conversion was determined by nuclear magnetic resonance (^{1}H NMR-BRUKER 400 MHz). Chloroform and tetramethylsilane were used as a solvent and internal standard, respectively, in ^{1}H NMR analysis. ^{1}H NMR spectra of biodiesel are shown in Fig. 2. The peak at 3.66 δ ppm is obtained for biodiesel, and it shows the presence of the methyl ester (CH_3COOR) group. However, the presence of methanol (CH_3OH) confirmed through the peak appeared at 2.7 δ ppm. Gas chromatography and mass spectrometer (Perkin Elmer Clarus 500) was used to find out the biodiesel composition, and fatty acid methyl ester composition is shown in Table 1. The capillary column (5% Phenyl 95% dimethyl polysiloxane)

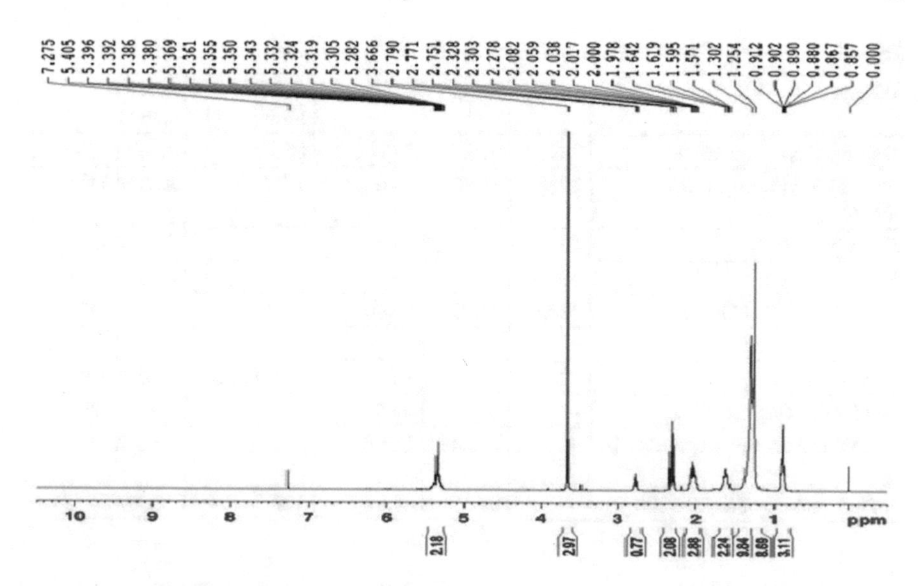

Fig. 2 ^{1}H NMR spectra of biodiesel

Table 1 List of fatty acid composition in biodiesel

Compounds name	Molecular formula	Molecular weight (g/mol)	Retention time (min)	Peak area (%)
Methyl octanoic acid	$C_9H_{18}O_2$	158	8.74	0.083
Methyl laurate acid	$C_{13}H_{26}O_2$	214	18.15	0.153
Methyl oleate acid	$C_{19}H_{36}O_2$	296	33.84	10.01
Methyl palmitate acid	$C_{17}H_{32}O_2$	268	28.45	0.013
Methyl palmitate acid	$C_{17}H_{34}O_2$	270	29.32	24.61
Methyl stearic acid	$C_{18}H_{36}O_2$	284	33.62	3.033
Methyl linoleate acid	$C_{19}H_{34}O_2$	294	34.18	50.04
Methyl stearate acid	$C_{19}H_{38}O_2$	298	34.38	7.804
Methyl icosanoate acid	$C_{21}H_{42}O_2$	326	38.24	1.662
Methyl behenate acid	$C_{23}H_{46}O_2$	354	42.42	2.074
Methyl tetracosanoic acid	$C_{25}H_{50}O_2$	382	46.52	0.450

(30 m × 0.25 mm × 0.25 μm) was employed in gas chromatography. About 1 μl of biodiesel was injected into the sample holder which was soluble with hexane. The helium gas was used as a carrier gas which was supplied at the flow rate of

Table 2 Physical and chemical properties of diesel, WFO, and biodiesel

Property	Diesel	WFO	Biodiesel	ASTM limit	ASTM standards
Density (kg/m^3) at 15 °C	829	916.7	876	860–890	ASTM 1298
Kinematic viscosity (cSt) at 40 °C	2.39	38.53	4.76	1.9–6.0	ASTM D445
Flash point (°C)	46	327	160	>130	ASTM D93
Fire point (°C)	54	332	165	–	ASTM D93
Calorific value (MJ/kg)	42.14	38.97	39.665	–	ASTM D240
Cloud point (°C)	0	14	5	–	ASTM D2500
Pour point (°C)	−15	3	−7	−15 to 10	ASTM D97
Acid value (mg KOH/g)	–	6	0.68	<0.8	ASTM D974
Saponification value (mg KOH/g)	–	211.41	182.68	–	ASTM D5558
Iodine value g of Iodine/100 g	–	97.38	49.48	120	ASTMD5554

1 ml/min. Initially, the oven temperature was maintained at 60 °C for 1 min and then it increased to a temperature of 280 °C. The gas chromatography and mass spectrometer (GC-MS) run time was 54.5 min, and the result was stored in GC-MS library.

Table 2 shows the physical and chemical properties of diesel and biodiesel. From Table 2, it noticed that WFO has the viscosity of 38.53 cSt, which is 9 times higher than the biodiesel and 12 times higher than the diesel. Higher viscosity leads to disturb the injection pattern and reduced the evaporation rate fuels. High injection pressure is proposed for biodiesel fuel for better atomization due to higher viscosity. The flash point of biodiesel was found to be greater than 100 °C, making it safe for handling and storage. Biodiesel has poor cloud and pour points, and it is difficult to use in cold atmosphere condition. Presence of oxygen in molecular structure leads to reduce the calorific value of fuel. Acid value is a direct measure of free fatty acid present in biodiesel. Biodiesel showed the lowest free fatty acid (FFA) of 0.68 mg KOH/g. Free fatty acid may be an indication of water in the biodiesel. Carbon residue of biodiesel was 0.02 wt%. The carbon residue is a measure of residual carbon remains after combustion. Iodine value represents the amount of saturated fatty acid present in fuel. Biodiesel has the iodine value of 49.48 g of Iodine/100 g.

2.3 *Engine Setup and Measurements*

The engine test was carried out on a common rail direct injection diesel engine with a rated power of 3.7 kW at constant speed of 1500 rpm. The experiment was conducted based on the International Standard Organization 8178 test cycles with D2 mode. Experimental design matrix was given by design of experiment as shown in Table 3. The performance, emission and combustion characteristics of diesel, B20, B40, B60, B80, and B100 were recorded at 100% rated load condition with injection pressure of 500 bar and main injection timing of 17, 19, 21, 23, and 25° bTDC and post-injection timing varied from −5° bTDC, 0° TDC, and 5° aTDC, respectively. The split injection parameters such as main injection fuel quantity, post-injection fuel quantity, and dwell period are the percentage of 90%, 80% and 75% and 10%, 20% and 25% and 5, 10, 15 and 20° bTDC, respectively. From the overall 100% fuel

Table 3 Experimental design matrix

Run no.	Main injection fuel quantity (%)	Post-injection fuel quantity (%)	Dwell period (°CA)	Main injection timing (°CA bTDC)	Post-injection timing (°CA)
1	80	20	5	19	2
2	80	20	10	25	3
3	90	10	5	25	1
4	75	25	10	25	−1
5	90	10	5	19	−1
6	75	25	10	23	−5
7	90	10	10	21	−3
8	75	25	10	19	3
9	75	25	5	21	5
10	90	10	5	21	−1
11	75	25	10	21	5
12	90	10	15	19	4
13	80	20	10	17	5
14	75	25	10	19	0
15	80	20	20	25	−5
16	90	10	15	23	0
17	75	25	10	23	3
18	80	20	10	21	5
19	80	20	15	19	0
20	80	20	10	21	5
21	90	10	15	25	3
22	75	25	10	17	5
23	90	10	15	17	3
24	90	10	10	21	5
25	80	20	10	17	5
26	80	20	10	21	−5
27	80	20	5	23	−1
28	90	10	20	17	0
29	90	10	10	25	−3
30	90	10	15	23	−5

quantity, the main injection fuel quantity is sprayed at 90, 80, and 75% and the post-injection fuel quantity is injected at 10, 20, and 25%, respectively. CRDI system has provision to modify the injection timing, injection pressure, injection duration, and fuel quantity. The operating injection pressure of CRDI system is varied up to 1200 bar. CRDI system attained above 600 bar could not control the injection duration and engine speed which resulted in heavy vibration and sound. In order to safeguard the system, the injection pressure was fixed up to 600 bar. CRDI system was connected to the engine and electronic control unit (ECU). ECU received the signal from engine, and it sends to data receiver or computer.

The test engine was started with the help of crank lever. The engine speed was allowed to achieve steady state by adjusting the mass flow rate or potentiometer. SAE 20–40 oil was used as lubricating oil in diesel engine. A differential airflow sensor, standard burette, and rotameter were employed to measure the flow rate of air, fuel flow rate, and flow rate of cooling water, respectively. The quartz transducer (Make- Kistler, Model- 6613 °C) was employed to measure the cylinder pressure in the range of 0–100 bar. The Kuebler Sendix 5000 model crank angle encoder is employed for crank angle measurement. The crank angle data is transferred to high-speed data acquisition system (HDAQ). The high-speed data acquisition system is connected with CRDI driver. Both CRDI driver and HDAQ are connected with engine software through Ethernet cable. The injection angle is varied by setting injection timing and injection duration through engine software. The common rail pressure sensor measures the rail pressure, and the pressure relief valve is used for maintaining the pressure within the limit. The solenoid sensor controls the amount of fuel injection and injection timing. The fuel injection duration is controlled by potentiometer. The eddy current dynamometer is connected with engine to load.

Exhaust gas recirculation technique was employed to minimize the exhaust gas emissions coming out from engines with minimal compromise in performance. The counter flow water-cooled heat exchanger was used to cool down the exhaust gas temperature and the gases sent back to the surge tank to mix with fresh air. Regulating value and orifice meters are used to send required amount of exhaust gas into the surge tank. Performance and emissions characteristics for B20 and B40 blends were tested with different EGR rates of 5, 10, 15, 20, 25, and 30% on a CRDI diesel engine, and obtained results were compared with diesel. The measurements were taken thrice to ensure repeatability of the data, and an average of all three readings was considered in the present study. The engine and dynamometer specification are given in Table 4. Figures 3, 4, and 5 show the schematic view of engine setup, CRDI system, and EGR setup, respectively. The carbon monoxide (CO), carbon dioxide (CO_2), oxygen (O_2), nitric oxide (NO), and unburned hydrocarbon (UBHC) were measured by AVL Digas 444 five gas analyzer. The CO, CO_2, and O_2 were measured in terms of percentage volume (%vol.). The NO and UBHC concentrations were measured in terms of parts per million (ppm). The AVL smoke meter 415SE was used to measure the filter smoke number. The specification of AVL 444 five gas analyzer and AVL smoke meter is given in Table 5.

Table 4 Specification of single-cylinder diesel engine and eddy current dynamometer

Make	KIRLOSKAR
Model	AV1
No. of cylinders	1
No. of strokes	4
Bore	80 mm
Stroke	110 mm
Displacement volume	553 cc
Compression ratio	16.5:1
Rated power output	3.7 kW at 1500 rpm
Cooling medium	Water
Fuel injection timing	23° bTDC
Fuel injection pressure	200 bar
Nozzle hole diameter	0.3 mm
No. of nozzle holes	3
Piston bowl	Hemispherical
Temperature measurement unit	K-type (Chromel-Alumel) (0–1250 °C), digital indicator
Pressure sensor	Make, model: Kistler, model—6613C, Range: 0–100 bar, Frequency range: (0.016–20,000 Hz)
Crank angle encoder	Make, Model: Kuebler, Sendix5000, Vibration resistance: 100 m/s, 10–2000 Hz, working temperature range: (−40 to 85 °C)
Eddy current dynamometer	Make: Power Mag Maximum power: 5.5 kW Maximum speed: 3000 rpm Operating voltage: 230 V, 50 Hz Operating temperature: (0–70 °C) Cooling: air cooled, self-ventilation Weight measurement: linear (load cell) Output: PC interface with a load cell indicator

3 Results and Discussion

Figures 6 and 7 represent the comparison of nitric oxide (NO) emission with respect to main injection fuel quantity, post-injection fuel quantity, and main injection timing for tested fuels. From Fig. 6, it can be seen that the NO emission increased with increase in main injection fuel quantity from 75 to 90%. It is concluded that from the experimental results, that the B20, B40, B80, B100, and diesel fuels produce maximum NO emission of 1406, 1302, 1203, 1135, and 1490 ppm, respectively at 90% (Main injection) - 10 (Dwell period) – 10% (Post injection), 90%-15-10%, 90%-10-10%, 80%-15-20% diesel-90%-15-10%. The nitric oxide formation is due to higher in-cylinder temperature, pressure, and residence time.

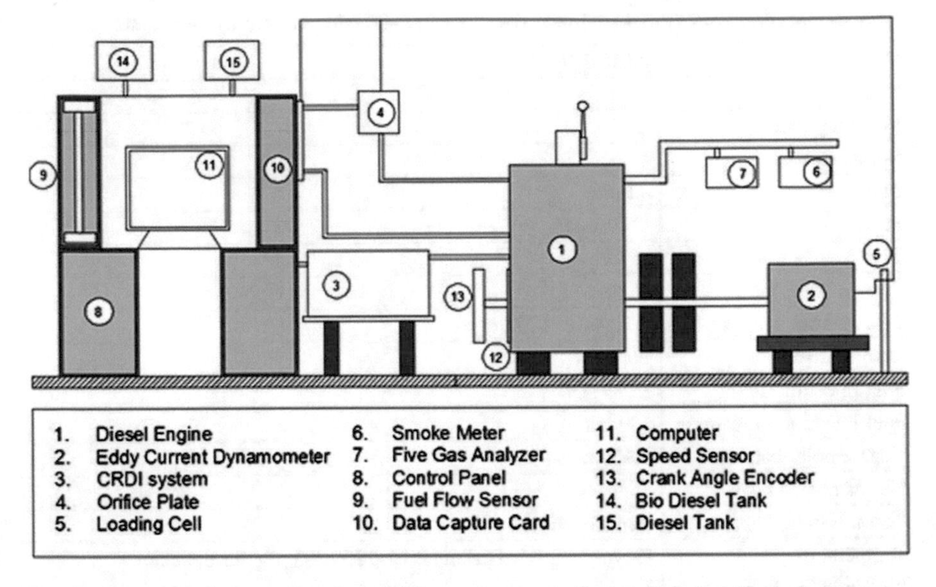

Fig. 3 Schematic diagram of engine experimental setup. 1. Fuel tank, 2. fuel filters, 3. fuel pump, 4. high-pressure pump, 5. pressure-regulating value, 6. distribution line (rail), 7. rail pressure sensor, 8. injector, 9. pressure gauge, 10. ECU controller

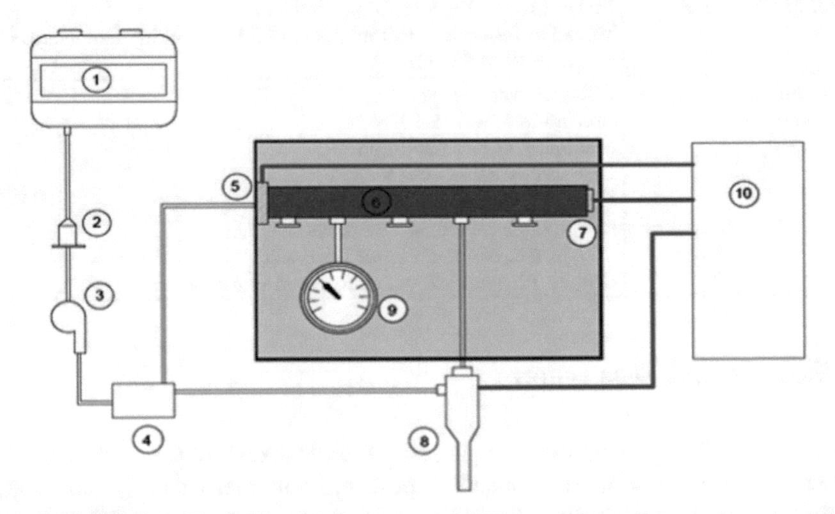

1. Fuel tank 2. Fuel filters 3.Fuel pump 4. High pressure pump 5.pressure regulating value 6. Distribution line (rail) 7. Rail pressure sensor 8. Injector 9. Pressure gauge 10. ECU controller

Fig. 4 Schematic diagram of CRDI system. 1. Engine block, 2. engine air inlet, 3. engine exhaust, 4. heat exchanger, 5. water inlet to heat exchanger, 6. water outlet from heat exchanger, 7. flow control value, 8. electronic control system, 9. manometer, 10. surge tank, 11. fresh air inlet to surge tank

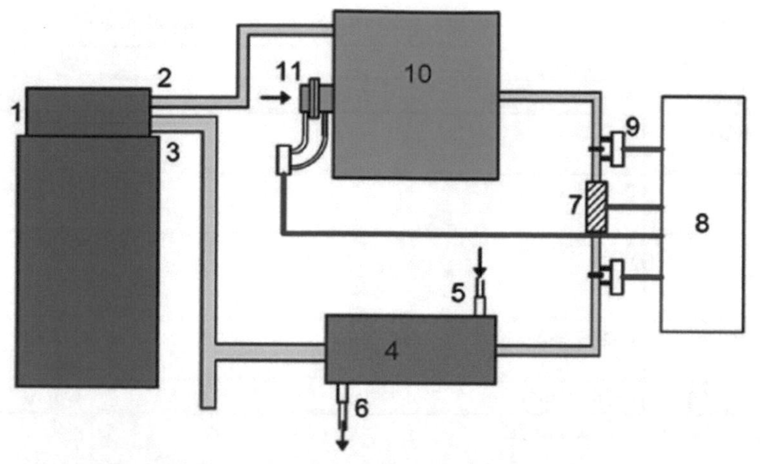

1. Engine block 2. Engine air inlet 3. Engine exhaust
4. Heat exchanger 5. Water inlet to heat exchanger 6. Water outlet from heat exchanger 7. Flow control value 8. Electronic control system 9. Mano meter
10. Surge tank 11. Fresh air inlet to surge tank

Fig. 5 Schematic diagram of exhaust gas recirculation setup

The higher percentage of main injection fuel quantity and higher dwell period lead to more premixed combustion and it results in higher NO emission formation [23, 24].

Figure 7 shows the variation of nitric oxide (NO) emission with respect to main injection fuel quantity and main injection timing for various fuels. From Fig. 7, it noticed that the NO emission is increased with increase in main injection timing from 17° bTDC to 25° bTDC and increased with increase in main injection fuel quantity from 75 to 90%. The minimum NO emission of 645, 638, 603, and 582 ppm was identified, respectively, for B20, B40, B80, and B100 fuels with the conditions of 75%-10-25%-17, 75%-20-25%-17, 75%-20-25%-17, and 75%-10-25%-17. Retarded fuel injection timing reduced the in-cylinder pressure due to more fuels burned after TDC. Lower cylinder pressure leads to lower in-cylinder temperature with minimum NO emission [25]. B100 has the minimum nitric oxide emission because of lower iodine value. Lower iodine value indicates more saturated fatty acid present in the fuel. Saturated fatty acid enhances the fuel properties such as cetane number and cloud and pour points and calorific value, and it reduces the NO formation [26, 27].

Figures 8 and 9 illustrate the contour plot of carbon monoxide (CO) emission with respect to split injection for tested fuels. In Fig. 8, it is observed that the CO emission is increased with increase in main injection fuel quantity from 75 to 90%. This is due to minimum quantity of fuel with more delay period that causes complete combustion. At full-load condition with higher injection pressure of 500 bar, the CO emission was not significant change. The minimum CO emission

Table 5 Specification of AVL Digas 444 five gas analyzer and AVL Smoke meter 415SE

AVL Digas 444 five gas analyzer		
Measured quality	Measuring range	Resolution
Carbon monoxide (CO)	0–10% vol.	0.01% vol.
Carbon dioxide (CO_2)	0–20% vol.	0.1% vol.
Unburned hydrocarbon (UBHC)	0–20,000 ppm	2 ppm
Oxygen (O_2)	0–22% vol.	0.01% vol.
Nitric oxide (NO)	0–5000 ppm	1 ppm
Temperature: −30 to 125 °C	−30 to 125 °C	
Engine speed	400–6000 rpm	
AVL Smoke meter 415SE		
Measurement range	0–10 FSN	
Detection limit	0.002 FSN or ~0.02 mg/m^3	
Resolution	0.001 FSN or 0.01 mg/m^3	
Exhaust pressure ranges	(−300*) −100 to 400 Mbar (−500*) −200 to 750 Mbar	
Maximum exhaust temperature	600 °C with standard 340 mm sample probe (800 °C with 780 mm long sample probe)	
Interfaces	RS232 interfaces with AK protocol digital via instrument controller 4210	
Power supply	100–115 VAC or 230 VAC, 50/60 Hz	
Sample flow	10 L/min	
Repeatability	Standard deviation 1 $\sigma \leq \pm$ (0.005 FSN + 3% of the measured value @ 10 s intake time)	

of 0.02, 0.02, 0.01, and 0.01% vol. was obtained for B20, B40, B80, and B100 fuels with condition of 80%-25-20%, 75%-10-25%, 75%-10-25%, and 80%-20-20%, respectively. The minimum CO emission occurred due to better atomization, evaporation, and air-fuel mixing at high injection pressure which leads to complete combustion with lesser emission [28].

Figure 9 shows the variation of carbon monoxide (CO) emission with respect to main injection fuel quantity and main injection timing. Figure 9 shows that the CO emission decreased with advancement of main injection timing from 17° bTDC to 25° bTDC. The maximum CO emission of 0.04, 0.04, 0.04, and 0.03% vol. was obtained for B20, B40, B80, and B100 fuel with the case of 75%-10-25%-19, 75%-10-25%-17, 80%-10-20%-17, and 80%-10-20-17%. The reduction of CO emission for biodiesel is due to the presence of oxygen and higher cetane number which resulted complete combustion. Advanced injection time enhanced the evaporation

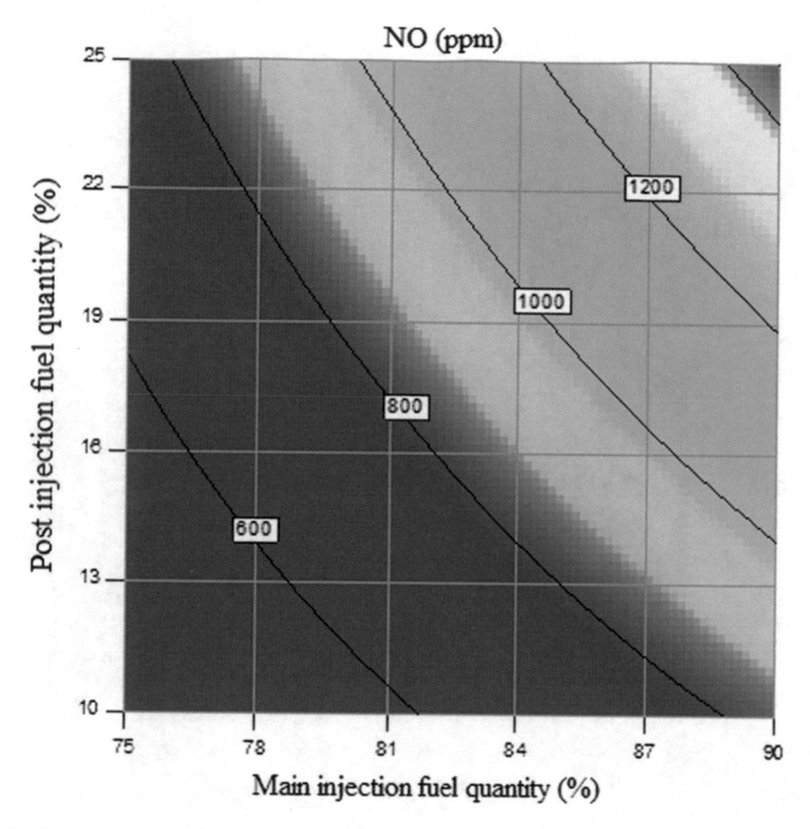

Fig. 6 Comparison of NO emission for tested fuel

rate of fuel which resulted in complete combustion with lower emission. However, retarded injection timing reduced the ignition delay period which results in incomplete combustion with more exhaust emissions [29].

Figures 10 and 11 show the contour plot of unburned hydrocarbon (UBHC) emission with respect to split injection for tested fuels. As in Fig. 10, at full-load condition, the UBHC emission increased with increase in main injection fuel quantity from 75 to 90%. The UBHC emission is indication of incomplete combustion or partial combustion of fuels. The biodiesel has the minimum UBHC emission compared to other biodiesel blends such as B80, B40, B20, and diesel due to excess oxygen present in biodiesel with higher cetane number results in complete combustion with lower emission [30]. The minimum UBHC emission of 13, 12, 11, and 10 ppm was obtained for B20, B40, B80, and B100 fuel with following condition 80%-20-20%-25, 75%-10-25%-23, 80%-10-20%-25, and 75%-5-25%-25. It can be concluded from the above value that the minimum UBHC emission is obtained with advanced injection timing irrespective of main injection fuel quantity and post-injection fuel quantity. Advanced injection timing improves the spray characteristic which leads to complete combustion [23].

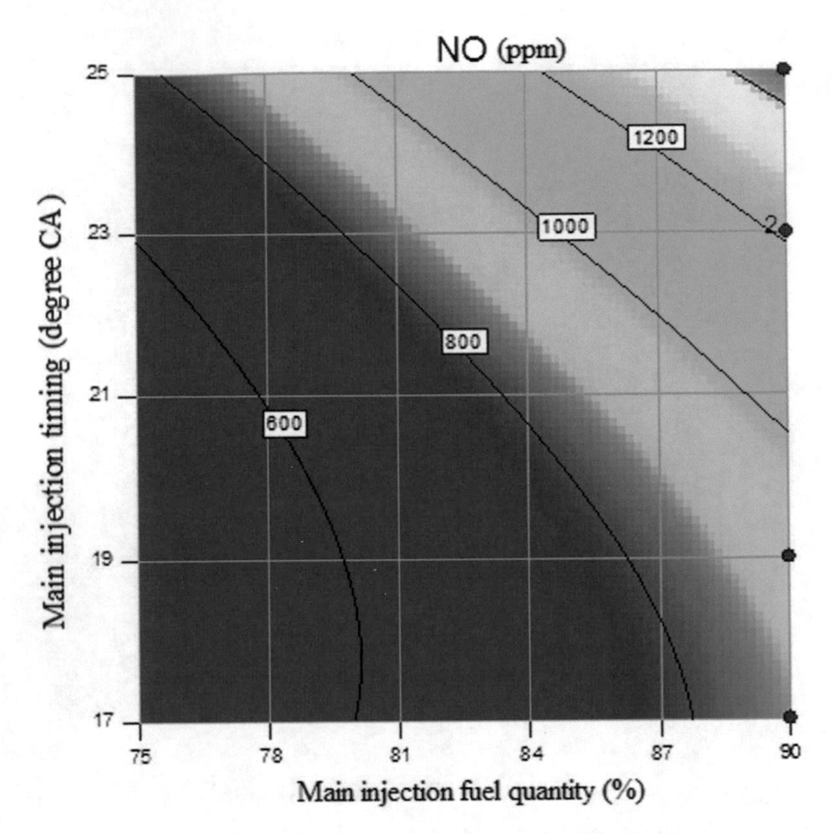

Fig. 7 Variation of NO emission at different injection timings and main injection

The variation of unburned hydrocarbon (UBHC) emission with respect to main injection timing and main injection fuel quantity for tested fuel is shown in Fig. 11. From Fig. 11, it is inferred that the UBHC emission decreased with increase in main injection timing from 17° bTDC to 25° bTDC and it decreased with increase in main injection fuel quantity from 75 to 90%. The minimum UBHC emission is occurred at advanced injection timing with main injection fuel quantity from 75 to 80%. The B20, B40, B80, and B100 have the highest UBHC emission of 25, 23, 19, and 16 ppm with the condition of 80%-15-20%-17, 90%-20-10%-17, 80%-10-20%-19, and 75%-10-25%. The higher injection pressure, longer dwell period and fixed fuel injection quantity with advanced injection timing enhanced the premixed combustion and it leads to lower unburned hydrocarbon emission [31].

Figures 12 and 13 represent the filter smoke number (FSN) with respect to main injection fuel quantity, post-injection fuel quantity, and main injection timing for test fuels. In Fig. 12, it is noticed that the FSN value decreased with decrease in the main injection fuel quantity from 90 to 75% and it decreased with increase in the main injection timing from 17° bTDC to 25° bTDC. The FSN emission is an indication of combustion quality. The minimum FSN value of 2.203, 1.864, 1.729,

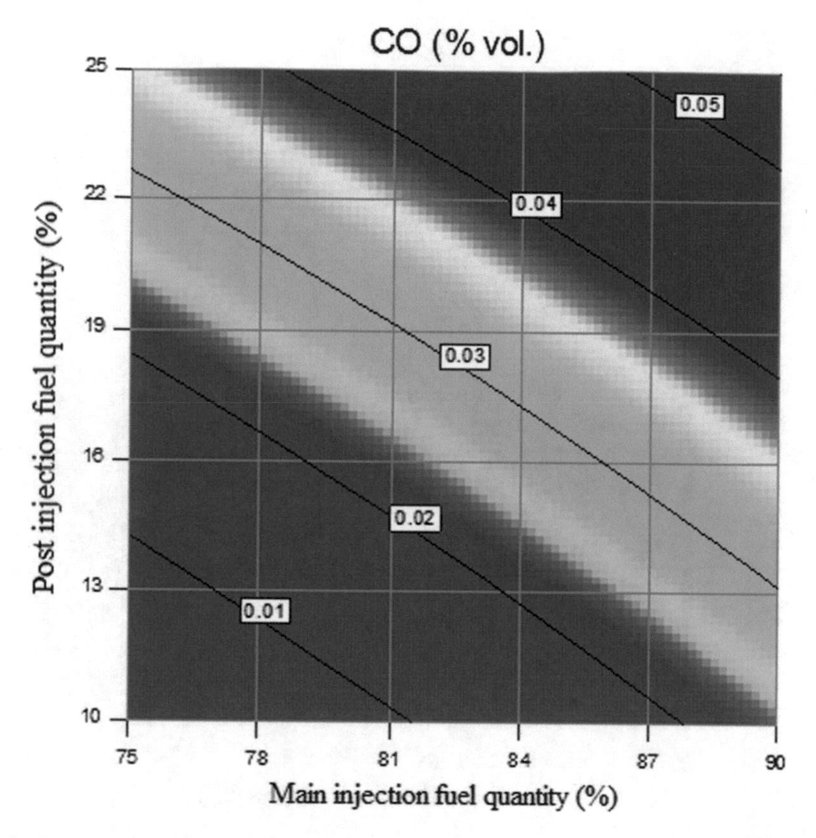

Fig. 8 Contour plot of CO emission for tested fuel

and 1.634 was obtained for B20, B40, B80, and B100, respectively, with the percentage of main injection fuel quantity, dwell period, and post-injection fuel quantity such as 75%-5-25%-25, 80%-20-20%-25, 75%-5-25%-25, and 80%-10-20%-25. The B100 fuel has the minimum FSN value compared to other tested fuels. The biodiesel fuel with higher injection pressure improve the combustion efficiency and lower the FSN emission [12]. The experimental data reveals that the 75 to 80% main injection fuel quantity with advanced main injection timing produced minimum emission.

Figure 13 shows the contour plot of filter smoke number (FSN) with respect to main injection fuel quantity and main injection timing. The results unveiled that the FSN value decreased with increase in the main injection timing from retarded to advancement. Advanced injection timing increases the delay period which improves the evaporation and mixing rate of air and fuel which results in better combustion. Reduction of soot is achieved by introducing main injection. The maximum FSN value was obtained for B20, B40, B80, and B100 at full-load

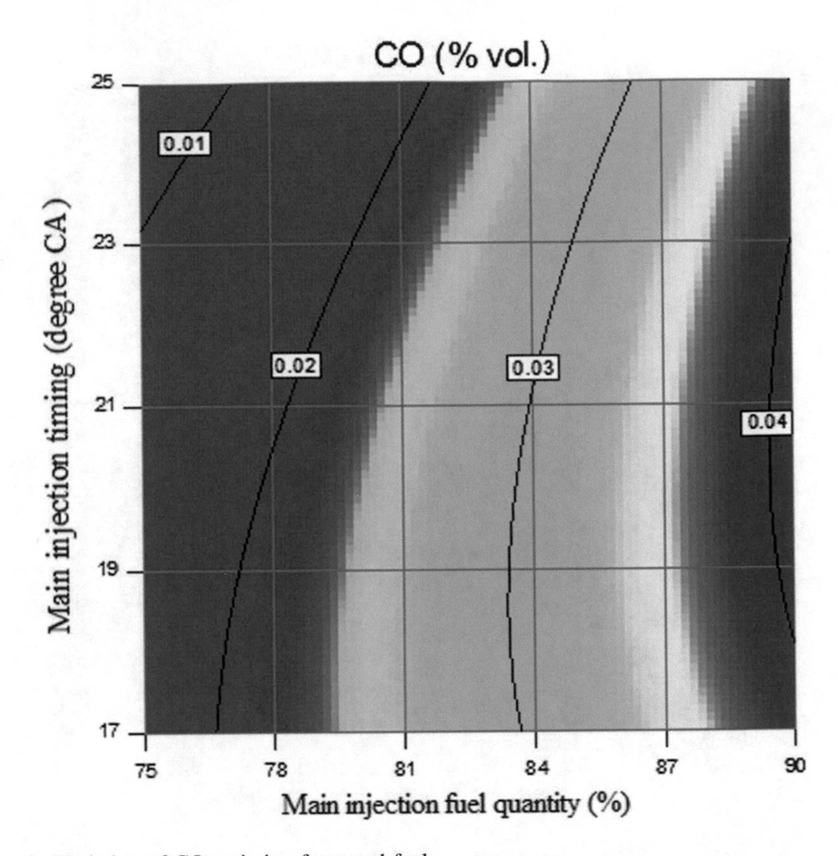

Fig. 9 Variation of CO emission for tested fuel

condition such as 3.167, 2.854, 2.428, and 2.384, respectively. The conditions are 90%-10-10%-17, 90%-20-10%-17, 80%-10-20%-17, and 80%-15-20%-17 [32].

Figures 14 and 15 illustrate the brake specific fuel consumption (BSFC) and brake thermal efficiency (BTE) with respect to split injection for tested fuels. In Fig. 14, it noticed that the BSFC increased with increase in main injection fuel quantity from 75 to 90%. This is due to more accumulation of fuels with lesser delay period resulted in incomplete combustion with more fuel consumption, at full-load condition. The biodiesel blends have higher BSFC compared to diesel due to its properties such as lower calorific value and higher viscosity. For the same power output, the engine consumed more biodiesel fuel to maintain speed of engine [10]. The maximum BSFC of 0.265, 0.268, 0.291, and 0.323 kg/kWh was obtained for B20, B40, B80, and B100 with the condition of 80%-10-20%-21, 80%-10-20%-25, 80%-10-20%-25, and 90%-10-10%-21. The minimum BSFC of 0.230, 0.232, 0.252, and 0.264 kg/kWh was obtained for B20, B40, B80, and B100 fuels with the case of 80%-10-20%-17, 75%-10-25%-19, 75%-10-25%-17, and 80%-10-20%-17. The above experimental results clearly show that the maximum BSFC was occurred at advanced main injection timing rather than the retarded injection timing. This is

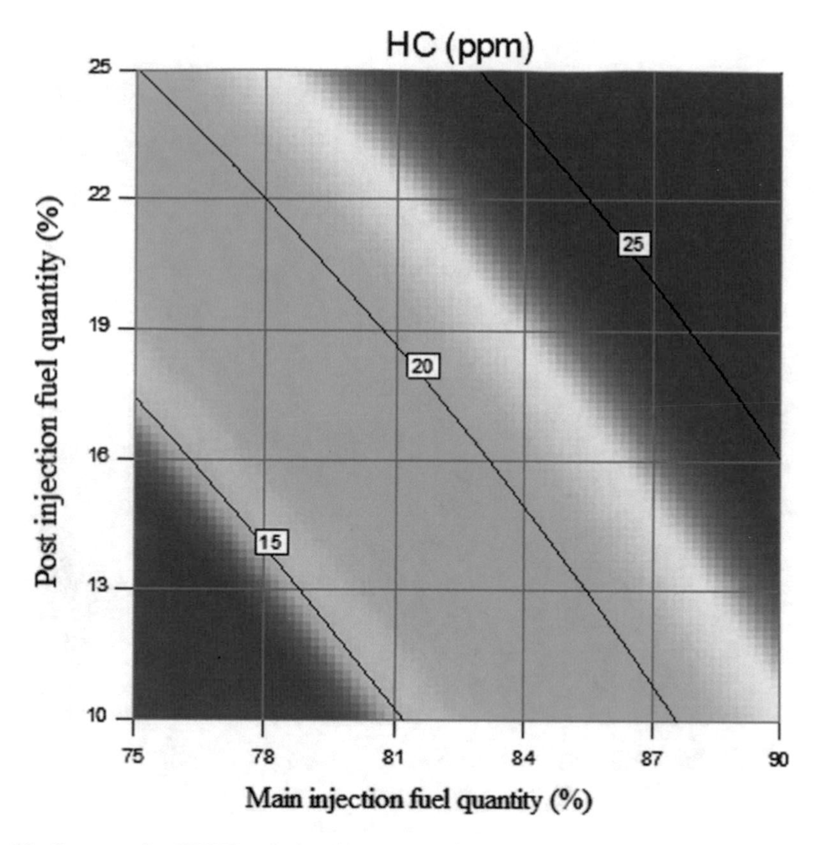

Fig. 10 Contour plot UBHC emission for tested fuel

because advance injection timing enhanced the delay period resulted in more accumulation of fuels.

Figure 15 shows that the BTE of engine increased with decrease in main injection fuel quantity and it increased with retarded main injection timing from 25 to 17° bTDC. Main injection fuel quantity with advanced main injection timing increased the delay period, and it leads to more fuels burned at premixed combustion rather than controlled combustion. Retarded injection timing reduced the delay period which results in lower accumulation fuels with complete combustion. The combustion process extended up to power stroke that leads to better thermal efficiency [33]. The maximum BTE was obtained for B20 blends compared to other blends due to higher calorific value and lower viscosity. The fuel properties enhanced the evaporation rate. The maximum BTE of 37.46, 37.32, 36.23, and 34.33% was obtained for B20, B40, B80, and B100 fuel with the case of 75%-10-25%-23, 80%-10-20%-17, 80%-10-20%-21, and 80%-20-20%-19. The minimum BTE of 32.41, 32.40, 30.35, and 28.08% was obtained for B20, B40, B80, and B100 fuel.

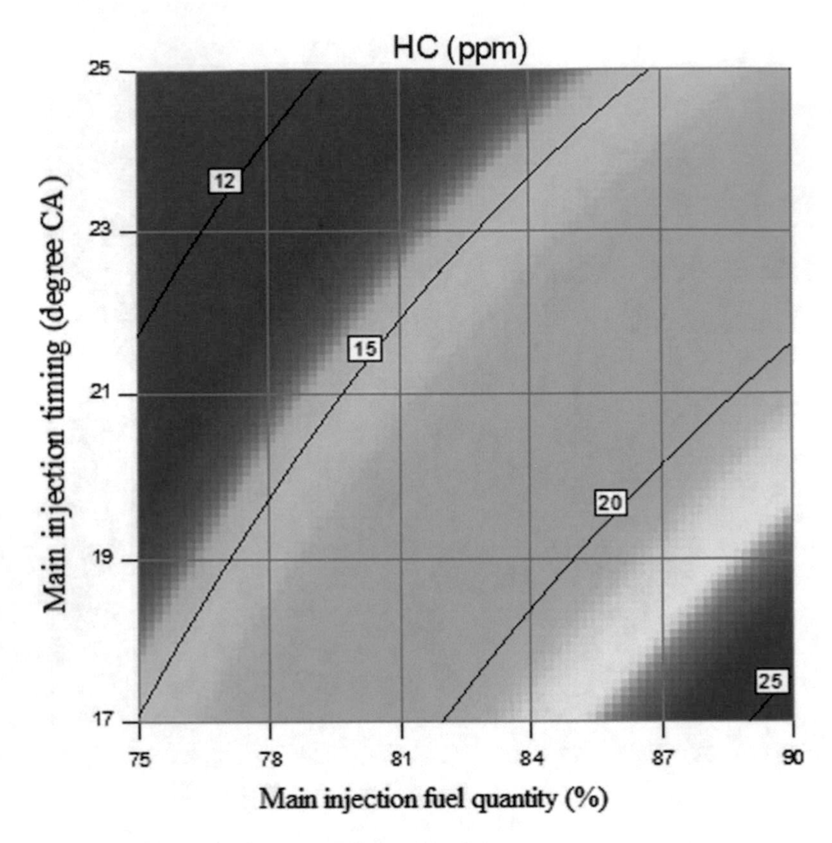

Fig. 11 Effect of UBHC emission at different injections

Figure 16 represents the variation of cylinder gas peak pressure (CGPP) with respect to split injection for different biodiesel blends. In-cylinder gas peak pressure increased with increase in main injection fuel quantity due to higher fuel consumption. Biodiesel blends have lesser peak pressure due to its physical properties such as lesser calorific value, more viscosity, and higher density. The fuel CGPP increased due to better combustion at high injection pressure with advanced main injection timing [34]. By advancing the main injection timing, the CGPP increased because advancement of injection will give more time for mixing of fuel with air resulted in more premixed combustion than the controlled combustion; thus, high gas temperatures and higher gas pressure is obtained [35]. The minimum CGPP of 60.89, 60.59, 60.39, and 60.11 bar was obtained for B20, B40, B80, and B100, respectively, with the case of 90%-20-10%-17, 90%-10-10%-17, 90%-15-10%-17, and 80%-10-20%-17. The retarded main injection timing has reduced the delay period which leads to poor vaporization resulted in lower cylinder peak pressure. The maximum CGPP of 63.83, 63.81, 63.78, and 63.75 bar was achieved for B20, B40, B80, and B100 fuel with condition of 80%-10-20%-25, 90%-15-10%-25,

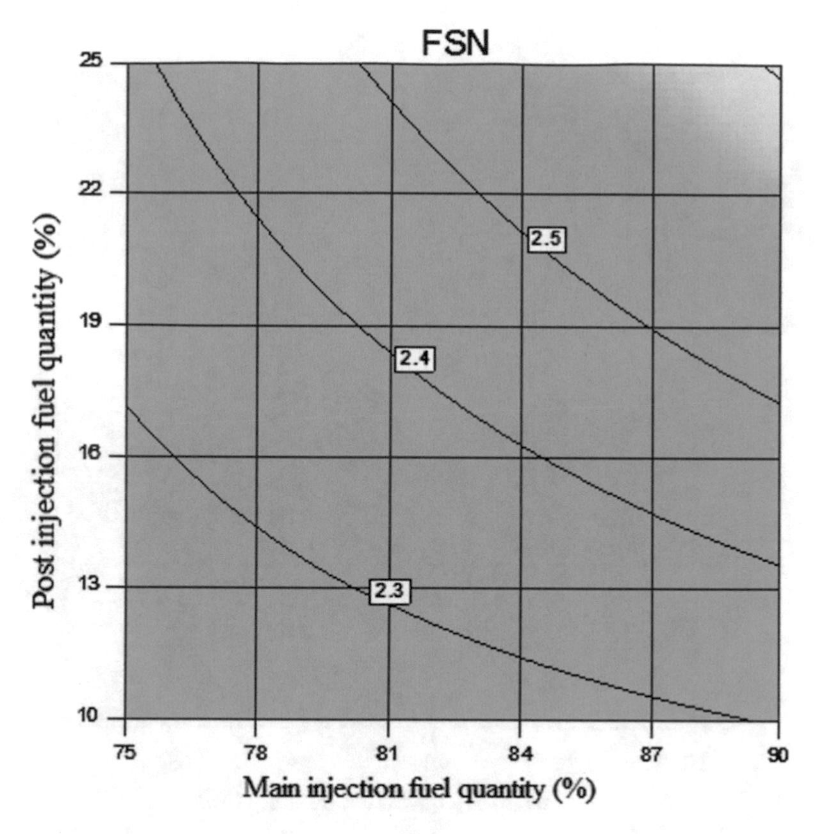

Fig. 12 Contour plot of FSN emission for tested fuel

80%-10-20%-25, and 80%-10-20%-25. The maximum CGPP was attained at advanced main injection timing with main injection fuel quantity.

Figure 17 shows the variation of ignition delay (ID) with respect to main injection fuel quantity, post-injection fuel quantity, and main injection timing for different biodiesel blends. The ignition delay period is calculated from heat release rate curve. Ignition delay period is the period between start of injection and start of combustion. Ignition delay consists of physical and chemical delays. Physical delay depends on fuel properties like viscosity and density, and chemical delay depends on volatility of fuel and latent heat of evaporation. By increasing biodiesel blends, the ignition delay decreased due to high cetane number than the diesel. Biodiesel has higher ignition quality and lesser auto-ignition temperature as observed by Atadashi et al. [36]. At full-load condition, the ignition delay period decreased due to high in-cylinder pressure and temperatures which results in faster combustion. From Fig. 17, the effect of injection pressure and injection timing is negligible at low-load condition. At higher load, ignition delay period is reduced due to better atomization of the fuel at high injection pressure with retarded main injection timing as observed by Sayin et al. [29]. Advanced injection timing increases the

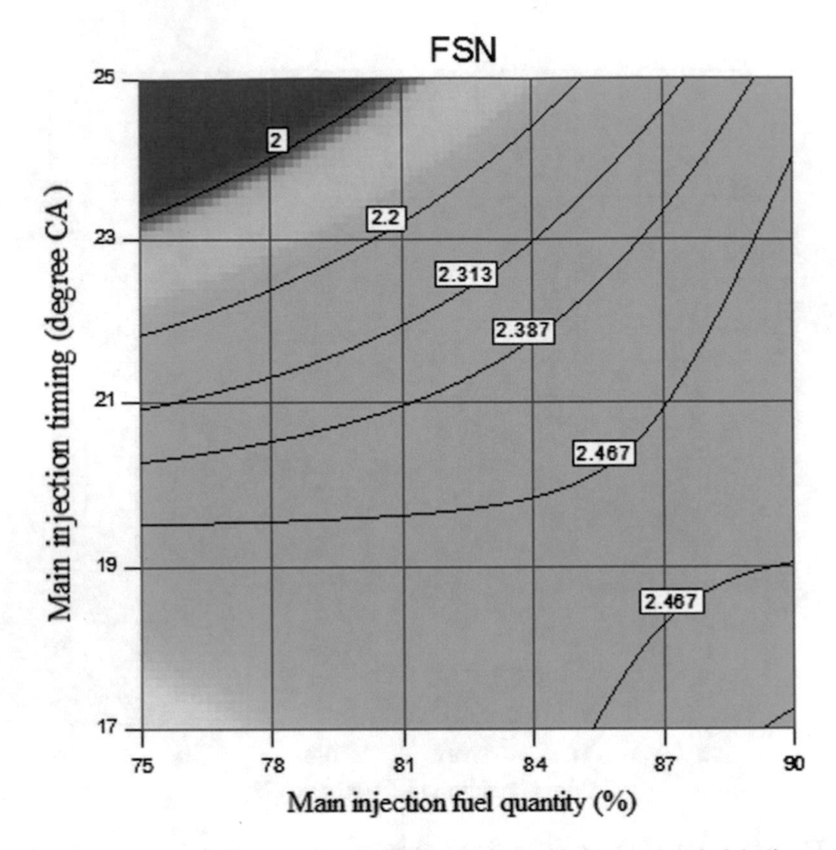

Fig. 13 Contour plot of FSN emission for different injection timings and main injection

delay period and it is due to low temperature and pressure inside the cylinder. The minimum ignition delay period of 11 °CA was obtained for B20, B40, B80, and B100 fuel. The maximum delay period of 16, 15.5, 15, and 15 °CA was found for the biodiesel blends with the conditions of 90%-5-10%-25, 80%-10-20%-25, 90%-20-10%-25, and 80%-10-20%-25.

Figure 18 represents the variation of brake specific fuel consumption (BSFC) with respect to exhaust gas recirculation. The fuel consumption slightly increased with increase in exhaust gas recirculation rate from 0 to 30%. This is due to reduction in oxygen content, increase in specific heat capacity of flue gases and lower calorific value of biodiesel [17]. From Fig. 18, it noticed that the maximum BSFC of 0.2506 kg/kWh was obtained for B40 blend at full-load condition with 30% EGR. The minimum BSFC of 0.2456 was obtained for B20 blend with 5% EGR. The maximum fuel consumption occurs for biodiesel blends due to less calorific value.

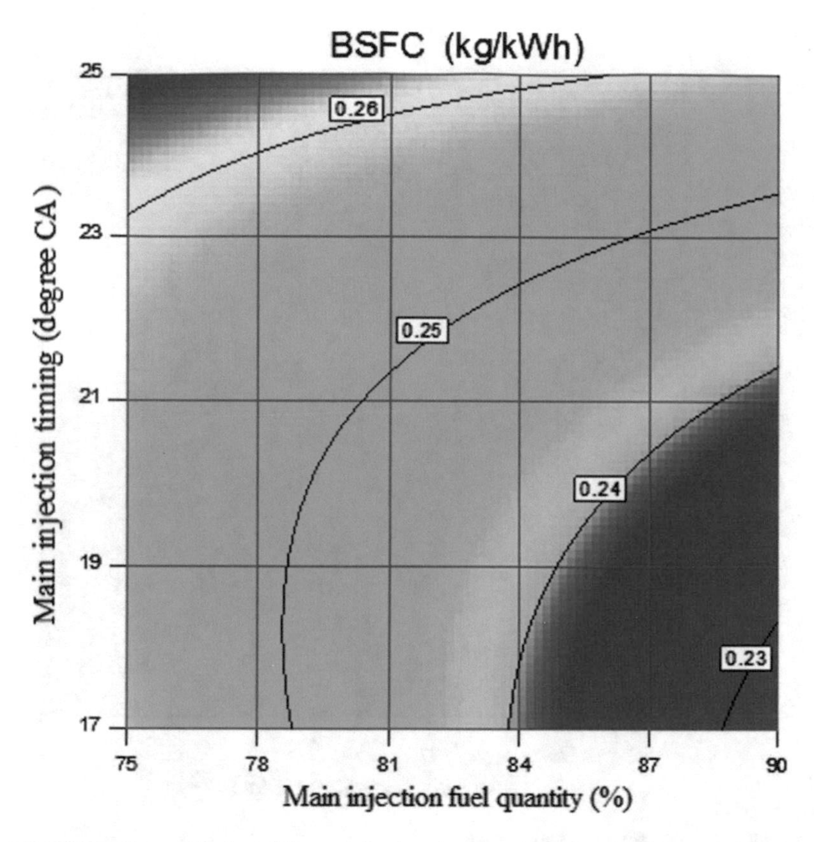

Fig. 14 BSFC of tested fuel at different injection timings

Figure 19 represents the variation of brake thermal efficiency (BTE) with respect to percentage of exhaust gas recirculation. From Fig. 19, it is observed that the brake thermal efficiency decreased with increase in EGR percentage. This is due to lesser oxygen concentration and lower in-cylinder pressure, temperature that leads to incomplete combustion [17, 37]. Maximum BTE of 35.46% was obtained for B20 blend with 5% EGR at full-load condition. Minimum BTE was obtained for B40 blend at 30% EGR, which is 2% lesser than B20 blends with 5% EGR and 3% lesser than diesel. Brake thermal efficiency mainly depends on caloric value, latent heat of evaporation, and lower viscosity. Biodiesel blends obtained lesser BTE with EGR due to dilution effect, thermal effect and lower caloric value and lower content of oxygen in fuels [20, 38].

Figure 20 represents the variation of carbon monoxide (CO) concentration with respect to percentage of exhaust gas recirculation. The CO emission increased with increase in exhaust gas recirculation percentage. This is due to less availability of oxygen in cylinder. By increasing the amount of exhaust gas recirculation, the amount of oxygen gets reduced. Due to lesser oxygen density, the fuels goes partial pyrolysis and produces more CO and UBHC [17]. For biodiesel blends, the CO

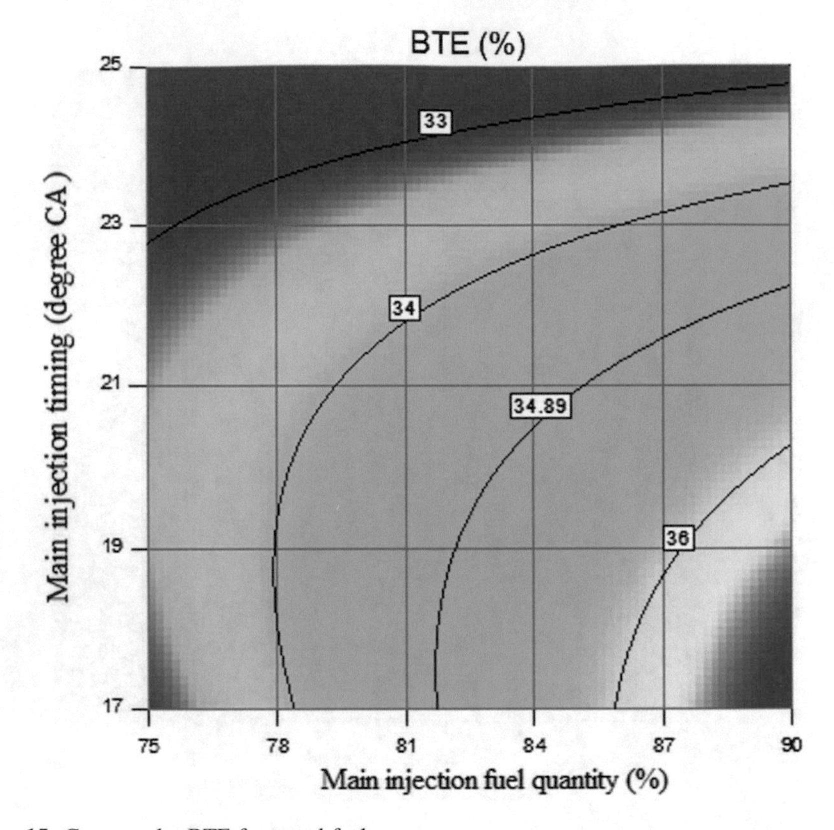

Fig. 15 Contour plot BTE for tested fuel

emissions are less compared to diesel. CO emission for biodiesel at 15% EGR is almost equal to diesel. This is due to more availability of oxygen in biodiesel. Up to 15% EGR, the effect on CO emission is considerable. Thermal effect reduces the flame temperature by quenching effect, and its controlled combustion process causes more CO emission. [38].

Figure 21 represents the variation of unburned hydrocarbon (UBHC) concentration with respect to percentage of exhaust gas recirculation. From Fig. 21, it is noticed that the UBHC emission increased with increase in exhaust gas recirculation percentage. At full load with 30% EGR, B20 and B40 blends have the UBHC emission of 23 and 19 ppm, respectively. B40 has obtained the minimum emission which is 42% lesser than diesel and 17% lesser than the B20 blends. This is due to more oxygen availability in biodiesel blends. Less availability of oxygen and partial combustion of fuel lead to more UBHC emission [17, 39]. For biodiesel blends, the UBHC emissions are less compared to diesel. UBHC emission for biodiesel at 10% EGR is almost equal to diesel fuel, and this is due to more availability of oxygen in biodiesel.

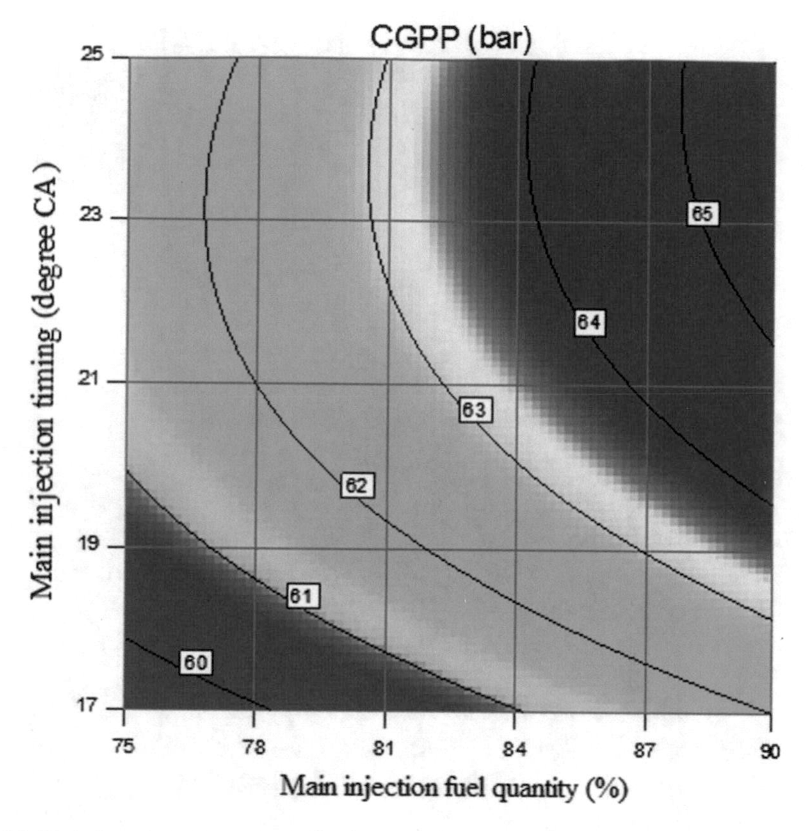

Fig. 16 Variation of CGPP for tested fuel

Figure 22 represents the variation of nitric oxide emissions (NO) with respect to percentage of exhaust gas recirculation. From Fig. 22, it is observed that the NO emission decreased with increase in EGR percentage. The higher percentage of EGR rate dilute the fresh intake air charge and reduced the oxygen concentration which suppressed the in-cylinder pressure and temperature resulted in lower NO emission formation [39]. Recirculation of exhaust gas increased the CO_2 concentration and heating capacity of charge mixture, which lead to decrease the gas temperature inside the cylinder. Because higher specific heat capacity of CO_2 consumed more energy resulted in lesser in-cylinder temperature [17]. By recirculation of 15% EGR, the NO emission decreased by 51% for diesel and 55% for B20 and B40 blends. At full load with 30% EGR, B20 and B40 blends have the NO emission of 356 ppm and 301 ppm, respectively. B40 blend obtained the minimum NO emission compared to B20. This is 39% lesser than diesel and 16% lesser than the B20 blends. Biodiesel blends have obtained the minimum NO emission due to lower iodine value [26, 27].

Figure 23 represents the variation of filter smoke number (FSN) with respect to percentage of exhaust gas recirculation. From Fig. 23, it is noticed that the filter

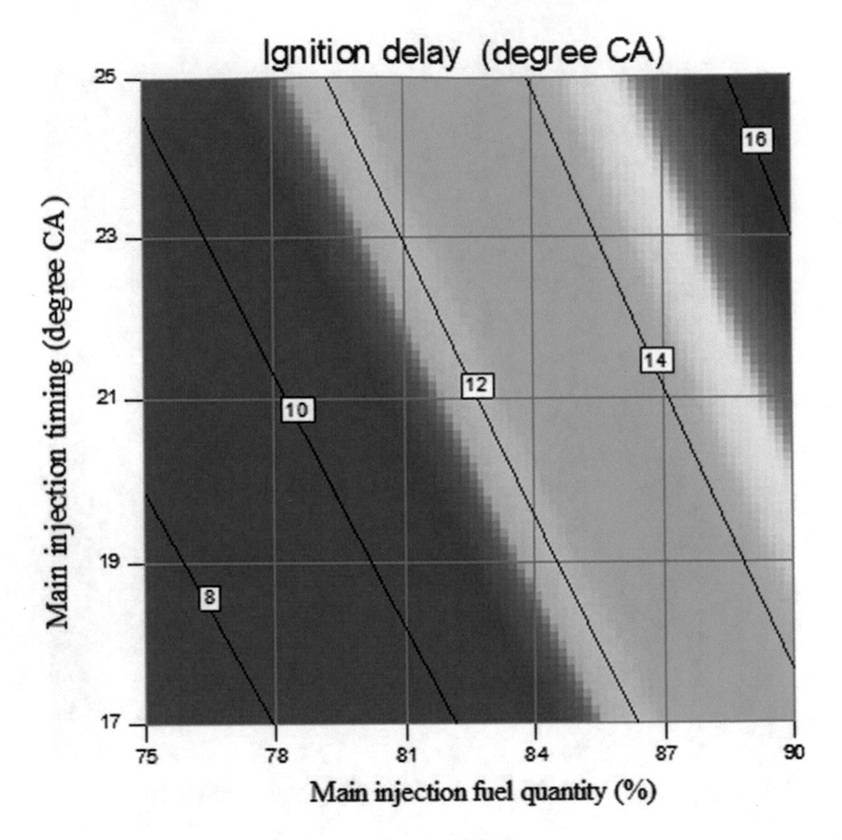

Fig. 17 Variation of ignition delay period for tested fuel

smoke number increased with increase in EGR percentage due to lesser availability of oxygen and dilution of intake gas as observed by Rajesh kumar et at. [37]. At 30% EGR, B20 and B40 blends have obtained the maximum FSN emission of 2.47 and 2.20. The B40 blend has the minimum FSN value than the B20 and diesel fuel, which is 35% lesser than the diesel and 10% lesser than the B20 blend. At 15% EGR at full-load condition, B20 and B40 have the FSN emission of 1.72 and 1.50, respectively. B40 blend has the minimum FSN value than the B20 and diesel fuel, which is 48% lesser than diesel and 11% lesser than B20. About 15% EGR is the maximum effect on reduction of FSN value compared to 30% EGR. The increase in FSN value is due to incomplete combustion caused by lesser oxygen availability in cylinder [39].

Figure 24 illustrates the variation of in-cylinder gas peak pressure (CGPP) with respect to percentage of exhaust gas recirculation. From Fig. 24, it noticed that the peak cylinder pressure decreased with increase in EGR percentage. Peak cylinder pressure mainly depends on premixed combustion with advanced injection timing. Advanced injection timing increased the ignition delay period. Due to better mixing of fuel with air and more accumulation of fuel leads to complete combustion

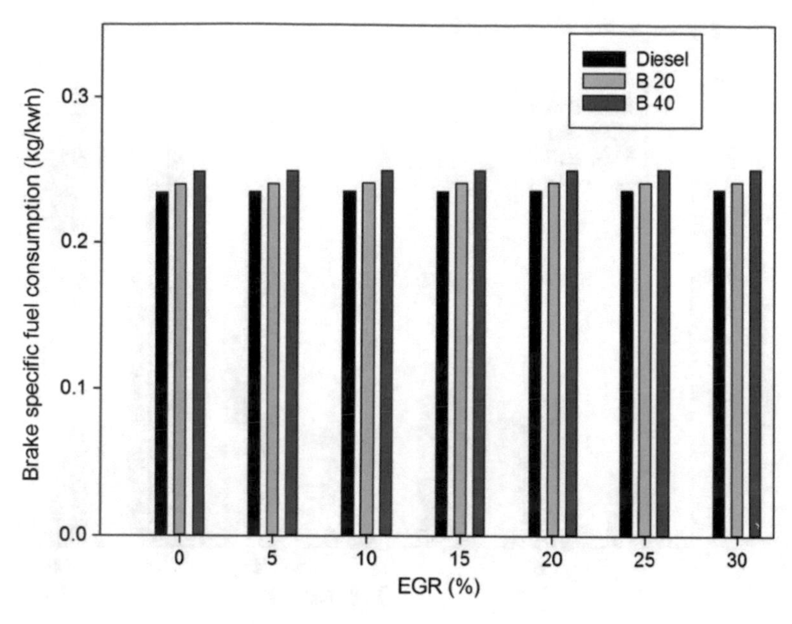

Fig. 18 Effect of EGR on engine BSFC

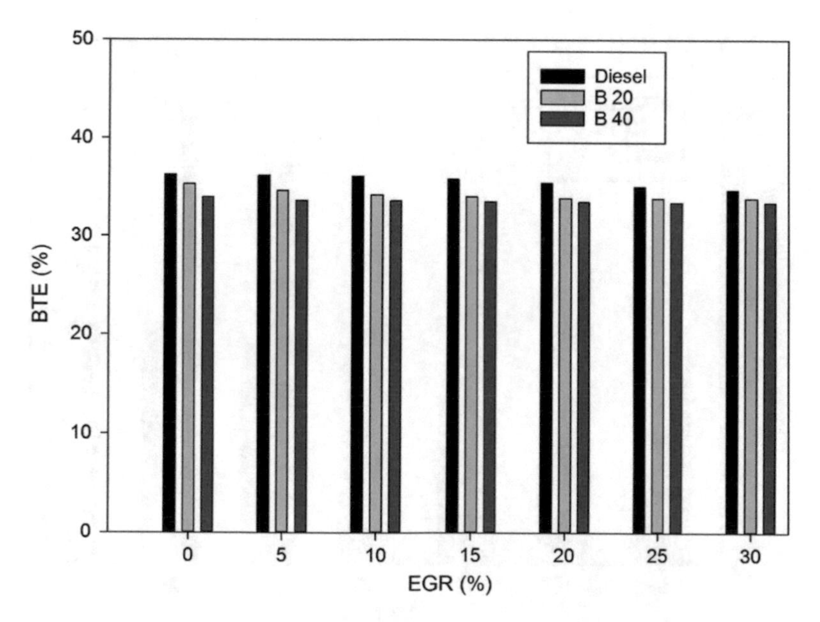

Fig. 19 Comparison of BTE for different fuels at different EGR rates

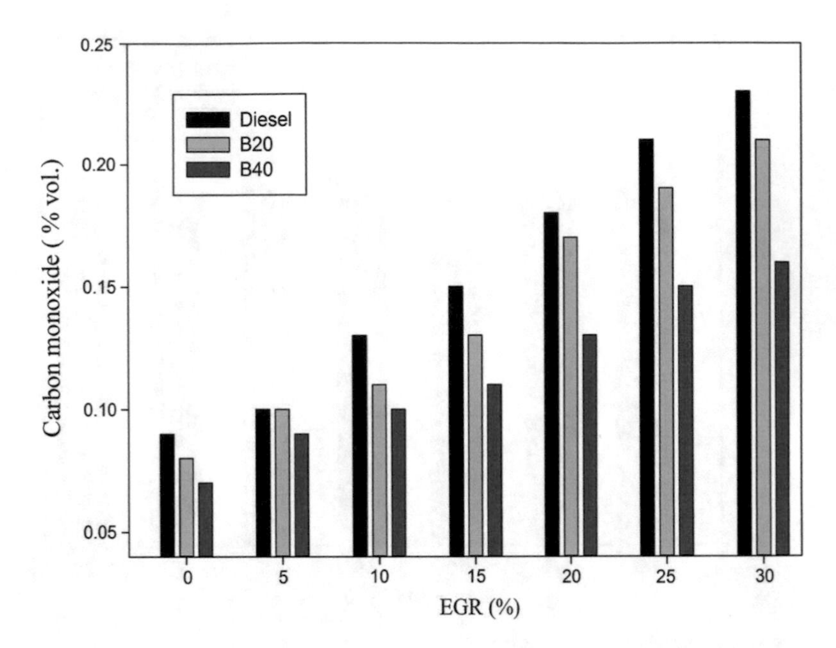

Fig. 20 Comparison of CO emission at different EGR rates

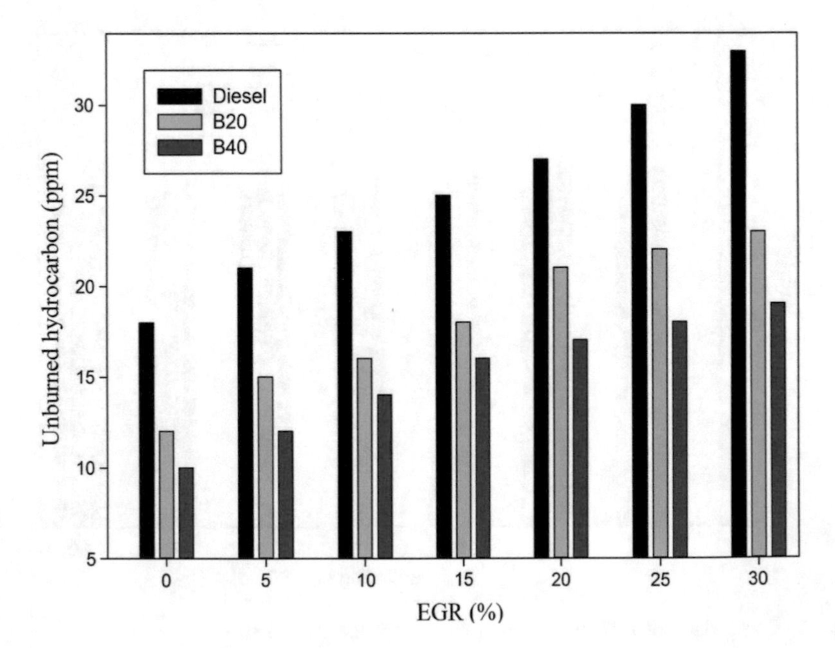

Fig. 21 Variation of UBHC emission with EGR

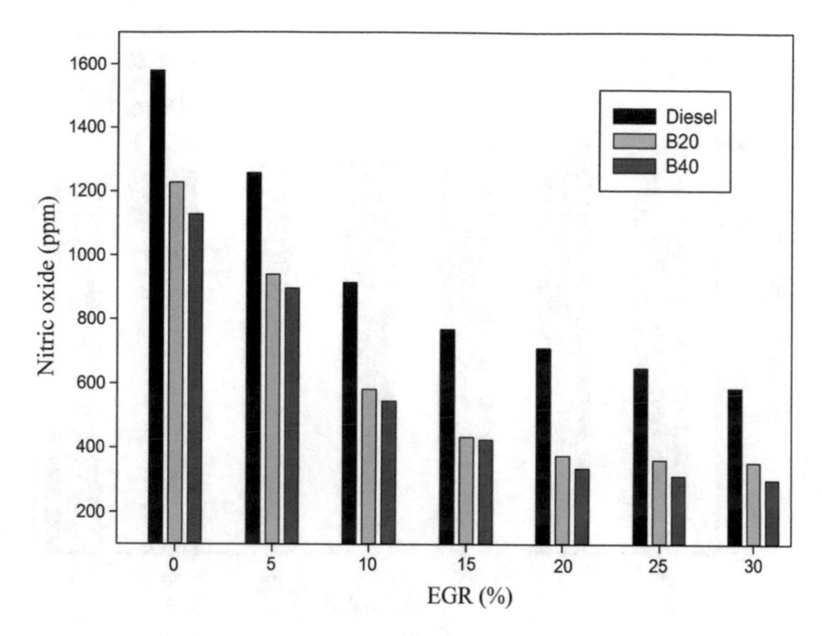

Fig. 22 Comparison of NO emission at different EGR rates

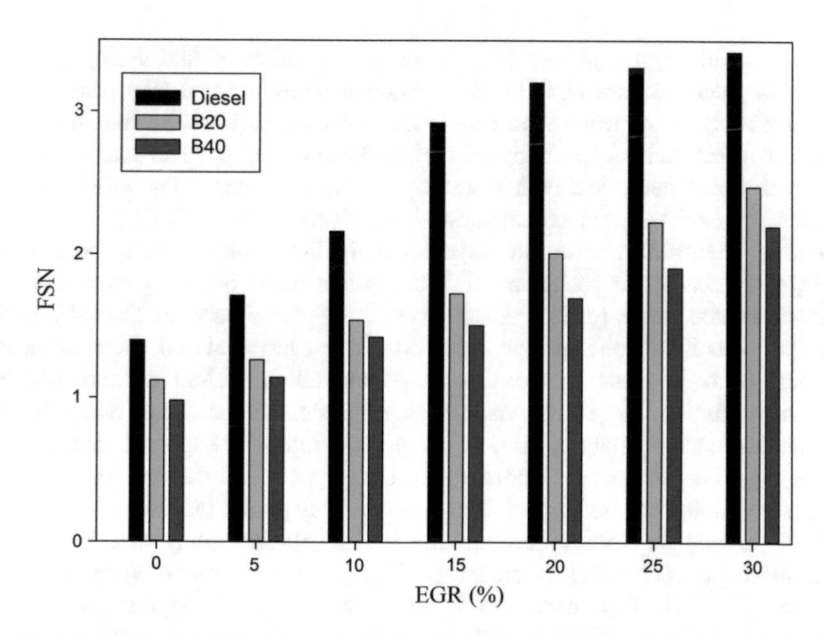

Fig. 23 Comparison of FSN emission at different EGR rates

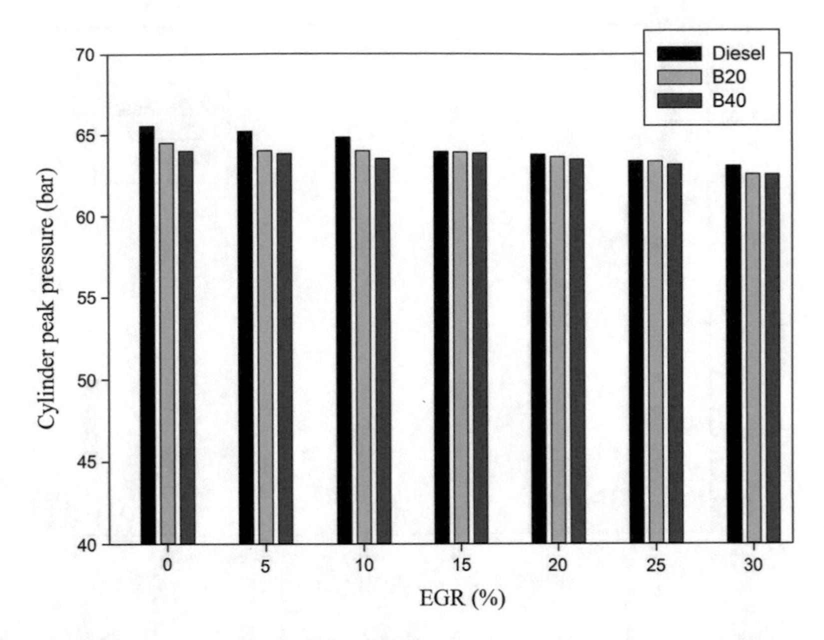

Fig. 24 Variation of CGPP at different EGR rates

associated with high peak pressure. More accumulation of fuel during premixed combustion stage causes rapid burning which leads to high cylinder peak pressure [37]. Less oxygen concentration results in poor combustion. The maximum peak pressure of 65.5 bar occurs for diesel fuel at 0% EGR supply. The peak pressure of biodiesel blends decreased with increase in EGR percentage. The minimum peak pressure of 62.5 bar was obtained for B40 blends with 30% EGR at full-load condition. Figure 25 shows the variation of ignition delay with respect to percentage of exhaust gas recirculation. The ignition delay period is measured from heat release rate curve. Ignition delay period is the period between start of injection and start of combustion. Ignition delay consists of physical and chemical delays. Physical delay depends on atomization, vaporization, fuel-air mixture and also depends on fuel properties like viscosity and density. The chemical delay depends on combustion temperature, pressure, and swirl ratio. The ignition delay period decreased with increase in biodiesel percentage in blend due to better ignition quality and higher cetane number. The ignition delay period increased with increase in EGR percentage. Combustion duration is also getting slightly reduced with rise in exhaust gas recirculation percentage. This is due to less oxygen density in cylinder [37, 39]. The maximum ignition delay of 16.2 degrees crank angle occurred for B20 at 30% EGR. This is one degree less than the diesel at full-load condition.

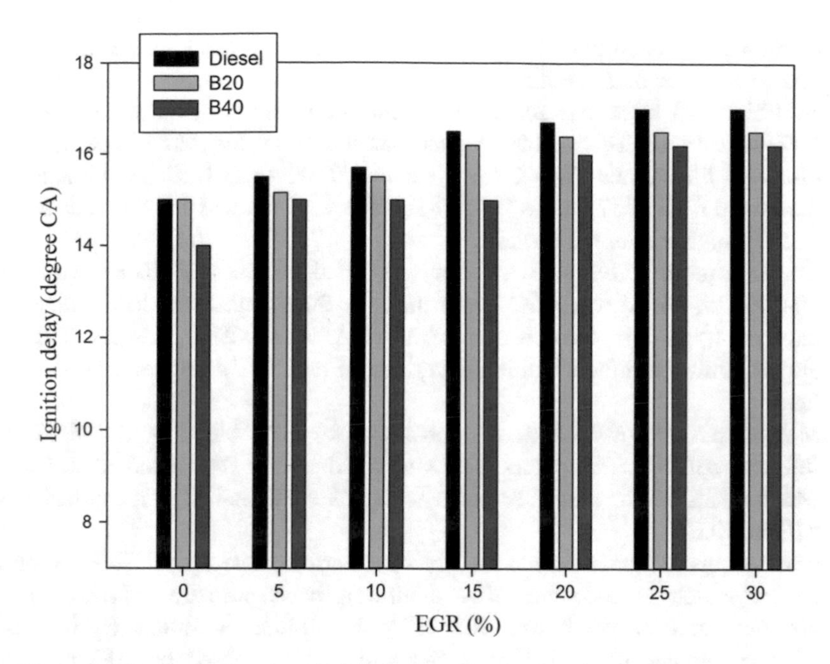

Fig. 25 Variation of ignition delay with different EGR rates

4 Conclusions

This research work was carried out using sodium methoxide for biodiesel extraction from WFO. Performance and emission and combustion characteristics of CRDI diesel engine test were carried out based on the D2 cycle. Response surface methodology-based central composite design was employed for optimization of engine operating variables. Exhaust gas recirculation (EGR) technique has been implemented for the simultaneous reduction of NO and soot emissions. The results show that biodiesel-diesel blends have the lowest emission of CO, UBHC, and FSN compared to diesel fuel. Up to 60% biodiesel blend has a prominent alternative fuel properties for next generation without any engine modification. Higher injection pressure and split injection strategies are advised for higher viscosity fuel for better atomization and better performance. Minimum NO emission was achieved through EGR technique. The main conclusions are summarized below:

- The lowest NO emission of 582 ppm was obtained for B100 fuel at full load with the condition of 75%-10-25%-17, which is 59% lesser than diesel. The lowest CO emission of 0.01% vol. was obtained in B80 blend with condition of 75%-10-25%. This is 0.02% vol. lower than the diesel fuel. B80 has obtained the lowest UBHC of 10 ppm with case of 75%-5-25%-25, which was 65% lesser than the diesel. The lowest FSN emission of 1.729 was obtained for B80 blend with the condition of 75%-5-25%-25, which was 6% higher than the B100 and 51% lower than the diesel.

- B20 biodiesel blend has achieved the minimum BSFC of 0.230 kg/kWh with 80%-10-20%-17. This was 5% higher than the diesel fuel, 20% lower than B80 biodiesel blend, and 28% lower than the B100 fuel. B20 has obtained the maximum BTE of 37.46% with 75%-10-25%-23. This was 1% lower than diesel and 6% higher than the diesel.
- The maximum CGPP of 63.83 bar was obtained for B20 blend with 80%-10-20%-25, which was 1.5% lower than the diesel fuel. B20 has obtained the maximum ignition delay period of 16 °CA with 90%-5-10%-25. This was almost similar to diesel ignition delay period and 5 °CA higher than the B100 fuel.
- Maximum BTE of 35.28% was obtained for B20 blend with 5% EGR at full-load condition. Minimum BTE was obtained for B40 blend at 30% EGR, which is 2% lesser than B20 blend with 5% EGR and 3% lesser than diesel without EGR.
- Exhaust gas recirculation has an effect on performance up to 5% EGR on the engine which is allowable. The minimum peak pressure of 62.5 bar was obtained for B40 blends with 30% EGR at full-load condition. By increasing EGR percentage up to 15%, the NO emission decreased by 50% and other emissions are increased. The CO and HC emissions for biodiesel at 10% EGR are considerable, and those are almost equal to diesel emission without EGR.

References

1. Tan KT, Lee KT, Mohamed AR (2011) Potential of waste palm cooking oil for catalyst free biodiesel production. Energy 36:2085–2088
2. Zhang Y, Dube MA, McLean DD, Kates M (2003) Biodiesel production from waste cooking oil, Economic assessment and sensitivity analysis. Biores Technol 90:229–240
3. Gui MM, Lee KT, Bhatia S (2008) Feasibility of edible oil vs. non edible oil vs. waste cooking oil as biodiesel feedstock. Energy 33:1646–1653
4. Kulkarni MG, Dalai AK (2006) Waste cooking oils an economical source for biodiesel. Ind Eng Chem Res 45:2901–2913
5. Chen Y, Xiao B, Chang J, Wang X (2009) Synthesis of biodiesel from waste cooking oil using immobilized lipase in a fixed bed reactor. Energy Convers Manag 50:668–673
6. Saka S, Kusdiana D (2001) Biodiesel fuel from rapeseed oil as preparation in supercritical methanol. Fuel 80:225–231
7. Freedman BE, Pryde H, Mounts TL (1984) Variables affecting the yields of fatty esters from transesterified vegetable oils. JAOCS 61:1638–1643
8. Vicente G (1998) Application of the factorial design of experiments and response surface methodology to optimize biodiesel production. Ind Crop Prod. 8:29–35
9. Hamamre ZA, Yamin J (2014) Parametric study of the alkali catalyzed transesterification of waste frying oil for Biodiesel production. Energy Convers Manag 79:246–254
10. Imtenan S, AshrafurRahman SM, Masjuki HH, Varman M, Kalam MA (2015) Effect of dynamic injection pressure on performance, emission and combustion characteristics of a compression ignition engine. Renew Sustain Energy Rev 52:1205–1211

11. Agarwal AK, Dhar A, Srivastava DK, Maurya RK, Akhilendra Singh AP (2013) Effect of fuel injection pressure on diesel particulate size and number distribution in a CRDI single cylinder research engine. Fuel 107:84–89
12. Yehliu K, Boehman AL, Armas O (2010) Emissions from different alternative diesel fuels operating with single and split fuel injection. Fuel 89:423–437
13. Qi D, Leick M, Liu Y, Lee CF (2011) Effect of EGR and injection timing on combustion and emission characteristics of split injection strategy DI-diesel engine fuelled with biodiesel. Fuel 90:1884–1891
14. Choi CY, Reitz RD (1999) An experimental study on the effects of oxygenated fuel blends and multiple injection strategies on DI diesel engine emissions. Fuel 78:1303–1317
15. Park SH, Yoon SH, Lee CS (2011) Effects of multiple-injection strategies on overall spray behavior, combustion and emissions reduction characteristics of biodiesel fuel. Appl Energy 88:88–98
16. Rajesh Kumar B, Saravanan S, Rana D, Anish V, Nagendran A (2016) Combined effect of injection timing and exhaust gas recirculation on performance and emissions of a DI diesel engine fuelled with next generation advanced biofuel- diesel blends using response surface methodology. Energy Convers Manag 123:470–486
17. Yasina MHM, Mamata R, Yusopa AF, Daing Idrisa DMN, Yusaf T, Rasulc M, Najafid G (2017) Study of a diesel engine performance with exhaust gas recirculation (EGR) system fuelled with palm biodiesel. Energy Procedia 110:26–31
18. Tornatore C, Marchitto L, Mazzei A, Valentino G, Corcione FE, Merola SS (2012) Effect of butanol blend on in-cylinder combustion process part 2: compression ignition engine. J Kones Powertrain Transp 18:473–483
19. Valentino G, Corcione FE, Iannuzzi SE (2012) Effects of gasoline-diesel and n-butanol-diesel blends on performance and emissions of an automotive direct-injection diesel engine. Int J Engine Res 13:199–215
20. Can Ozer, Ozturk Erkan, Solmaz Hamit, Aksoy Fatih, Cinar Can, Yucesu H Serdar (2016) Combined effects of soybean biodiesel fuel addition and EGR application on the combustion and exhaust emissions in a diesel engine. Appl Therm Eng 95:115–124
21. Gerpen JV (2005) Biodiesel processing and production. Fuel Process Technol 86:1097–1107
22. Berrios M, Skelton RL (2008) Comparison of purification methods for biodiesel. Chem Eng J 144:459–465
23. Kannan GR, Karvembu R, Anand R (2011) Effect of metal based additives on performance emission and combustion characteristics of diesel engine fuelled with biodiesel. Appl Energy 88:3694–3703
24. Ehleskog R (2006) Experimental and numerical investigation of split injections at low load in an HCCI diesel engine equipped with a piezo injector. SAE Paper No. 2006-01-3433
25. Kannan GR, Anand R (2012) Effect of injection pressure and injection timing on DI diesel engine fuelled with biodiesel from waste cooking oil. Biomass Bioenerg 46:343–352
26. McCormick R, Tennant C, Hayes R, Black S (2005) Regulated emissions from biodiesel tested in heavy-duty engines meeting 2004 emission standards. SAE Technical Paper, 2005-01-2200
27. Babu D, Anand R (2017) Effect of biodiesel-diesel-n-pentanol and biodiesel-diesel-n-hexanol blends on diesel engine emission and combustion characteristics. Energy 133:761–776
28. Maheswari N, Balaji C, Ramesh A (2011) A nonlinear regression based multi-objective optimization of parameters based on experimental data from an IC engine fuelled with biodiesel blends. Biomass Bioenerg 35:2171–2183
29. Sayin C, Gumus M, Canakci M (2012) Effect of fuel injection pressure on the injection, combustion and performance characteristics of a DI diesel engine fuelled with canola oil methyl esters-diesel fuel blends. Biomass Bioenerg 46:435–446
30. Gnanasekaran S, Ilangkumaran M (2016) Influence of injection timing on performance, emission and combustion characteristics of a DI diesel engine running on fish oil biodiesel. Energy 116:1218–1229

31. Han D, Li K, Duan Y, Lin H, Huang Z (2017) Numerical study on fuel physical effects on the split injection processes on a common rail injection system. Energy Convers Manag 134:47–58
32. Fayad MA, Tsolakis A, Fernández-Rodríguez D, Herreros JM, Martos FJ, Lapuerta M (2017) Manipulating modern diesel engine particulate emission characteristics through butanol fuel blending and fuel injection strategies for efficient diesel oxidation catalysts. Appl Energy 190:490–500
33. Zhu R, Miao H, Wang X, Huang Z (2013) Effects of fuel constituents and injection timing on combustion and emission characteristics of a compression-ignition engine fuelled with diesel-DMM blends. In: Proceedings of the Combustion Institute, vol 34, pp 3013–3020
34. Agarwal AK, Srivastava DK, Dhar A, Maurya RK, Shukla PC, Singh AP (2013) Effect of fuel injection timing and pressure on combustion, emissions and performance characteristics of a single cylinder diesel engine. Fuel 111:374–383
35. Puhan S, Jegan R, Balasubbramanian K, Nagarajan G (2009) Effect of injection pressure on performance, emission and combustion characteristics of high linolenic linseed oil methyl ester in a DI diesel engine. Renew Energy 34:1227–1233
36. Atadashi IM, Aroua MK, Abdul Aziz A (2010) High quality biodiesel and its diesel engine application: A review. Renew Sustain Energy Rev 14:1999–2008
37. Rajesh Kumar B, Saravanan S, Rana D, Anish V, Nagendran A (2016) Effect of a sustainable biofuel—n-octanol—on the combustion, performance and emissions of a DI diesel engine under naturally aspirated and exhaust gas recirculation (EGR) modes. Energy Convers Manag 118:275–286
38. Poures MVD, Sathiyagnanam AP, Rana, Rajesh Kumar DB, Saravanan S (2017) Hexanol as a sustainable biofuel in DI diesel engines and its effect on combustion and emissions under the influence of injection timing and exhaust gas recirculation (EGR). Appl Thermal Eng 113:1505–1513
39. Rajesh kumar B, Saravanan S (2015) Effect of exhaust gas recirculation (EGR) on performance and emissions of a constant speed DI diesel engine fueled with pentanol/diesel blends. Fuel 160:217–226

Biodiesel Soot Characteristics

Joonsik Hwang, Choongsik Bae, Chetankumar Patel, Tarun Gupta and Avinash Kumar Agarwal

Abstract Diesel engines are widely utilized as a power source in various applications, such as passenger and commercial vehicles, electricity generation, marine transportation, and construction equipment. Despite several advantages for instance, low operating cost, high durability, and high power-to-weight ratio, diesel engines have the inherent disadvantage of high NO_x and soot emissions. In terms of environmental concerns, oxygenated fuels such as biodiesel have been considered as a promising alternative fuel for diesel engines. It is well known that the oxygen content of biodiesel suppresses the formation of soot precursors in diffusion flames. Many researchers are now trying to examine the soot formation with biodiesels. In this chapter, the effect of biodiesel on soot particles in compression ignition engines is discussed. The morphological characteristics of soot particles were analyzed with transmission electron microscope (TEM). The chemical composition was determined by the elemental analysis (EA) and thermogravimetric analysis (TGA). From the previous studies, it emerged that soot particles from biodiesel were composed of smaller primary particles compared to those in case of diesel. The nano-structure analysis of biodiesel soot particles showed that biodiesel origin soot is preferable

---

J. Hwang · C. Bae (✉)
Department of Mechanical Engineering, Korea Advanced Institute of Science and Technology, 291 Daehak-Ro, Yuseong-Gu, Daejeon 305-701, Republic of Korea
e-mail: csbae@kaist.ac.kr

J. Hwang
e-mail: beluga0410@gmail.com

C. Patel · A. K. Agarwal
Department of Mechanical Engineering, Indian Institute of Technology Kanpur, Kanpur, India
e-mail: chetanpatel.iitk@gmail.com

A. K. Agarwal
e-mail: akag@iitk.ac.in

T. Gupta
Department of Civil Engineering, Indian Institute of Technology Kanpur, Kanpur, India
e-mail: tarun@iitk.ac.in

N. Sharma et al. (eds.), *Air Pollution and Control*, Energy, Environment, and Sustainability, https://doi.org/10.1007/978-981-10-7185-0_3

for oxidation due to its unstable graphene structure. On the other hand, biodiesel soot particles were composed of lesser carbonaceous materials. They contained lower carbon-containing species but higher hydrogen, oxygen, and nitrogen in the soot.

Keywords Biodiesel · Soot particles · Primary particles · Transmission electron microscopy (TEM) · Elemental analysis (EA) · Thermogravimetric analysis (TGA)

1 Introduction

1.1 Soot Formation in Compression Ignition (CI) Engine

The soot formation process can be described into six different steps: pyrolysis, nucleation, surface growth, coalescence, agglomeration, and oxidation (Fig. 1). At first, under high-temperature conditions, fuel molecules break into smaller simpler molecules without undergoing oxidation. This process depends on the level of in-cylinder temperature and concentration of air–fuel mixture. In this step, soot precursors and building blocks are produced and they are also consumed by oxidation. Thus, the competition between pyrolysis and oxidation determines total amount of soot precursors. Pyrolysis can be promoted by hydrogen (H), oxygen (O), and hydroxyl (OH) radicals [2]. Unsaturated hydrocarbons, polycyclic aromatic hydrocarbons (PAHs) are produced by pyrolysis of many different fuels.

Secondly, in the nucleation process, particles are produced from gaseous reactants. The “nuclei” in Fig. 1 represents the smallest solid particles having a diameter of 1.5–2 nm. This process occurs at a temperature range of 1300–1600 K and the nuclei formed are not directly related to the soot particle mass but it affects the surface growth [2]. The formation of large PAHs and consequently nucleation can happen through three major processes; (1) formation of chain and ring structures, for example, combination of acetylene molecules to a benzene ring, (2) dehydrogenation of aromatic species and formation of polycyclic molecules, finally (3) breakup and re-cyclization of rings at higher temperatures [3]. Third, the surface growth is the process, in which mass is added to the soot nuclei. The boundary between nucleation and surface growth processes is not very clear so they may be considered as simultaneous processes. The temperature of the soot surface remains hot enough to accept gaseous hydrocarbons hence the soot nuclei mass increases.

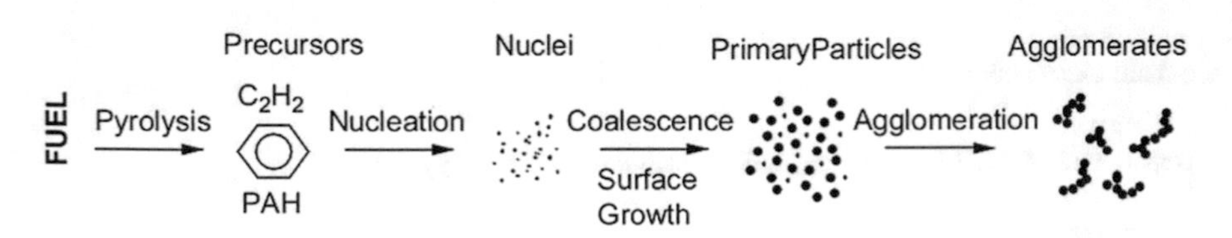

Fig. 1 Process of soot formation [1]

This process is continued until the nuclei moves to lower temperature region, where the concentration of hydrocarbons is low. The residence time in surface growth stage significantly affects the soot volume fraction. Fourth, the coalescence and agglomeration processes take place. During the coalescence process, total soot mass remains constant, however, the soot particle number decreases. A number of particles stick together and form a larger chain-like structure in the agglomeration process. Finally, oxidation process converts some of the hydrocarbons or carbon to combustion products. The oxidation process takes place in two ways. Oxygen is attached to the surface, and then it is desorbed from the attached fuel component as reaction product. This process significantly affects soot particles above 800 K temperature [4]. OH is a dominant soot oxidizer under fuel-rich or stoichiometric conditions, while the O_2 is responsible for soot oxidization under fuel-lean conditions [5].

1.2 Conceptual Model of CI Engine Soot Formation Process

Diesel spray combustion is dominated by diffusion flames. Kosaka et al. proposed a conceptual model of soot formation process in a diesel engine [6] (Fig. 2).

At first, fuel spray from the injector nozzle penetrates high temperature region in the cylinder and starts to evaporate. In this situation, the soot precursors are mainly formed in the middle of spray flames, where the fuel–air mixture is very rich in

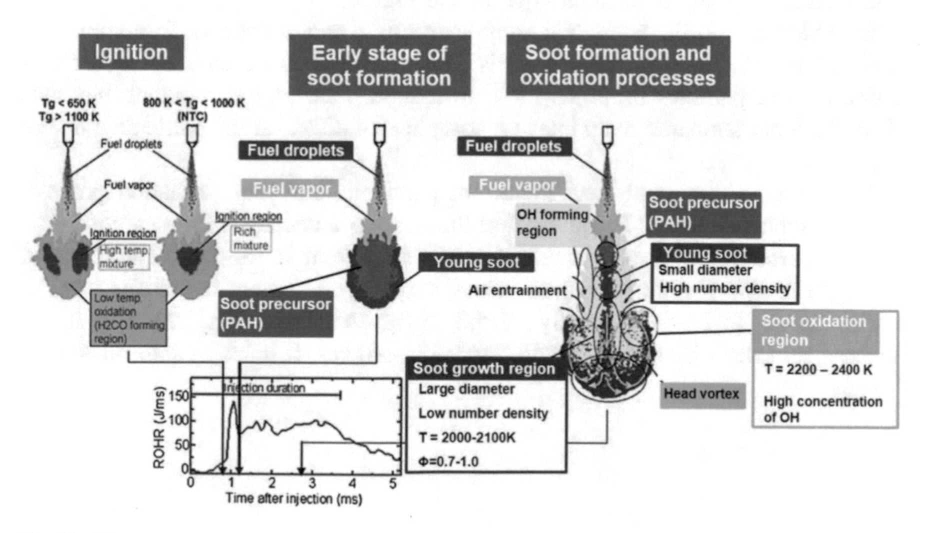

Fig. 2 Conceptual model of soot formation and oxidation processes [6]

conventional diesel combustion. The precursors are turned into particulate matter by surface growth as flame develops. As the particulate matter gets toward the end of flames, they enter into oxidation zone because of head vortex. A portion of soot particles may escape the in-flame oxidation zone and can be partially oxidized. These soot particles further oxidize in the hot environment of the combustion chamber (late-cycle oxidation) [7]. This late-cycle oxidation is terminated as the piston goes down toward the end of power cycle. Soot particles which survive the late-cycle oxidation are released into the exhaust flow as engine-out emissions. The characterization of engine-out soot particles is important because it determines the design criteria for after-treatment systems such as diesel particulate filters (DPF). Adverse impact of engine-out emissions to human beings and environment is affected by the soot characteristics. Therefore, the morphological and compositional characteristics of biodiesel soot particles should be identified and compared with the conventional fuels.

1.3 Effect of In-Cylinder Parameters on Soot Formation Process

In-cylinder temperature is the most significant parameter in soot formation and oxidation process. The soot formation and oxidation rates increase with temperature. The total soot production rate reduces by increasing in-cylinder temperature due to faster oxidation in premixed flames. On the other hand, soot formation increases monotonically with increasing in-cylinder temperature because of enhanced fuel pyrolysis in the non-premixed region.

Stoichiometry in the spray and combustion field can affect soot formation process. In general, oxygen addition leads to soot reduction because of higher oxidation of soot particles in all stages of formation. Fuel oxygen content may also affect the local temperature and have a complicated effect on the particulate matter formation.

Fuel composition is also an important parameter for soot formation process. From previous studies, it is known that the sooting tendency decreases with lowering of carbon content and increasing oxygen content in the fuel. The effect of hydrogen is also similar to that of increasing oxygen content in the fuel. Sulfur content in the fuel is not directly related to the soot formation, however, it can increase soot mass by attaching with the soot particles. It helps in soot nucleation though.

2 Experimental Method

2.1 Soot Characterization

Particulate matters (PM) from the exhaust were sampled on a cold glass surface. These surface were kept cold by low temperature coolant. At the same time, PM emissions were also collected on a lacey carbon transmission electron microscopy (TEM) grid for morphological investigations. Figure 3 shows a schematic of the single-cylinder engine setup.

The sampled particulates were utilized to perform *thermogravimetric analysis* (TGA) and elemental analysis (EA). The samples were kept in a constant temperature and humidity oven for 24 h at 278 K in order to remove moisture content before its inspection. TGA analyzer (TG 209 F3; Tarsus) was utilized for the TGA of soot samples under an air atmosphere, up to 1173 K. The temperature was raised at a ramp rate of 10 K per minute. Elemental composition of the sample was analyzed by elemental analysis. The most common form of elemental analysis was accomplished by combustion analysis. In this method, soot sample was burned in an excess of oxygen and various traps, which collected the combustion products such as carbon dioxide, water, sulfur oxide, and nitric oxide. These combustion products mass were used to calculate the composition of the unknown soot sample. In this research, soot particle components broken into carbon, hydrogen, nitrogen, sulfur, and oxygen (by weight) were classified by the elemental analyzer (EA 1110; FISONS). Soot morphology of the soot particles collected on TEM grid was visualized by the high-resolution transmission electron microscope (HRTEM) (JEM 3010; JEOL).

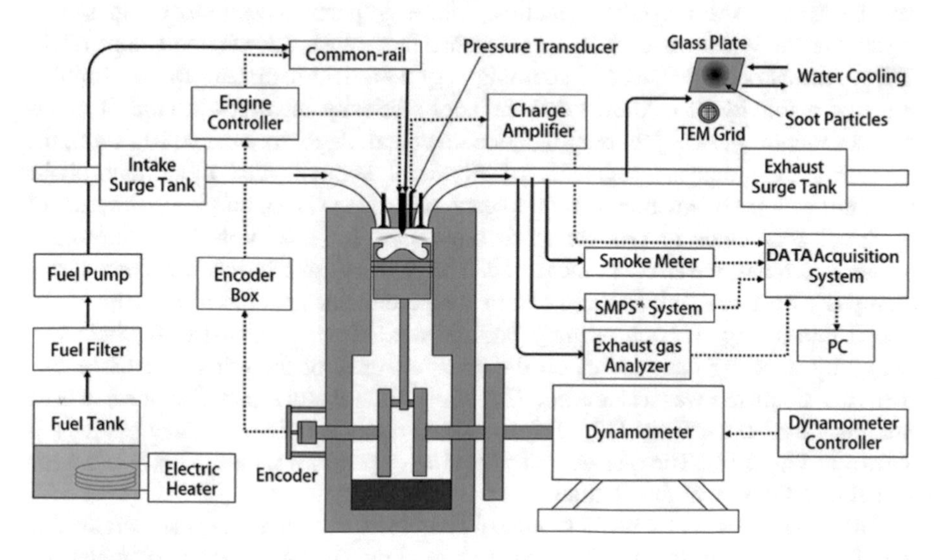

Fig. 3 A schematic diagram of engine test setup

2.2 Experimental Conditions

Soot inception experiment was performed in a 1 L single-cylinder CI engine. The bore and stroke were 100 and 125 mm. The engine was equipped with a common-rail direct injection (CRDI) fuel injection system. The injection parameters including injection pressure, injection timing, and injection quantity were controlled by an engine controller (ZB-9013; Zenobalti). Conventional diesel and biodiesel produced from waste cooking oil (WCO) were injected by a solenoid injector. Engine speed and load were kept at 1200 rpm and 0.75 MPa IMEP, respectively. The fuels were injected at −5 CAD a TDC at an injection pressure of 80 MPa.

3 Experimental Results

3.1 Morphological Characteristics

The TEM images for agglomerated soot particles are shown in Fig. 4 [8]. The TEM images were taken in six different regions. It can be seen that the soot agglomerates are composed of many circular shape primary particles showing chain or graph like structures. In general, soot particles consist of 100–500 primary particles. The TEM images were taken in six different regions. It is difficult to compare the macroscopic morphology between two different fuels from this figure; however, biodiesel soot particles are seen to be less compact than diesel, thus showing smaller fractal dimensions [9]. Inside primary particles, nano-size short individual graphene layers can be seen in the magnified pictures. These graphene layers are composed of crystallite stacks shown as dark lines because they block the electron beams of the TEM. Formation of this nano-structure is very important parameter for determining the soot reactivity. It is known that the soot reactivity increases in case of amorphous structure showing shorter graphene layer and bigger tortuosity [10]. From the literature, the graphene segments of biodiesel soot showed more amorphous structure with no orientation relative to each other [11]. The amorphous structure of biodiesel soot tends to provide more edge sites for oxidation. To characterize primary particles, the size was measured. The primary particle size was defined as a diameter of a circle, which can contain whole circular area described with white dotted line in Fig. 4. Each primary particle was selected manually. A circle was drawn by selecting three points on the circumference of the primary particle, and then the diameter was calculated. Generally, diesel soot particles have a size ranging from 20 to 50 nm [12]. The measured diameter of the primary particles is shown in Fig. 5 [8]. The diameter of the primary particles ranges from 9 to 54 nm and showed a normal distribution.

The mean diameter of WCO biodiesel and diesel soot particles is shown as 19.9 and 23.7 nm, respectively. The results regarding smaller primary diameter for

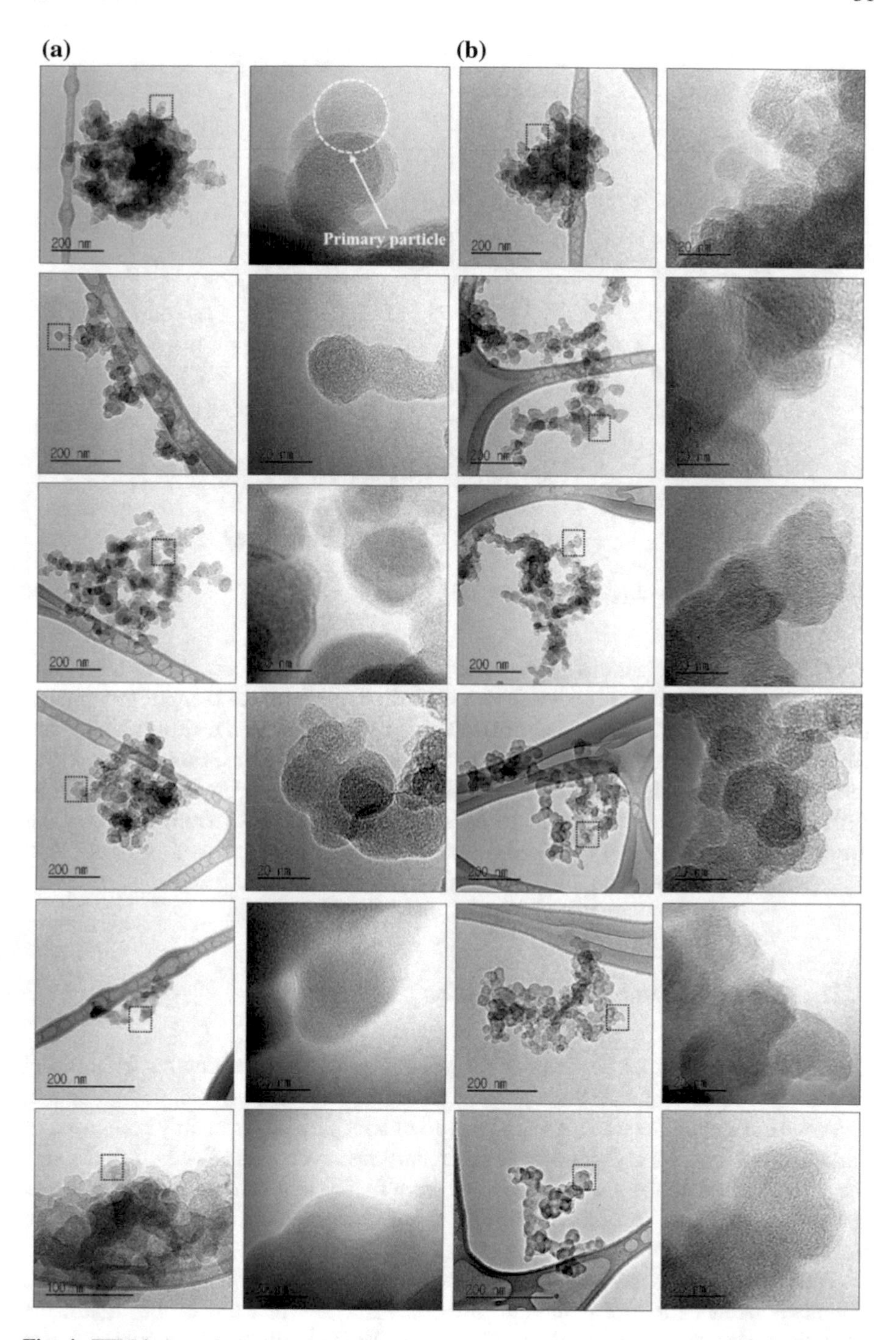

Fig. 4 TEM images for **a** waste cooking oil biodiesel and **b** conventional diesel [8]

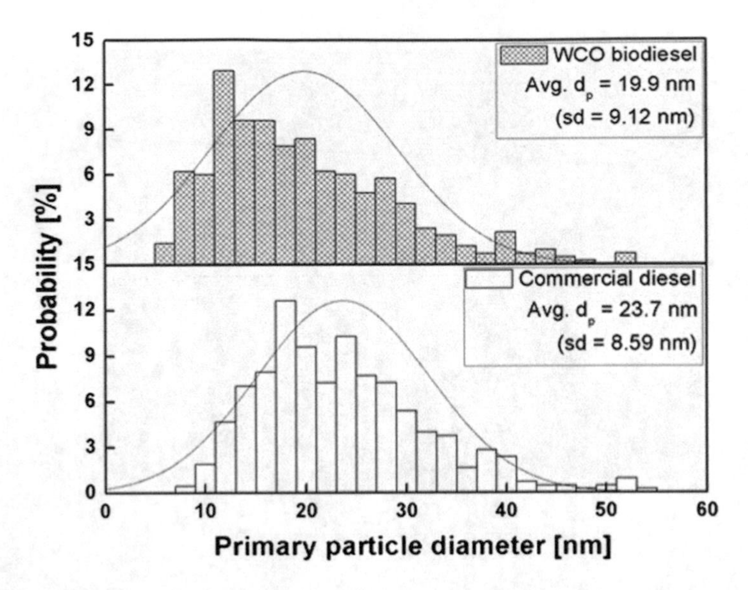

Fig. 5 Size distribution of primary particles [8]

WCO biodiesel is consistent with previous studies [13]. The smaller primary particle diameter from WCO biodiesel soot is attributed to surface oxygen groups that promoted faster oxidation of soot particles. The lower contents of aromatics and sulfur also resulted in lower surface growth rate of soot particles. From the previous results of Kook et al., primary particle diameter from non-aromatic fuels such as Fisher-Tropsch diesel was found to be smaller than the ones from conventional aromatics-rich test fuels [14]. The primary particle size increased at lower fuel injection pressure under high load condition [15].

3.2 Compositional Characteristics

Oxidation curve of soot particles is shown in Fig. 6. The particulates from diesel engines can be divided into three groups; solid fraction (carbon), sulfate particulate, and soluble organic fraction (SOF). The core of soot particles is mainly composed of carbonaceous species and condensed hydrocarbons layer is absorbed on to the surface. The hydrocarbons content originates from fuel or lubricant oil primarily. These species are the main constituent of SOF. It has been reported that the oxidation of volatile organic fraction (VOF) starts from ~473 K and ends ~693 K [16].

The unique graphene structure, chemical composition, and burning mode are dominant factors for soot oxidation process. As it can be seen in the figure, the oxidation process is terminated at 550 °C with WCO biodiesel. Meanwhile, diesel soot showed a termination temperature of 650 °C. It implies that the oxidation

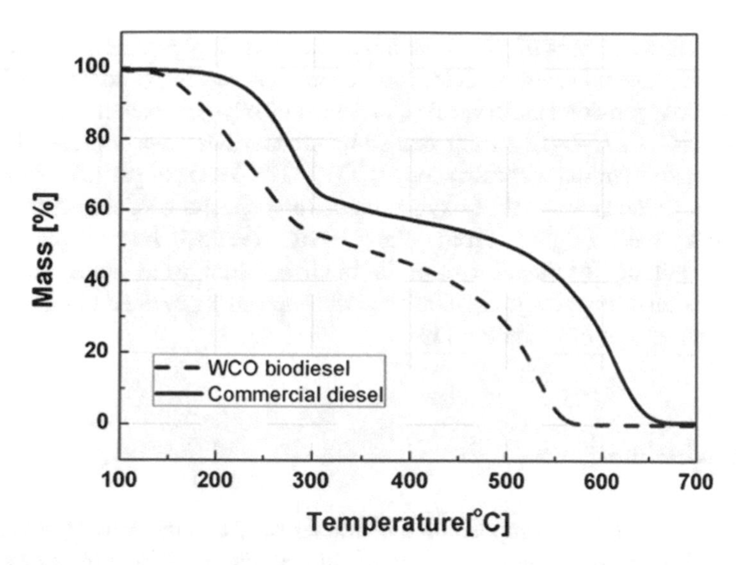

Fig. 6 Oxidation curve of soot particles [8]

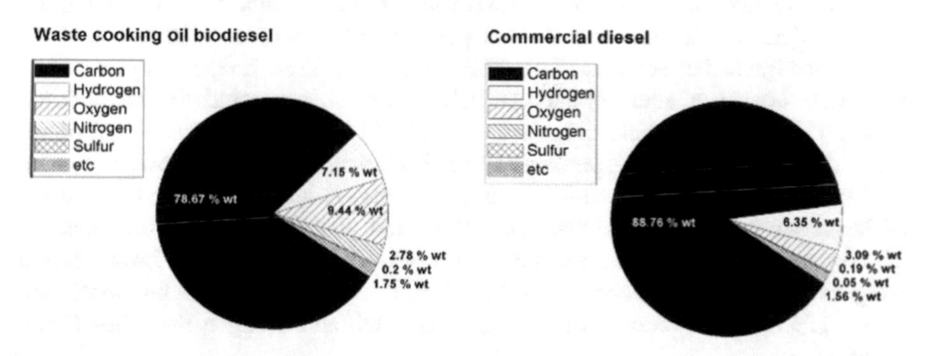

Fig. 7 Elemental analysis of soot from biodiesel and diesel [8]

process of biodiesel soot particles was much faster than that of diesel. The WCO biodiesel showed much larger mass reduction in the early stage of oxidation. The main reason is that the WCO biodiesel soot is composed of less species but more volatile substances. Another possible reason is that oxidation of biodiesel soot occurred not only on the surface but also inside the soot because of its unique oxidation process [17]. This oxidation process led to more drastic structural deformation and resulted in faster oxidation than diesel soot.

The results from elemental analysis are shown in Fig. 7. The graph presents relative weight fractions of particulates from WCO biodiesel and diesel.

The EA results show that WCO biodiesel has lower carbon weight fraction than diesel soot. This result is consistent with the TGA results, showing that the WCO biodiesel soot is less carbonaceous than diesel one. The hydrogen, oxygen, and nitrogen weight fractions of WCO biodiesel soot were higher than diesel soot. The amount of hydrogen in soot particles can be varied according to soot maturity [15]. In terms of oxygen-containing species, presence of oxygen-containing groups such as C=O, C–O–C, C–O–H during oxidation process has been reported based on Fourier transform infrared spectroscopy (FTIR) [18]. The soot particles from WCO biodiesel have larger amount of oxygen-containing species because of intrinsically oxygenated fuel origin. From the FTIR results, larger presence of oxygen-containing functional groups in biodiesel soot were confirmed, so the devolatilization of oxygen group bonded at the edge sites provided higher reactivity during initial stages of oxidation [19].

4 Conclusions

The morphological and compositional characteristics of soot particles from waste cooking oil (WCO) biodiesel and conventional diesel were investigated in a single-cylinder diesel engine. The morphology of soot particles was analyzed by transmission electron microscope (TEM) images. At the same time, the compositional analysis of the soot particle was performed by thermogravimetric analysis (TGA) and elemental analysis (EA). In the results of TEM image analysis, WCO biodiesel showed smaller primary particle diameters compared to conventional diesel. The primary particle diameters of WCO biodiesel and conventional diesel were 19.9 and 23.3 nm, respectively. The TGA analysis results showed that the WCO biodiesel soot particles underwent faster mass reduction because of presence of larger amount of volatile fraction (VOF) and absorbed hydrocarbon species. The EA results revealed that soot particles from WCO biodiesel had lower carbon weight fraction (less carbonaceous) than that of diesel soot. WCO biodiesel soot also exhibited larger amount of hydrogen, oxygen, and nitrogen than diesel soot particles.

References

1. Tree DR, Svensson KI (2007) Soot processes in compression ignition engines. Prog Energy Combust Sci 33(3):272–309
2. Smith OI (1981) Fundamentals of soot formation in flames with application to diesel engine particulate emissions. Prog Energy Combust Sci 7(4):275–291
3. Bryce D et al (1999) Investigating the effect of oxygenated and aromatic compounds in fuel by comparing laser soot measurements in laminar diffusion flames with diesel-engine emissions. J Inst Energy 72(493):150–156
4. Glassman I (1996) Combustion. Academic Press, San Diego
5. Bartok W, Sarofim AF (1991) Fossil Fuel Combustion: A Source Book 1st Edition. William Bartok, Adel F Sarofim, Wiley, New York
6. Kosaka H, Aizawa T, Kamimoto T (2005) Two-dimensional imaging of ignition and soot formation processes in a diesel flame. Int J Engine Res 6(1):21–42

7. Gallo Y et al (2017) Investigation of late-cycle soot oxidation using laser extinction and in-cylinder gas sampling at varying inlet oxygen concentrations in diesel engines. Fuel 193:308–314
8. Hwang J, Jung Y, Bae C (2015) Comprehensive assessment of soot particles from waste cooking oil biodiesel and diesel in a compression ignition engine. SAE Int J Fuels Lubr 8 (2015-01-0809):290–297
9. Aizawa T et al (2012) Transmission electron microscopy of soot particles directly sampled in diesel spray flame-A comparison between US# 2 and biodiesel soot. SAE Int J Fuels Lubr 5 (2012-01-0695):665–673
10. Wal V, Randy L, Tomasek AJ (2003) Soot oxidation: dependence upon initial nanostructure. Combust Flame 134(1):1–9
11. Hwang J, Jung Y, Bae C (2015) Spray and combustion of waste cooking oil biodiesel in a compression-ignition engine. Int J Engine Res 16(5):664–679
12. Neer A, Koylu Umit O (2006) Effect of operating conditions on the size, morphology, and concentration of submicrometer particulates emitted from a diesel engine. Combust Flame 146(1):142–154
13. Merchan-Merchan W, Sanmiguel SG, McCollam S (2012) Analysis of soot particles derived from biodiesels and diesel fuel air-flames. Fuel 102:525–535
14. Kook S, Pickett LM (2012) Soot volume fraction and morphology of conventional, fischer-tropsch, coal-derived, and surrogate fuel at diesel conditions. SAE Int J Fuels Lubr 5 (2012-01-0678):647–664
15. Jung Y, Hwang J, Bae C (2016) Assessment of particulate matter in exhaust gas for biodiesel and diesel under conventional and low temperature combustion in a compression ignition engine. Fuel 165:413–424
16. Stratakis GA, Stamatelos AM (2003) Thermogravimetric analysis of soot emitted by a modern diesel engine run on catalyst-doped fuel. Combust Flame 132(1):157–169
17. Song J et al (2006) Examination of the oxidation behavior of biodiesel soot. Combust Flame 146(4):589–604
18. Stanmore BR, Brihac JF, Gilot P (2001) The oxidation of soot: a review of experiments, mechanisms and models. Carbon 39:2247–2268
19. Song J, Alam M, Boehman AL (2007) Impact of alternative fuels on soot properties and DPF regeneration. Combust Sci Technol 179(9):1991–2037

Techniques to Control Emissions from a Diesel Engine

Pravesh Chandra Shukla, Tarun Gupta and Avinash Kumar Agarwal

Abstract Diesel particulate and NO_x emission cause several serious health problems; therefore, it is necessary to reduce these emissions from the tailpipe. In the past decades, significant technological advancements have been made in the field of engine emission control. In modern diesel engines, smarter electronic fuel injection strategies are being employed. Control of engine emissions can be put under two baskets: (1) active control techniques, and (2) passive control techniques. Active control techniques are those which restrict the formation of the pollutants in the combustion chamber itself. Passive control techniques refer to after-treatment devices. Active control techniques include advancement in the combustion chamber design, use smarter electronic fuel injection system, exhaust gas recirculation, high-pressure multi-fuel injection with precise injection timing, homogenous charge compression ignition, etc. which if used properly restrict the formation of the pollutants. Some other in-cylinder technologies are also effective in reducing the pollutant emission. Although active control techniques are able to reduce the emission up to some extent, but in order to meet the modern emission regulations, passive techniques are also required in addition to active techniques. Passive control technique involves after-treatment devices like diesel oxidation control, diesel particulate trap, NO_x absorber, selective catalytic reduction.

Keywords Diesel oxidation catalyst · Diesel particulate filter · NO_x storage-reduction catalysts (NSR) · Selective catalytic reduction (SCR)

P. C. Shukla · T. Gupta
Department of Civil Engineering, Indian Institute of Technology Kanpur, Kanpur 208016, India

A. K. Agarwal (✉)
Engine Research Laboratory, Department of Mechanical Engineering, Indian Institute of Technology Kanpur, Kanpur 208016, India
e-mail: akag@iitk.ac.in

N. Sharma et al. (eds.), *Air Pollution and Control*, Energy, Environment, and Sustainability, https://doi.org/10.1007/978-981-10-7185-0_4

List of Abbreviations

NSR	NO_x storage-reduction catalysts
ULSD	Ultra-low sulfur diesel fuel
DPF	Diesel particulate filter
EGR	Exhaust gas recirculation
SCR	Selective catalytic reduction
SOF	Soluble organic fraction
USCR	Urea-selective catalytic reduction
LSD	Low sulfur diesel

1 Introduction

Diesel particulate and NO_x emission led to deleterious effects on human health [1, 2]. This has led to several ongoing advance research around the world to come up with alternatives to current combustion strategies. In the past decades, several technological advancements have been carried out in the field of engine emission control. There are mainly two broad categories under which all the engine emission control techniques can be classified [3]:

1. Active control techniques
2. Passive control techniques

Active control techniques are the techniques which restrict the formation of the pollutants in the combustion chamber itself. Passive control techniques are the techniques which are used to control the pollutants before they are emitted into the ambient atmosphere from the tailpipes. Active control techniques include advancement in the combustion chamber design, use of modern electronic fuel injection system, exhaust gas recirculation, high-pressure multi-fuel injection with precise injection timing, homogenous charge compression ignition, etc. [4–6]. Some other in-cylinder technologies are also effective in reducing the pollutant emission.

Passive control technique involves after-treatment devices like diesel oxidation control, diesel particulate trap, NO_x absorber, selective catalytic reduction. Following sections give a brief explanation about the active and passive control techniques for NO_x and particulate control.

2 Active Control

In 1970s, there was a lack in oil supply from Middle East countries to the US and other industrialized countries [7]. This crisis motivated the engine manufacturers to shift toward the diesel engines. 1980s and 1990s are the transitional decades for diesel engine development. By 1991, newer and stricter emission regulations were implemented in US and Europe. To meet these emission regulations, engine manufacturers were forced to implement the new engine technologies which helped in reduction of pollutants. The active technologies by which NO_x and particulate emission can be controlled effectively are described in detail below [8].

2.1 NO_x Control

Injection timing retard, intake charge cooling, exhaust gas recirculation, etc. are some of very effective development to control in-cylinder combustion and have led to lower NO_x emissions.

Injection Timing Retard: Several studies have been carried out to investigate the effect of injection timing on the net emission of NO_x from the engine. Figure 1 shows the effect of injection timing on NO_x emission from one of those studies [9]. By retarding the injection timing, peak cylinder pressure reduces consequently the peak cylinder temperature also reduces. The lower peak cylinder temperature significantly reduces the production of NO_x.

Cooling the Intake Air: Cooling of intake air has relatively lower effect as with injection timing retardation [9]. By cooling the intake air, heat capacity of air increases in the cylinder. This leads to lower peak cylinder pressure and lower peak temperature which helps in overall less production of NO_x. Figures 2 and 3 show the effect of intake manifold temperature on NO_x emission and in-cylinder pressure.

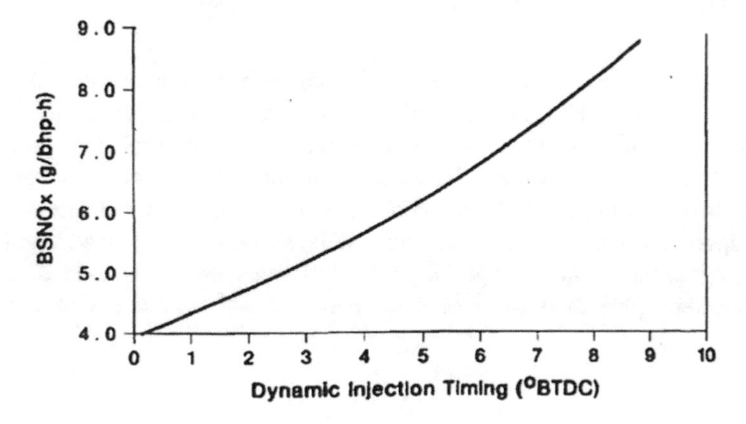

Fig. 1 Effect of injection timing and NO_x emissions [9]

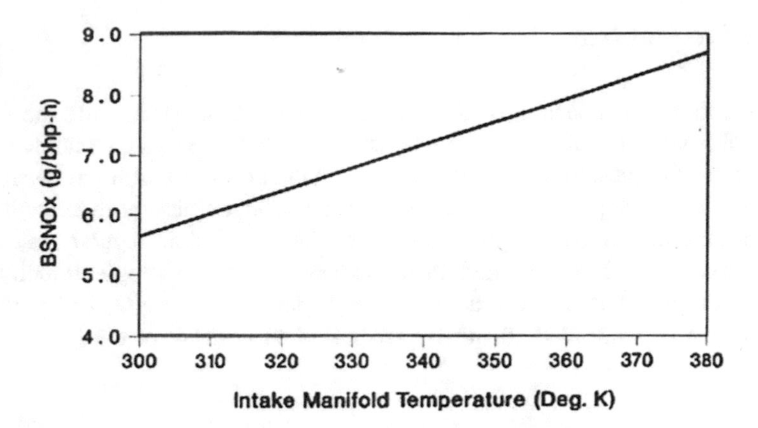

Fig. 2 Effect of intake manifold temperature on NO_x emissions [9]

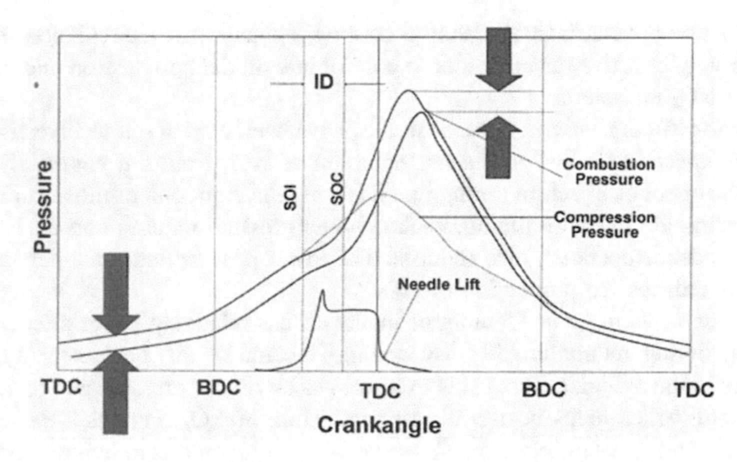

Fig. 3 Schematic representation showing how intake manifold temperature affects cylinder pressure [17]

Exhaust Gas Recirculation (EGR): Modern engines implement exhaust gas recirculation system. NO_x is produced during the combustion event at high temperatures. Automobiles use exhaust gas recirculation to control NO_x production. EGR controls NO_x by lowering the combustion temperature and reducing the net oxygen content in the combustion chamber. EGR is mostly used at higher vehicle/engine speeds with moderate acceleration. EGR systems have helped significantly in cleaning up the urban air, especially NO_x emissions emanating from IC engines. However, the application of EGR also incurs penalties in terms of particulate emission.

2.2 Particulate Control

Particulate matter emanating from diesel engines is mainly composed of carbonaceous and soluble organic fraction (SOF). SOF contributes about 40–50% of the total particulate depending upon the engine load [10]. Other components are sulfates, ash content, nitrates, etc. In addition, lubricating oil significantly contributes to particulate emission. For reducing the particulate emission from lubricating oil, certain engine design modifications have been adopted like optimum piston design, optimum bowl design, piston liner modifications. Lubricating oil formulation is also an important parameter and is a broad field for research. The central idea of this example is that particulate emission cannot be reduced effectively by changing a single engine parameter; it needs number of modifications simultaneously for optimum performance and minimized emission level. In broad sense, there are mainly a few systems to reduce the particulate emission by engine modification as reported by Majewaski et al. [8]: air management, lubricating oil consumption control, and variation in fuel injection parameters.

Air Management: In diesel engine, charge air is supplied in the intake stroke and it comes out in the exhaust stroke. The formation of particulate matter is dependent on how the air flows and mixes with the fuel droplets inside the combustion chamber. Intake temperature, pressure, flow rate, air cooling have significant influence on pollutant formation. Turbo charging, charge air cooling, optimum intake manifold design, etc. are the main modifications related to air management that can be employed in a diesel engine for reducing the particulate emission. Turbochargers are the devices which use the waste energy carried by the exhaust gases. These devices use the heat energy of the exhaust gas to compress the intake air. The density of the intake air increases and the engine's volumetric efficiency increases. Diesel engine typically shows high smoke emission at low engine speed with high engine load. To reduce the smoke emission in this condition, turbochargers are one of the good options. Turbocharger maximizes the air–fuel ratio at low engine speed with high engine load condition which leads to reduction of incomplete combustion and eventually reduces net smoke emission. In the previous section, it has already been explained that the charge air cooling has net effect on reducing the NO_x. It also reduces the particulate emission up to some extent. By cooling the charge air, intake air density increases and it results in better fuel consumption and optimum combustion. In the modern engines, air-to-air cooling is one of the most common ways to cool the intake charge air. Figure 4 shows the schematic of air-to-air cooling device [11].

Lubricating Oil Consumption Control: It has been demonstrated that the lubricating oil is a major contributor in overall particulate emission from diesel engines [12]. There are mainly three areas of lubricating oil consumption/accumulation namely: cylinder wall, intake and exhaust system, and lubricating oil formulation. Out of these three, cylinder walls are responsible for highest lube oil contribution to the particulate formation. At the cylinder walls, lube oil comes in direct contact with the extremely hot burned gases. In order to tackle this problem, engineers have

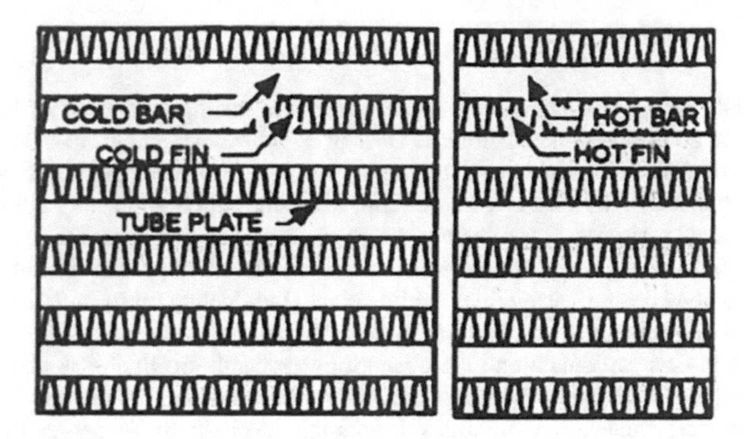

Fig. 4 Air-to-air heat exchanger [11]

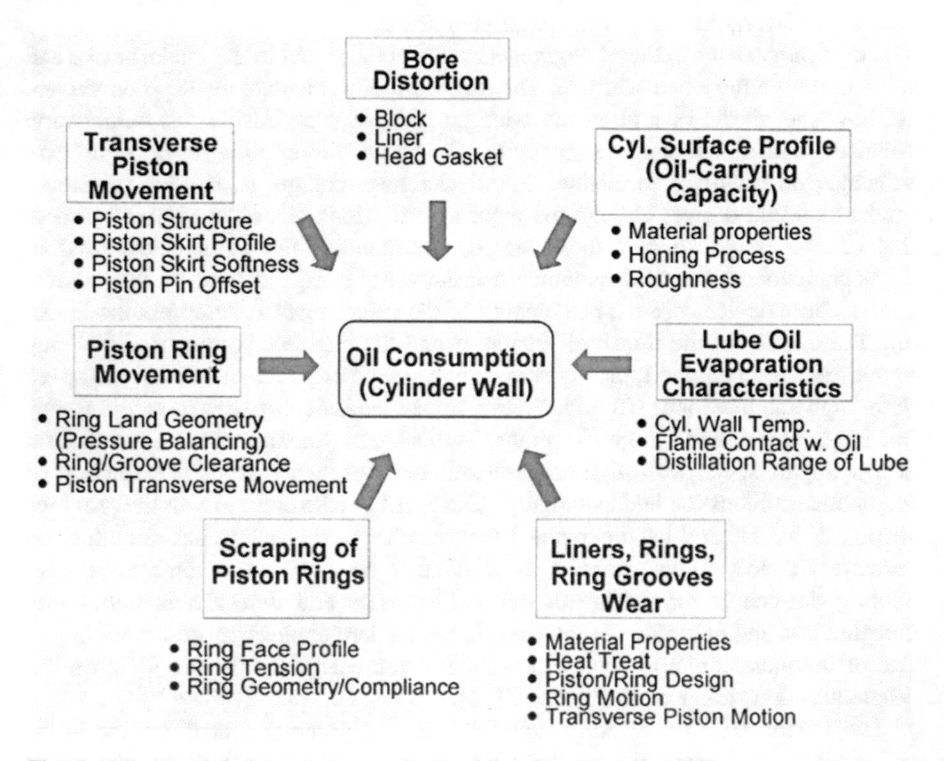

Fig. 5 Criteria and design parameters affecting lube oil consumption [13]

made modifications in piston design, piston rings, liners, and other engine parts. Figure 5 gives an overall idea about the engine design parameters to be modified in order to reduce the lube oil consumption [13].

Variation in Fuel Injection Parameters: Fuel injection system for diesel engine is considered to be the heart of the system. There are mainly four fuel injection parameters which control the fuel–air mixing and optimum combustion: injection timing, injection pressure, injection duration, and number of injection in one thermodynamic cycle.

Injection Timing: It has already been discussed that the retarded fuel injection timing leads to reduction in NO_x emission. Unfortunately, fuel consumption increases with retarded injection timing.

Injection Pressure: In the early decades of diesel fuel injection system development, some researchers have believed that the injection rate shaping is more effective than the injection pressure in reducing particulate emissions [9]. However, fuel injection pressure changes the fuel injection rate, spray behavior, injection duration, etc. Injection pressure and injection shaping are two interdependent parameters. Higher injection pressure results in better spray characteristics, and atomization leads to good fuel economy and lower smoke (Fig. 6) emissions.

Injection Duration: For a typical diesel engine, the fuel injection duration spans over less than 30 CAD. There should be optimum match between combustion system and injection duration.

Multiple Injections: New multi-injection technology is being used in the modern diesel vehicles. Multi-injection facilitates number of injection in a single

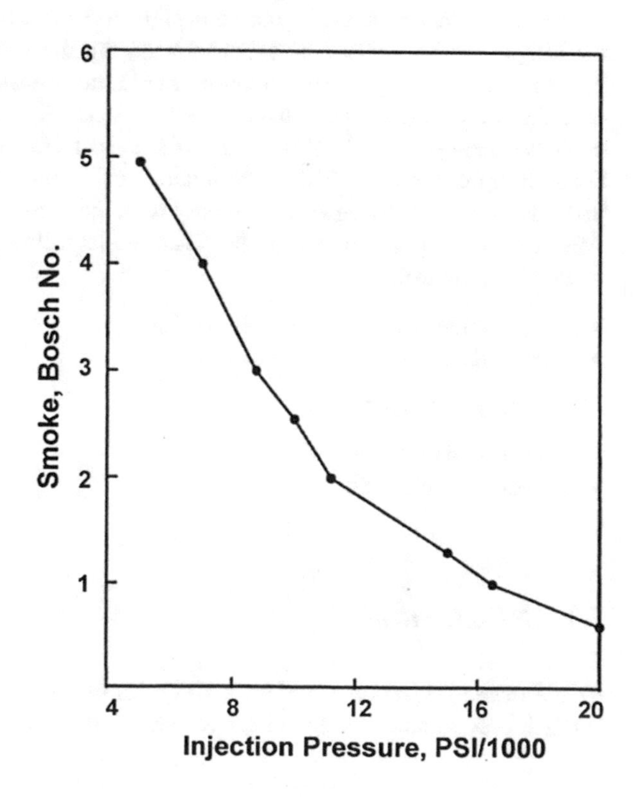

Fig. 6 Effect of injection pressure on smoke emissions (Osenga M, "EPA proposes stationary diesel emissions regulations" Industry news, Diesel progress North America edition, August 2005)

thermodynamic cycle. Apart from the main injection, there may be pre-injection and post-injection of fuel. In modern engine, all these injection parameters are electronically controlled very precisely and thus led to net reduction in PM emissions. To facilitate the above four fuel injection parameters in diesel engines, a sophisticated advanced common rail fuel injection system is required. Common rail fuel injection is fairly complex and usually costlier than the conventional one but in turn delivers the fuel at a very high pressure (it may be up to 2000 bar) and precisely in the combustion chamber. This optimum control in fuel injection strategy greatly helps in particulate emission reduction.

Electronic control unit is one of most important tools for controlling all the engine operating parameters. To meet the present and future emission regulations, there is a need to use the after-treatment devices in addition to the above-explained engine emissions control technologies. The following section explains various after-treatment technologies used for emission control.

3 Passive Control

During 1990s, the engine emission control was limited only to using a combination of several in-cylinder technologies like electronic fuel injection control, exhaust gas recirculation, modification in engine design, use of turbochargers. For controlling the NO_x emission, EGR is an effective way, but the volume of recirculated gas is limited as it in turn gives rise to higher particulate emissions. Oxidation catalyst is used for controlling the particulate emission. Later, it has been realized that the use of above-mentioned technologies is not able to fulfill the future emission regulations comprehensively. The combination of after-treatment technologies with in-cylinder control strategies is necessary with the present stringent emission norms. After-treatment technologies can be classified as follows:

For NO_x control:

- NO_x storage-reduction catalysts (NSR)
- Urea-selective catalytic reduction (SCR)

For particulate control:

- Diesel oxidation control
- Diesel particulate filter

3.1 NO_x Control

NO_x Storage-Reduction Catalysts (NSR): Unlike the gasoline engine, three way catalytic converters are not effective for diesel exhaust for NO_x control. NO_x

reduction requires a reducing atmosphere, whereas CO and HC conversion require oxidizing atmosphere. In the gasoline exhaust, reducing as well as oxidizing atmosphere is available for reducing NO_x and oxidizing HC and CO conversion, but in diesel exhaust, only oxidation atmosphere is available with ample availability of oxygen. Therefore, a different kind of catalytic reduction is required for NO_x control in diesel exhaust which provides a reducing environment. NO_x storage-reduction catalyst provides the required environment for NO_x reduction. The NO_x reduction in this device follows two steps: (i) NO_x storage during the engine lean mixture operation (ii) release and reduction of NO_x during the rich mixture engine operation or by adding additional hydrocarbon.

Figure 7 shows an artistic picture for NO_x conversion mechanism for the two stated steps. Generally, platinum catalyst is used on metal oxide supports like Al_2O_3, SiO_2, TiO_2 [14, 15] with BaO doping on it. Platinum catalyst leads the conversion of NO into NO_2 which gets absorbed in the support as nitrate. When the engine operates on relatively richer mixture (even for short duration), these nitrates get desorbed from the support and then reduced to N_2 in the presence of hydrocarbons, which provide a reducing environment. This conversion is difficult in low load engine and high load engine operations due to very little HC emission and inactivity of the catalyst. There is always an optimum exhaust temperature for which NO_x conversion efficiency is maximum which depends upon HC/NO_x ratio. Higher will be the HC/NO_x ratio, higher will be the value of temperature at which maximum NO_x conversion occurs.

Figure 8 shows the effect of HC/NO_x ratio on the value of optimum temperature for maximum NO_x conversion. When very little hydrocarbon is available for NO_x conversion, additional hydrocarbon can be added through post-injection. Common rail injection system facilitates a precise post-injection for increasing HC in the exhaust stream. Generally, post-injection is done at 90–200 CAD after the main injection. However, the fuel quantity injected during post-injection is kept around 2% of the fuel quantity injected during the main injection [14]. Zeolites are also incorporated in this system in order to absorb the excess hydrocarbons and release it when required [16]. It is recommended to use low sulfur fuel with these catalysts otherwise it leads to deactivation of the NSR catalysts.

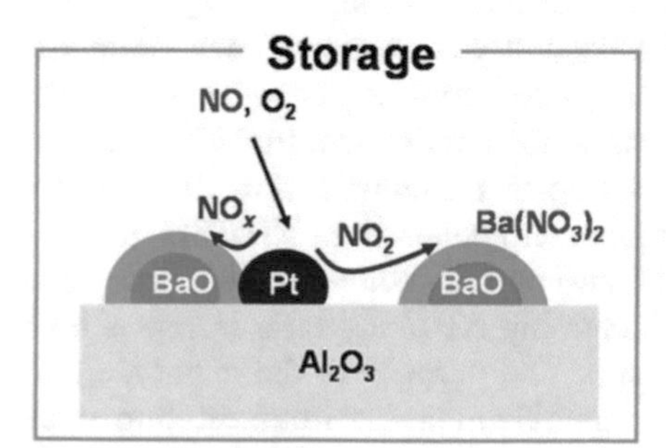

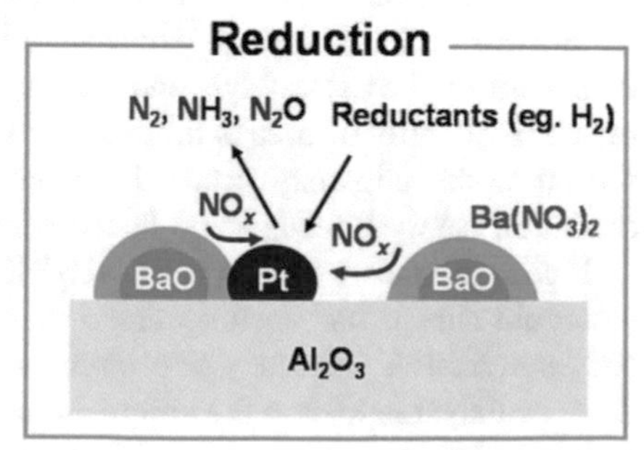

Fig. 7 NO_x storage-reduction catalysts (NSR) [18]

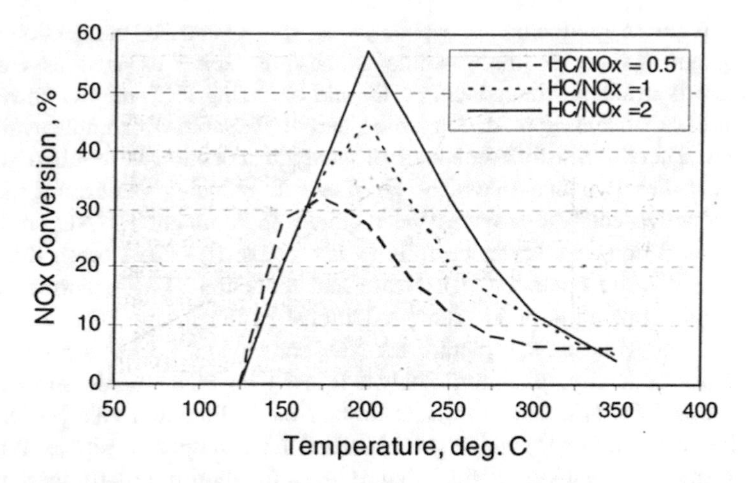

Fig. 8 Effect of HC/NO_x ratio and temperature on conversion efficiency of NSR catalyst in a laboratory model gas bench test. Adapted from [14]

Urea-Selective Catalytic Reduction (SCR): This control technique uses ammonia for providing a reducing environment for NO_x conversion. Urea is used as a source of ammonia with 30–40% concentration in water solution. SCR technology is not new to the automotive industry. It is being used since 1980s in different industries like thermal power plants for pollution control.

Figure 9 shows the schematic arrangement of SCR system. Ammonia is a good reducing agent for NO_x reduction in the catalytic environment. SCR system uses ceramic honeycomb substrate with catalytic coating. Mixture of vanadium, tungsten, and titanium oxide is used for catalytic coating over the substrate. There are mainly three sections in SCR system as follows: hydrolysis catalyst section, SCR catalyst section, and oxidation catalyst section. Urea injection is done at the upstream of SCR system. Urea injection quantity is also very important, and it is decided according to the NO_x concentration in the exhaust stream. NO_x detection sensors are fitted upstream of SCR which provide information about the NO_x concentration in the exhaust stream. Electronic control unit for urea injection injects a definite amount of urea in accordance to the requirement for NO_x reduction. Hydrolysis is the first stage where urea breaks and produces ammonia. Ammonia comes into the SCR catalyst and reduces NO_x into harmless gas N_2. Although the injection quantity of urea is limited according to the requirement for NO_x reduction but ammonia slip may occur. In order to stop this ammonia slip, an ammonia oxidation catalyst needs to be fitted. Ammonia slip depends on NH_3/NO_x ratio.

Figure 10 shows the effect of NH_3/NO_x ratio on NO_x conversion efficiency and ammonia slip. It has been observed that increasing NH_3/NO_x ratio increases both NO_x conversion efficiency and ammonia slip. The purpose of the ammonia oxidation catalyst is to stop the slip of excess ammonia in the environment. Sometime an oxidation catalyst is placed before the urea injection which helps in NO_2

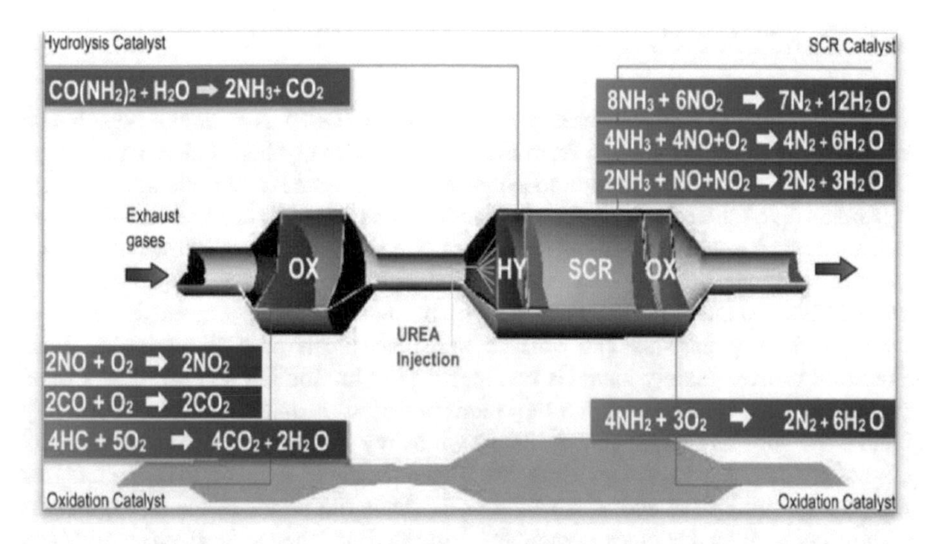

Fig. 9 Schematic layout of an advanced SCR catalyst system using pre-oxidation catalyst [19]

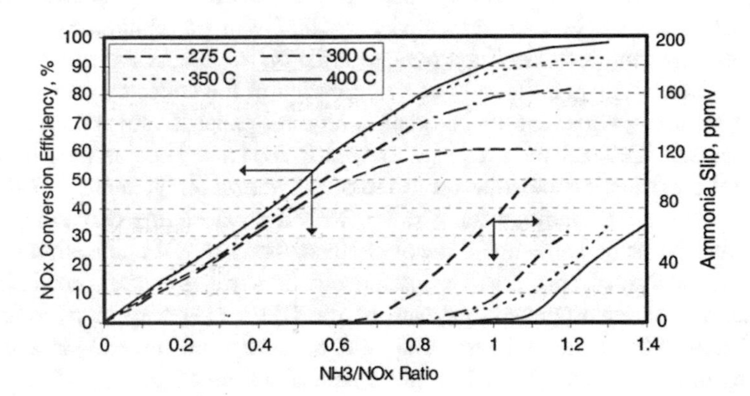

Fig. 10 NO_x conversion ammonia slip for a SCR catalyst as a function of NH_3/NO_x ratio on engine test bench. Adapted from [20]

formation in the exhaust stream. Urea-SCR system requires relatively lower space velocity for effective operation at around 10–30 k h^{-1}. This lower space velocity leads to a bigger size of the device; therefore, it is appropriate for heavy duty diesel vehicles and stationary NO_x emitter plants.

3.2 Particulate Control

Diesel Particulate Filter: Diesel particulate filter (DPF) is a device which can capture the particulate emitted from the tailpipe with very high efficiency.

Figure 11 shows the filtration mechanism in diesel particulate filters. Generally, it is made up of ceramic substrate with parallel channels. The channels are designed in such a manner that exhaust can enter from one side but it cannot flow out through the same channel. Exhaust has to pass through the porous walls between two consecutive channels. In this way, most of the particulate get trapped in the filter media and only gaseous and volatile substances can pass through the filter. Filtration process mainly includes two steps: (1) filtration (2) regeneration. Diesel particulate filter can capture the huge amount of soot quickly. It can effectively handle the huge amount of soot emitted by a heavy duty vehicle by capturing liters of soot per day. The continuous trapping of soot in diesel particulate filter increases the pressure difference across the filter and affects the performance of engine. At certain point, filter becomes completely saturated and hence ineffective and it is required to remove the collected particulate from these filters. The soot removal process from filter for continuous operation is known as regeneration process. Regeneration process can be performed periodically or on continuous basis. DPF for the modern engine is also incorporated with the engine management system

The particle removal is carried out by oxidation of the collected soot material in the filter channels, where the soot oxidation requires exhaust temperature should be more than 600 °C. There are two ways to oxidize the loaded soot in the filter: either increase the exhaust temperature up to the soot ignition temperature (>600 °C) or by lowering the soot ignition temperature to a normal engine operating exhaust temperature by using some noble metal catalyst. The catalyst application can be a fuel-borne catalyst or coated in the filter itself. Sometimes a diesel oxidation catalyst can also be incorporated upstream of the DPF in order to increase the concentration of NO_2 in the exhaust. It should be noted that the NO_2 is a stronger oxidizing agent than oxygen and oxidizes soot more effectively.

Figure 12 shows a general concept of filtration process in diesel particulate filter.

Filtration mechanism: Filtration can be of two types: depth filtration and surface filtration. In depth filtration, the mean pore diameter of the filter media is much larger than the mean diameter of the collected particle. In the surface filtration, pore

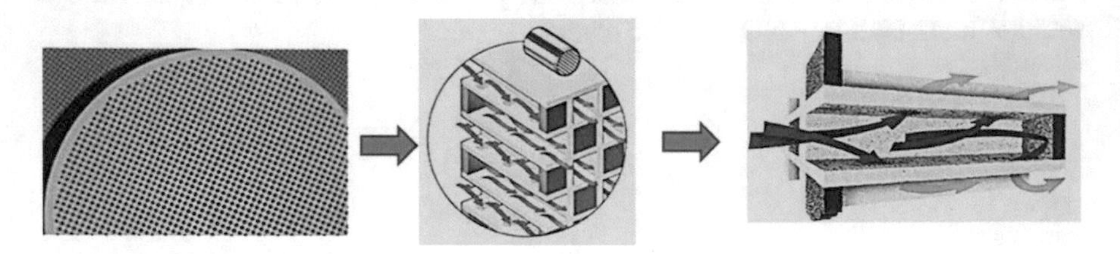

Fig. 11 Typical DPF and monolythic channel

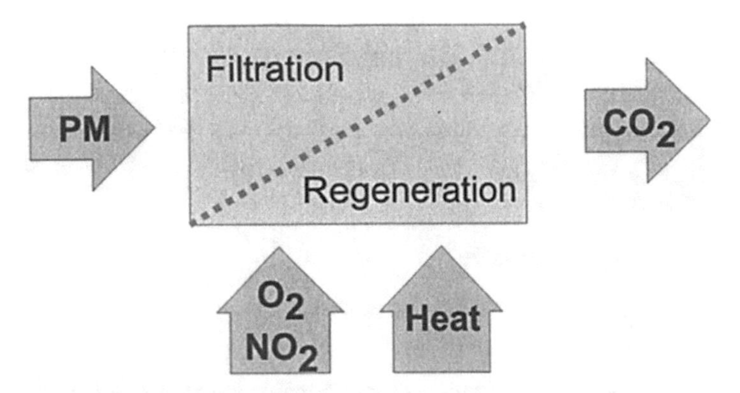

Fig. 12 Particulate filter with thermal regeneration [4]

diameter of the filter media is smaller than the mean diameter of the collected particle. Filtration mechanism in DPF is largely dominated by depth filtration.

Performance of DPF: Diesel particulate filter is an efficient particulate removal device, and its particle removal efficiency is the way to express performance.

Table 1 shows the reduction efficiency of individual components of the particulate by using DPF. DPF particle removal efficiency is defined as the ratio of collected mass of particulate in the filter to the total incoming mass of particulate into the filter. DPF removal efficiency may be as high as 90–98% depending upon the filter characteristics.

Figure 13 shows the effect of particulate filter on removing the total particulate mass. It shows that there is huge reduction in total particulate reduction, but it has high sulfate particulates. Catalyst present in the DPF promotes the formation of SO_3 which leads to the formation of sulfate particulates in the presence of moisture. Therefore, it is recommended to use ultra-low sulfur fuel when using diesel oxidation catalyst. Although the particulate mass filtration efficiency of DPF is quite high, but its performance in reducing the particle number emission is poor. The two major components of diesel particulates are carbonaceous material and organic

Table 1 Typical filtration efficiency of diesel particulate filter (PM mass) [4]

Typical filtration efficiency of diesel particulate filters (PM mass)		
Fraction of diesel PM	Efficiency	Comments
Elemental carbon	95–99.9	Physical filtration
Organic fraction (SOF)	50–90	Conversion primarily by catalytic oxidation depends on temperature. Lower conversion in non-catalytic filters
Sulfate particulates	No filtration	Catalytic oxidation of SO_2 to SO_3 can increase SO_4, causing negative efficiency when high sulfur fuel is used
TPM	70–95	

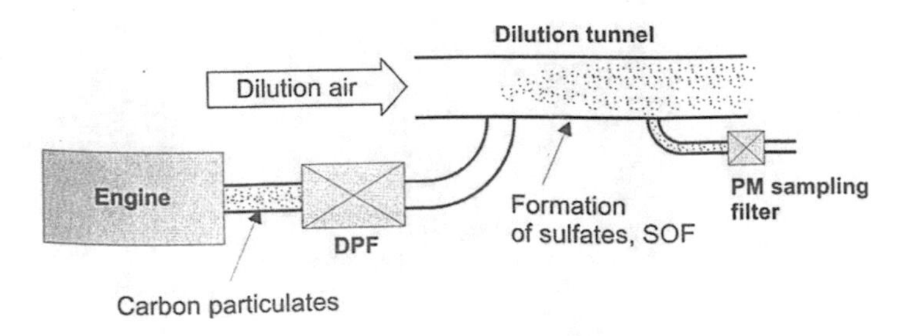

Fig. 13 Effect of liquid particulates downstream of the DPF [4]

material. A large portion of organic material remains in vapor phase at high exhaust temperature and can pass through the porous wall of the filter. Later, they nucleate due to condensation and form nuclei mode particles in huge number. This phenomenon increases the particle number in the downstream of DPF. Increased number emission is an important aspect with the use of DPF because future emission regulation will also regulate the particle number emission. Another aspect is that organic matters are largely responsible for the toxic potential of diesel exhaust. Therefore, there is a need to optimize the catalytic activity of DPF with all operating temperature range to resolve this problem. An easy solution to this problem is to employ a DPF and a DOC in tandem for them to effectively control these emissions.

Diesel Oxidation Catalyst: Diesel emissions are mainly composed of carbonaceous material and soluble organic fraction. Other components are sulfate and ash content. SOF part is the fraction which is mostly responsible for diesel exhaust toxicity. A big advantage with the SOF part is that it can be oxidized at relatively lower diesel exhaust temperature in the presence of a suitable catalyst. Diesel oxidation catalyst is a device which is very effective in removing the SOF fraction of diesel particulate in addition to eliminating CO and HC gaseous emissions. Diesel oxidation catalysts are generally made up of ceramic monoliths. Ceramic monoliths provide a strong adhesion for wash-coat material and offer lower cost comparative to metal monoliths. Wash-coating is done by Al_2O_3 as it provides a high surface area for the catalyst to adhere on. Sometimes SiO_2, TiO_2, or Zeolite materials are also used as wash-coating materials. DOC is very effective in removing the SOF part of the particulate but only in a certain temperature range. DOC provides its maximum SOF removal efficiency in the temperature range of 200–350 °C. It is less effective in removing organic fraction at temperature lower than 200 °C and encourages the formation of sulfate particulates at temperature more than 350 °C. Noble metals like platinum, palladium are used as catalyst in DOCs. One of the major problems with DOC is that it promotes sulfate formation at higher exhaust temperature. At higher exhaust temperature, catalyst increases the rate of SO_3 formation which leads to the formation of sulfate particulates in the presence of moisture. Therefore, it is recommended to use low sulfur diesel fuel

Table 2 Emission conversion efficiency of diesel oxidation catalysts [21, 22]

S. No.	Engine/vehicle type	Test cycle	Conversion efficiency			
			CO	HC	SOF	Total PM
1	Heavy duty	US HD	8.7	29	51.8	32.8
2	Passenger car	ECE15 + EUDC	95	90	–	35.0

(LSD) or ultra-low sulfur diesel fuel (ULSD). This limits the extent of formation of sulfate. Table 2 shows the emission conversion efficiency of a typical DOC for different species present in the engine exhaust.

References

1. McCormick RL, Ross JD, Graboski MS (1997) Effect of several oxygenates on regulated emissions from heavy-duty diesel engines. Environ Sci Technol 31:1144–1150
2. Mauzerall DL, Sultan B, Kim N, Bradford DF (2005) NO_x emissions from large point sources: variability in ozone production, resulting health damages and economic costs. Atmos Environ 39:2851–2866
3. Nuti M (2017) Emissions from two-stroke engines. Training 1998:08–15
4. Majewski WA, Khair MK (2006) Diesel emissions and their control. SAE Technical Paper
5. Yagi S, Miyaki K (1977) Exhaust gas recirculation for emissions control. Google Patents
6. Shi L, Cui Y, Deng K, Peng H, Chen Y (2006) Study of low emission homogeneous charge compression ignition (HCCI) engine using combined internal and external exhaust gas recirculation (EGR). Energy 31:2665–2676
7. Owen K, Coley T (1995) Automotive fuels reference book
8. Majewski WA, Khair MK (2006) Diesel emissions and their control. SAE International
9. Gill AP (1988) Design choices for 1990s low emission diesel engines ISSN. SAE Technical Paper
10. Pundir B (2007) Engine emissions: pollutant formation and advances in control technology. Alpha Science International, Limited
11. Merrion DF (1994) Diesel engine design for the 1990s: the fortieth L. Ray Buckendale lecture. Society of Automotive Engineers
12. Cartellieri WP, Wachter WF (1987) Status report on a preliminary survey of strategies to meet US-1991 HD diesel emission standards without exhaust gas aftertreatment. SAE Technical Paper
13. Zelenka P, Kriegler W, Herzog PL, Cartellieri WP (1990) Ways toward the clean heavy-duty diesel. SAE Technical Paper
14. Peters A, Langer H-J, Jokl B, Müller W, Klein H, Ostgathe K (1998) Catalytic NO_x reduction on a passenger car diesel common rail engine. SAE Technical Paper
15. Hodjati S, Semelle F, Moral N, Bert C, Rigaud M (2000) Impact of sulphur on the NO_x trap catalyst activity-poisoning and regeneration behaviour. SAE Technical Paper
16. Kitahara Y, Akama H, Kamikubo M, Shinzawa M (1998) Passive and active performance characteristics of NO_x catalysts for direct-injection diesel engines for passenger cars. SAE Technical Paper
17. Khair MK (1993) Diesel engine technology, SAE Seminar, Apr 1993
18. HU Z, Sun K, Xu B (2011) NO_x storage-reduction catalysis and structure-performance relationship of Pt-BaO catalyst. J Catal 32(1): 17–26
19. The Lubrizol Corporation, USA

20. Gieshoff J, Schäfer-Sindlinger A, Spurk P, Van Den Tillaart J, Garr G (2000) Improved SCR systems for heavy duty applications. SAE Technical Paper
21. Smedler G, Ahlström G, Fredholm S, Frost J, Lööf P, Marsh P, et al (1995) High performance diesel catalysts for Europe beyond 1996. SAE Technical Paper
22. Voss K, Yavuz B, Hirt C, Farrauto R (1994) Performance characteristics of a novel diesel oxidation catalyst. SAE Technical Paper

BS-III Diesel Vehicles in Imphal, India: An Emission Perspective

Thokchom Subhaschandra Singh, Tikendra Nath Verma, Prerana Nashine and Chitra Shijagurumayum

Abstract Air pollution is an inevitable factor in the modern era. Substances, in the form of particles or gases, which are subtended in the air and causing harm to the environment and living organisms both directly and indirectly, are termed as pollutants, and these pollutants lead to air pollution. The origin of air pollution is not restricted to only man made, but it may also be from natural causes like eruption of volcanoes or by forest fire. There are two divisions of air pollutants—primary and secondary. Primary pollutants are those which are emitted directly from a source like exhaust of automobile, industries, burning of fossil fuels. Some of them include hydrocarbons (HC), carbon dioxide (CO_2), carbon monoxide (CO), sulfur dioxide (SO_2), nitrogen oxide (NO_x), particulate matter. A secondary pollutant is a product of reaction among the primary pollutants or with water vapor and sunlight, and examples are sulfuric acid, ozone, peroxy-acyl-nitrate (PAN). The pollutants in the air cause severe adverse effect on the human health also, like asthma, bronchitis, and even lung cancer. Thus, it is important to measure, control, and check the pollution of air from time to time. The authors in this paper have attempted to study the emission characteristics of BS-III commercial diesel auto rickshaws operating in the city of Imphal, Manipur, India. A mobile flue gas analyzer (MFGA) was used for recording the data. The oxygen percentage in the exhaust ranged from 16.6 to 21.3% (vol.). The ambient temperature during recording of data varied from 24.8 to 27 °C. Higher NO_x

T. S. Singh (✉) · T. N. Verma
Department of Mechanical Engineering, National Institute of Technology Manipur, Imphal 795004, India
e-mail: th.subhas143@gmail.com

P. Nashine
Department of Mechanical Engineering, National Institute of Technology Rourkela, Rourkela 769008, India

C. Shijagurumayum
Department of Civil Engineering, Manipur Institute of Technology, Imphal 795001, India

N. Sharma et al. (eds.), *Air Pollution and Control*, Energy, Environment, and Sustainability, https://doi.org/10.1007/978-981-10-7185-0_5

values in some of the engine indicate that the engine has high combustion, which is due to improper cooling of the engine. Low CO_2 was observed in some vehicles which indicate that the engine has incomplete combustion.

Keywords MFGA · Emission · BS-III norms

1 Introduction

Emissions have posed a serious hazard in our environment and the living beings since the industrial era started. Various atmospheric pollutants such as polycyclic aromatic hydrocarbons were found to be toxic to the human health and cause harmful effects such as cancer, asthma, reproductive toxicity. These compounds were liberated as a result of incomplete combustion of fossil fuels. BS-III vehicles were once widely manufactured and sold vehicles in India. These types of vehicles were mostly used as commercial vehicles and a lot of them are still prevalent on the roads even though the official manufacturing of these types were stopped. These types of vehicles, when unregulated pose a high threat to the environment. Hence, it is the need of the hour to regulate and maintain these types of vehicles. On-road measurements were carried out to check the emission levels of various cities by many researchers. Another analysis was carried out to measure the organic pollutants from vehicles and its corresponding effects on the environment. The main source of polycyclic aromatic hydrocarbons (PAH) was found to be from vehicular exhaust gases, tire wear, motor lubricant oils, road surface wear, brake linings etc. About 2–6% of 5.8–29 kg emissions of PAH have ended up in the sewer system [1, 2]. The variance of fuel consumptions and carbon dioxide emissions from passenger cars in Europe from laboratory versus actual emissions was studied. It was observed that the laboratory values do not reflect the actual performance of the road vehicles since a difference of 30–40% was found. Since various factors such as driving behavior, vehicle configuration, and traffic conditions were found to be highly influential in evaluating the performance, negligence of such factors contributes high difference in the actual observed data and the laboratory collected data [3]. NH_3 and N_2O on-road emission measurements from a heavy duty EURO-V vehicle were performed. A portable emission measuring system, including heated exhaust lines, exhaust flow meter, exhaust gas analyzer, data logger to vehicle network, and a global positioning system, was used for the study (Fig. 1). The average N_2O and NH_3 ranged from 0.063 to 0.139 g/kWh and 0.9 to 5.7 PPM, respectively. NO_x emissions were reported to be higher than the specified limits of EURO-V for the study condition [4].

A study on introduction of emissions trading scheme (ETS) by allocating carbon quotas in the Chinese road transport sector was performed. Using a gray forecast model and trend exploration in a top-down approach, the CO_2 emission of the transport sector was studied. They have forecasted that the carbon emission by 6.6 billion tons by 2020 [5]. A comparative study of exhaust emissions of a

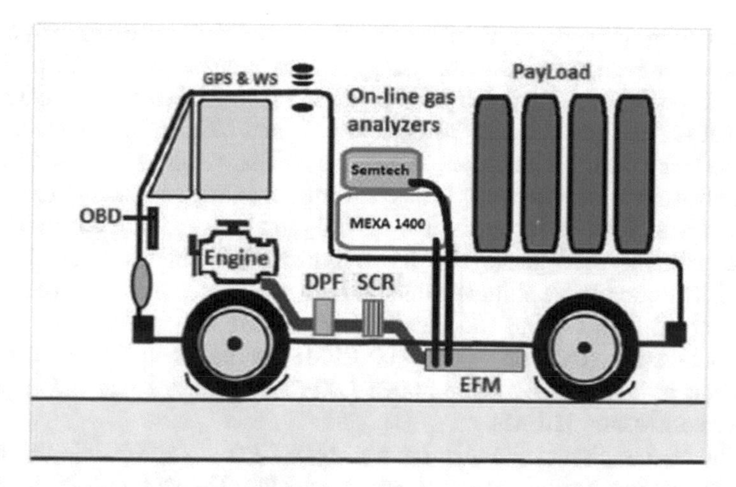

Fig. 1 On-road emission measurement [4]

non-road machine (tractor) using actual readings and laboratory data (similar engine on an engine dynamometer) was performed. Both the engines were equipped with oxidation catalyst (DOC) and selective catalytic reduction (SCR) system. The tests were performed using fossil diesel fuel with low sulfur content and hydro-treated vegetable oil (HVO). By replacing conventional fuel with HVO, the on-road NO_x emission was reduced by 20% and particle number 44%. A common trend was reported to be observed at the laboratory conditions also, although the emissions were lesser than on-road. It was concluded that during steady-state tests, the emissions were highly dependent on the engine load and engine speed on both the fuels [6]. The effects of exhaust gas recirculation (EGR) cooling on the diesel engine emissions were studied. Specially designed spiral fin exhaust was used for the study of chemical composition of the exhaust gas and reduction in pollutants. NO_x, CO_2, CO, and O_2 concentrations were studied. It was found that the reduction in the temperature of EGR has reduced the percentage of O_2, NO_x, and CO_2 while the percentage of CO was found to increase. Using water as heat transfer, fluid has shown a significant reduction in exhaust gas pollution and temperature, as compared to using air as the heat transfer media [7]. A downscaling methodology was applied based on commonly available spatial characteristics to study the on-road carbon dioxide emissions at a small scale (1×1 km) for Boston, MA. The study was tested using surface-level CO_2 emissions. Using impervious surface area and volume-weighted road density that might be scaled to any given resolution, regressions models were developed. The models were found to reflect a similar and strong relationship with the pre-existing inventory data. The approach was evaluated with independent on-road CO_2 observations to ensure the consistency of models. It shows that the improved spatial agreement of the models over the original inventories proved to be a viable basis for downscale study of other areas/cities and also country-level traffic volumes even with limited data [8]. The impact of driving style and road grade on exhaust gas emissions of passenger vehicles was measured by a portable emission measurement system. Two (2) diesel test vehicles of EURO-V and EURO-VI standards were taken for the study. It was found that severe driving led to increase in CO_2 and NO_x emissions while normal trips have significant less emission. There was no change in the CO and HC levels, which indicate that the driving style has less impact on the emission. The study was performed within 100 m segments, and the CO_2 and NO_x emissions showed a linear increase with the road grade, and larger emissions at higher road grade were obtained. The cumulative altitude gain and road grade (step increase from 0 to 5%) have increased the CO_2 and NO_x level by 65–81 and 85–115%, respectively [9]. Carbon monoxide (CO) of values 200 PPM and below indicates that proper combustion is taking place while above 600 PPM indicates that the engine has improper combustion [10–15].

The literatures discussed above have provided various methodologies in measuring exhaust gases from various types of vehicles. The author(s) in this chapter have emphasized the emissions taken from various BS-III vehicles using a portable gas analyzer, so that future researchers, engineers, and young students will have a glimpse of the hazardous exhaust from the vehicles.

2 Pollutant Formation and Adverse Affects on Living Organisms

It is a well-established fact that diesel engine exhaust is categorized as human carcinogens (GR-I). Even though the diesel engine is known to emit various chemicals, a few known pollutants were listed below, as the MFGA is limited to measure certain parameters only [16, 17].

- **Hydrocarbons (HC)**

Hydrocarbons or HC are present in the exhaust of the engine due to rich supply of fuel in the combustion chamber at much lower temperatures. Another possible reason is due to quenching of flames near the metallic walls of the combustion chamber. In most cases, the hydrocarbons get burned during expansion stroke if there are suitable exhaust temperature and oxygen availability. Due to less maintenance of the engine, the above-mentioned conditions were not satisfied, and hence, a large amount of HC will move out along with exhaust gases. HC have been reported to cause irritation of eyes and lungs in the living organisms.

- **Carbon dioxide (CO_2)**

Carbon dioxide formation in an engine is a result of complete combustion of the fuel supplied. A large amount of carbon dioxide in the vehicle exhaust indicates that the engine is running in good condition and the engine is capable of proper combustion of the supplied fuel. Even though presence of CO_2 shows good combustibility, deforestation and depleting of resources cause disturbance of natural CO_2 absorption. Thus, a large amount of CO_2 is suspended in the air which in turn increases global warming.

- **Carbon monoxide (CO)**

Carbon monoxide (CO) is an intermediate product which is formed due to incomplete combustion process. The presence of CO in the sample of air shows that the sample under test has not been properly oxidized during combustion. More formation is observed if the engine is running in rich mixture. As the exhaust gradually cools down, major part of CO mixed with oxygen converts into CO_2. CO is known to have strong affinity (as much as 200 times) to combine with hemoglobin of blood to form carboxyhemoglobin, which reduces the oxygen-carrying capacity of the blood tissues. It is also reported that high amounts of carboxyhemoglobin affect central nervous system and responsible for heart attacks.

- **Oxides of nitrogen (NO_x)**

The oxides of nitrogen were produced in small quantities, but in certain environment, it caused pollution. Prolonged inhalation of air exposed to NO_x has various health complications. NO_x is obtained due to the presence of high temperature during combustion, which is clearly explained in the following Eq. (1)

$$\begin{aligned} N_2 + O_2 &= 2NO \\ N_2 + 2H_2O &= 2NO + 2H_2 \end{aligned} \tag{1}$$

They are reported to cause occupational disease. An exposure to about 15 PPM of NO_x is estimated to cause eye and nasal irritation. Pulmonary discomfort and aggravation of diseases like asthma and bronchitis are found after exposure above 15–25 PPM.

- **Particulate matter (PM)**

In the combustion process of different hydrocarbons, solid particles like acetylene (C_2H_2) are formed through dehydrogenation, polymerization, and agglomeration etc. The solid molecules after simultaneous polymerization produce carbon particles which is the main constituent in PM. Due to continuous improvement in emission regulation, the PM (micrograms per cubic meter) has dropped from 109 (1995) to 11 (2009). PM gets absorbed in the lungs, and they act as carcinogen in the living organisms.

Other pollutants such as lead, SO_2, smog were reported to cause irritation of eyes, lungs, and reduction of visibility. The emission effect is highest among infants and elderly people. Those with chronic heart and lung disease are at higher risk.

3 Methods and Data Evaluation Methodology

The authors have collected raw data using a mobile flue gas analyzer (MFGA) [Testo 350] from the BS-III commercial diesel auto rickshaws operating in the city of Imphal, Manipur, India (24.8170°N latitude and 93.9368°E longitude as shown in Fig. 2). The specifications of the gas analyzer (Make: TESTO; Model: 350) are tabulated in Table 1.

An uncertainty table for the instrument is given in Table 2. Three (3) types of companies (vehicle manufacturer) were considered, and four (4) vehicles from each of the company were taken for collecting the data. Table 3 shows the necessary nomenclature used for the study. A dilution factor of $X - 1$ is taken for the study. The O_2 reference for the study is 3%. The data were taken during NO PASSENGER condition, and the vehicle was at idling condition 15 min prior to the data sampling.

Table 4 shows the permissible amount of pollutants which can be diluted in the air [16, 18, 19]. The increase in emission of pollutants in the air and deforestation caused various complications in the environment and living organisms.

The predefined procedure was followed during the course of the study. A tabulation sheet is prepared to record the data. The commercial vehicle is approached, and his (driver) consent is taken for taking the data. After taking the consent, the MFGA was turned ON and the probe was taken out. The instrument was kept 15 min before the taking the test. The probe was inserted onto the exhaust

Table 1 Specifications of MFGA [Testo 350]

S. No.	Feature	Value
1	Ambient temperature	23–113 °F [short term up to 176 °F by radiated heat, i:e., heat radiation from a hot exhaust channel]
2	Ambient pressure	240.87–441.60 H_2O
3	Ambient humidity	5…95%rF
4	Power supply and connectivity	Li-ion rechargeable battery, analyzer box, power supply [100 V AC/0.45 A–240 V AC/0.2 A (50–60 Hz)]; USB 2.0
5	Memory	250,000 readings
6	Weight	0.097 lbs.
7	Display	Graphic color display, 240 × 320 pixels
8	Dimensions	10 × 4.5 × 2.3 in.

Table 2 Uncertainties of the sensors from MFGA [Testo 350]

S. No.	Sensor	Range	Resolution	Uncertainty
1	O_2	0–25 vol.%	0.01 vol.%	±0.2%
2	CO	0–10,000 PPM	1 PPM	±2.0 PPM
3	NO_x	0–300 PPM	0.1 PPM	±2.0 PPM
4	CO_2	0–50 vol.%	0.01 vol.%	±0.3%

Table 3 Nomenclature of the study

S. No.	Vehicle manufacturer nomenclature	Vehicle nomenclature
1	Company-I (CI)	Vehicle-1 (V1)
		Vehicle-2 (V2)
		Vehicle-3 (V3)
		Vehicle-4 (V4)
2	Company-II (CII)	Vehicle-1 (V1)
		Vehicle-2 (V2)
		Vehicle-3 (V3)
		Vehicle-4 (V4)
3	Company-III (CIII)	Vehicle-1 (V1)
		Vehicle-2 (V2)
		Vehicle-3 (V3)
		Vehicle-4 (V4)

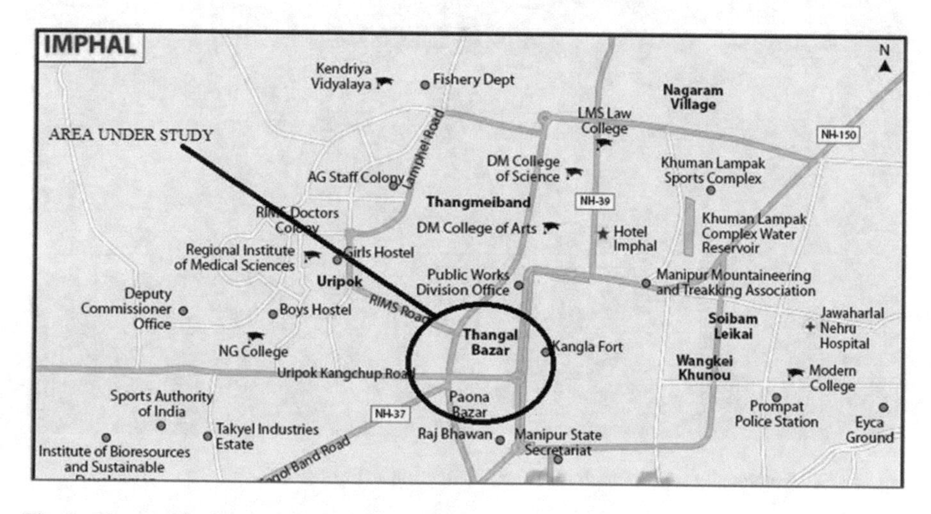

Fig. 2 Geographical map of Imphal city (Manipur) depicting the area under study [Image courtesy of www.mapsofindia.com]

Table 4 Permissible concentration of pollutants in ambient air

Pollutant	Permissible concentration [16]	Permissible concentration [18, 19]
Particulates	100–260 $\mu g/m^3$	60–100 $\mu g/m^3$
Carbon monoxide	40 PPM	2–4 $\mu g/m^3$
Sulfur dioxide	0.04–0.5 PPM	50–80 $\mu g/m^3$
Hydrocarbons	0.24 PPM	0.24 PPM
Nitrogen oxide	0.25 PPM	40–80 $\mu g/m^3$

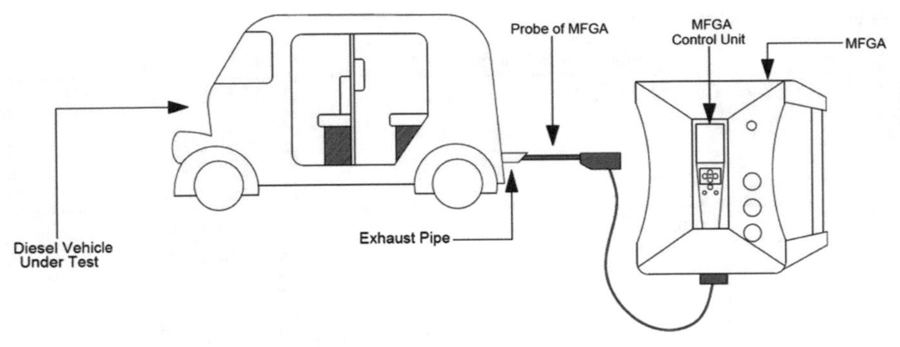

Fig. 3 Data sampling using MFGA

Fig. 4 Pictographically representation of data collection using MFGA

of the vehicle to be tested, and the data are recorded, as given in Fig. 3. Three (3) consecutive trials were performed for better accuracy of the data. The probe is taken out and is thoroughly cleaned before taking the next reading. The process continued for the remaining vehicles (Fig. 4).

4 Results and Discussion

The results were plotted, taking into account the best possible record obtained from the tabulated sheet. During recording of the data, the ambient temperature varied from 24.8 to 27 °C (as shown in Fig. 5). Since the ambient temperature is recorded

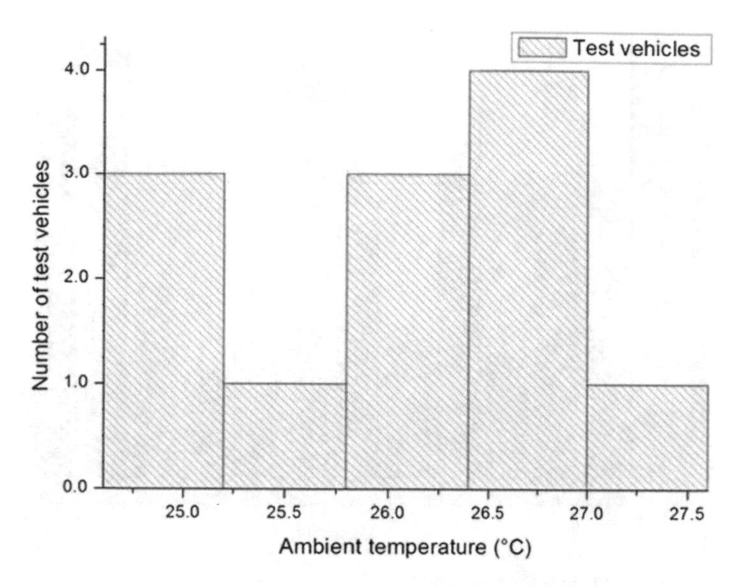

Fig. 5 Ambient air temperature during testing of the BS-III vehicles by MFGA

by the instrument, variation in the temperature from vehicle to vehicle is due to time difference to record the data. Efforts are being made by the crew and the supporting staff to ensure proper collection of data. The collected raw data were checked to make sure no junk data are fed during processing.

The carbon monoxide (CO) emissions from the samples varied between 157 and 680 PPM indicating high level of CO emission. Figure 6 shows the variation of CO emissions from the samples. It can be noted that except for few vehicles, most of the vehicles have CO emissions more than 350 PPM. Even though less CO emission is required, due to less maintenance by the driver/owner, insufficient supply of oxygen in the engine owing to improper cleaning of air filters resulted in high CO emissions. Few vehicles can be seen to emit CO above 550 PPM.

The carbon dioxide (CO_2) emission is expressed in terms of percentage of infra-red radiation (IR %) ranged from 0.48 to 4.781%, as shown in Fig. 7. The average CO_2 emission from the samples is about 0.5% which indicates that there is incomplete combustion during the operation of the engine consequently in emitting more CO than CO_2. This is quite common in poorly maintained diesel engines. Since diesel engines were operated through compression ignition, chances are there that there are missed cycles during the operation of the engine. This can be omitted by proper and regular checkup of engine and auxiliary components like air filter, fuel filter, fuel pump, and fuel delivery systems.

The emission of unburnt hydrocarbons (HC) from the exhaust in Fig. 8 varied from 1322 to 1584 PPM. In some vehicles, HC emissions were lower than 1400 PPM. Since HC are not actually a product of combustion, rather it is unburnt fuel, it is desired that the HC emissions are as low as possible. High emission of HC in the ambient air not only causes pollution to the environment, it also affects the

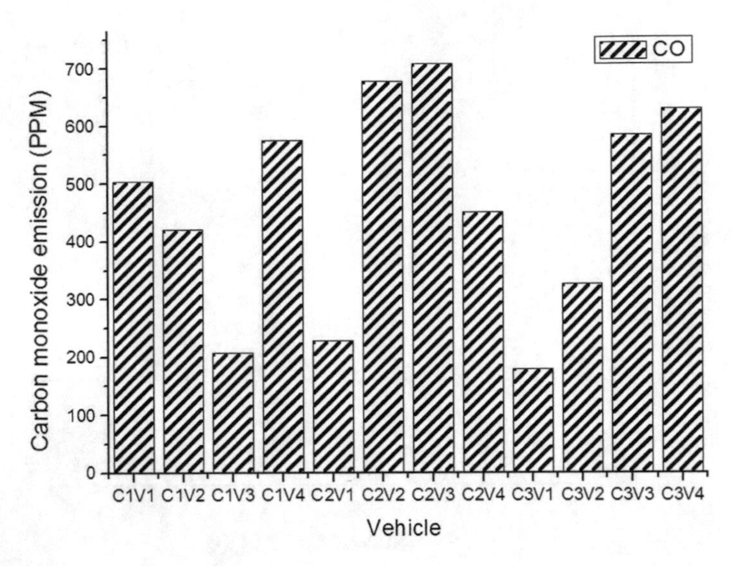

Fig. 6 Emission of CO from the test vehicles

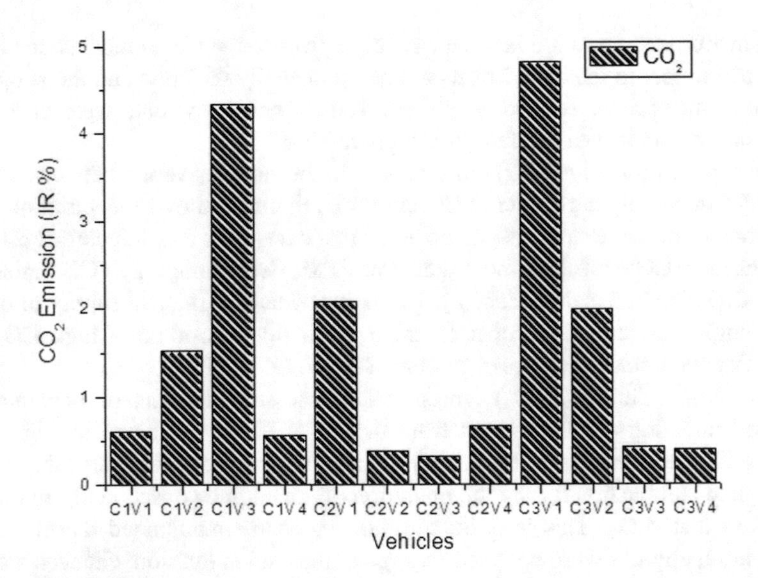

Fig. 7 Emission of CO_2 from the test vehicles

economy as it releases unburnt potential fuel to the air without utilizing the energy content of the fuel.

The oxides of nitrogen (NO_x) from the engine exhaust are found to vary from 14.6 to 44.3 PPM, as given in Fig. 9. The average NO_x emission is around 20 PPM. NO_x is formed when the temperature inside the engine is relatively higher (>1500 °C)

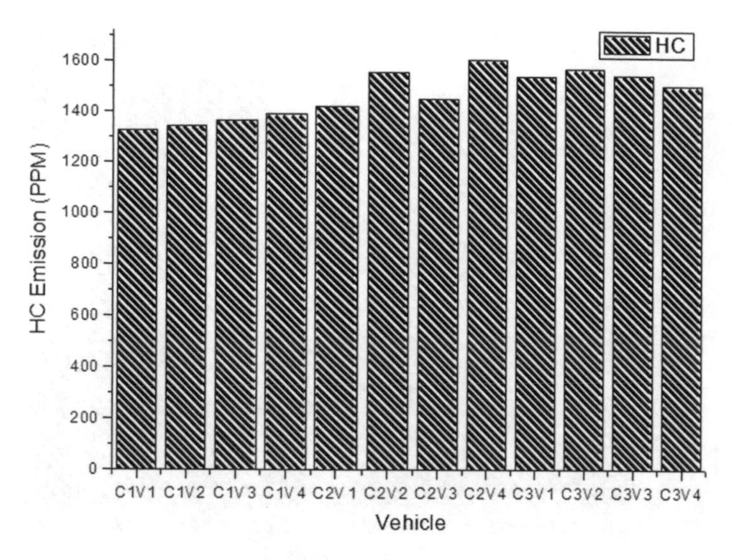

Fig. 8 Emission of HC from the test vehicles

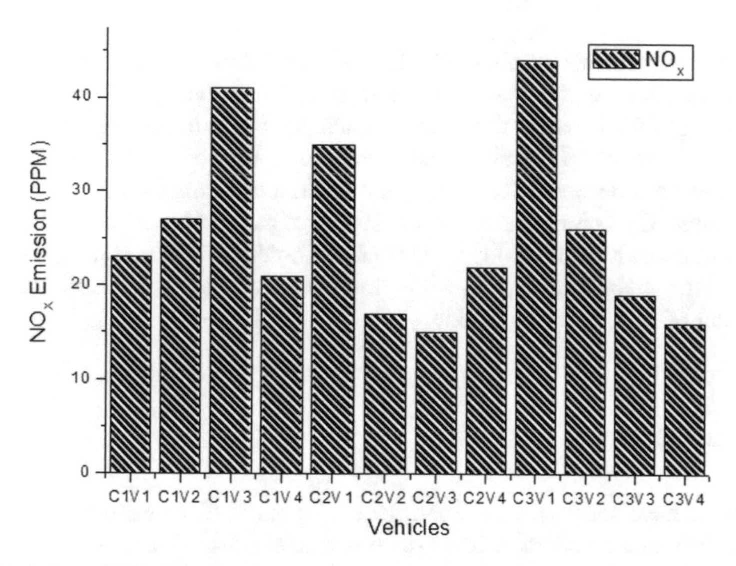

Fig. 9 Emission of NO_x from the test vehicles

than the operating conditions. The nitrogen in the air gets oxidized, and the oxides are formed. It is desirable, and most studied that the NO_x of a diesel engine must be as low as possible. Cooling systems and engine lubricant play an important role in NO_x emission. Improper cooling of the engine results in high temperature in the combustion chamber, and thus, NO_x formation takes place. Another important aspect

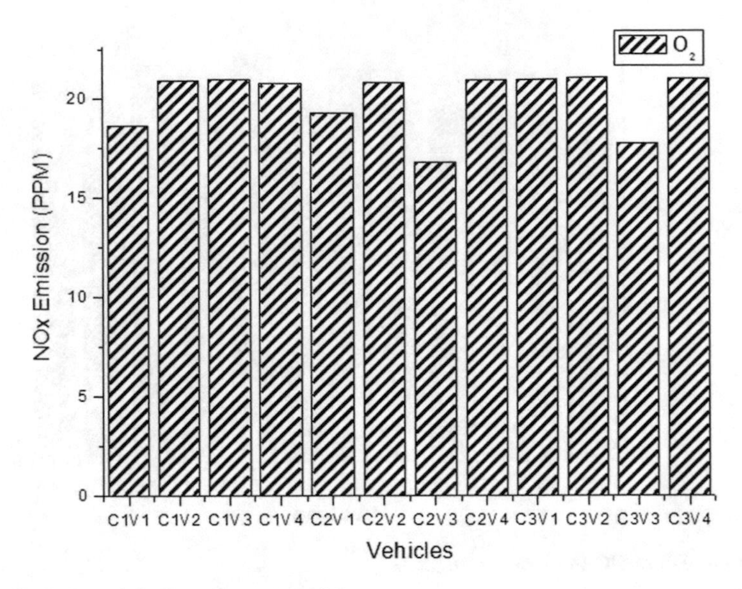

Fig. 10 Emission of O_2 from the test vehicles

is the stoichiometric air and fuel ratio during combustion. A total of 22–43% excess air in the combustion chamber also aids in NO_x formation.

The oxygen (O_2) content from the samples, as shown in Fig. 10, has an average value of around 20.7%, while some were found to have lower oxygen content (16.4%), and it is desired that the oxygen content during combustion must lay in the desired range. O_2 emission of about 17–20.6% of oxygen is desired during running conditions, and while idling, the O_2 emissions should be <10%. Since deficiency of oxygen in the engine can result in formation of CO, excess O_2 (20–25%) can aid the production of NO_x from the engine.

5 Conclusion

The authors have conducted an experiment using mobile flue gas analyzer (MFGA) to study the emission characteristics of commercial BS-III diesel vehicles operating in and around Imphal city, India. Twelve (12) vehicles, four (4) vehicles each from three different manufacturers, were taken for the study. The readings were taken on the spot from the vehicle stands, in the heart of Imphal city.

The exhaust emissions under study were CO, CO_2, HC, NO_x, and O_2. It was found that most of the vehicles under test have high soot and smoke formation from the exhaust. Poor maintenance and irregular check up of the engine and its auxiliary components might be the main reason behind the high emission of pollutants. It is desired that the authorities must also conduct special routine checks of the commercial vehicles to evaluate if the vehicular emissions are within the permissible range. This study by the author(s) is aimed at alerting the authorities of Imphal and other cities of India to take up precautionary measures in curbing emissions from poorly maintained and unregulated BS-III commercial vehicles running on the roads as failure to do so may result in deposition of high amount of the pollutants in the environment, and the air may become toxic, and the environment may suffer from severe pollution.

References

1. Keyte IJ, Albinet A, Harrison RM (2016) On-road traffic emissions of polycyclic aromatic hydrocarbons and their oxy- and nitro-derivative compounds measured in road tunnel environments. Sci Total Environ 566–567:1131–1142. doi:10.1016/j.scitotenv.2016.05.152
2. Markiewicz A, Björklund K, Eriksson E, Kalmykova Y, Strömvall AM, Siopi A (2017) Emissions of organic pollutants from traffic and roads: priority pollutants selection and substance flow analysis. Sci Total Environ 580: 1162–1174. doi:10.1016/j.scitotenv.2016.12.074
3. Fontaras G, Zacharof NG, Ciuffo B (2017) Fuel consumption and CO_2 emissions from passenger cars in Europe—Laboratory versus real-world emissions. Prog Combust Energy Sci 60:97–131. doi:10.1016/j.pecs.2016.12.004
4. Suarez-Bertoa R, Mendoza-Villafuerte P, Bonnel P, Lilova V, Hill L, Perujo A, Astorga C (2016) On-road measurement of NH_3 and N_2O emissions from a Euro V heavy-duty vehicle. Atmos Environ 139:167–175. doi:10.1016/j.atmosenv.2016.04.035
5. Han R, Yu BY, Tang BJ, Liao H, Wei YM (2017) Carbon emissions quotas in the Chinese road transport sector: a carbon trading perspective. Energy Policy 106:298–309. doi:10.1016/j.enpol.2017.03.071
6. Pirjola L, Rönkkö T, Saukko E, Parviainen H, Malinen A, Alanen J, Saveljeff H (2017) Exhaust emissions of non-road mobile machine: real-world and laboratory studies with diesel and HVO fuels. Fuel 202:154–164. doi:10.1016/j.fuel.2017.04.029
7. Abu-Hamdeh NH (2003) Effect of cooling the recirculated exhaust gases on diesel engine emissions. Energy Conver. Manag. 44(19). doi:10.1016/S0196-8904(03)00077-3
8. Brondfield MN, Hutyra LR, Gately CK, Raciti SM, Peterson SA. (2012). Modeling and validation of on-road CO_2 emissions inventories at the urban regional scale. Environ Pollut 170: 113–123. doi:10.1016/j.envpol.2012.06.003
9. Gallus J, Kirchner U, Vogt R, Benter T (2017) Impact of driving style and road grade on gaseous exhaust emissions of passenger vehicles measured by a portable emission measurement system (PEMS). Transp Res Part D Transp Environ 52:215–226. doi:10.1016/j.trd.2017.03.011
10. Degraeuwe B, Weiss M (2017) Does the new european driving cycle (NEDC) really fail to capture the NO_X emissions of diesel cars in Europe?. Environ Pollut 222:234–241. doi:10.1016/j.envpol.2016.12.050
11. Grote M, Williams I, Preston J, Kemp S (2016) Including congestion effects in urban road traffic CO_2 emissions modelling: do local government authorities have the right options? Transp Res Part D Transp Environ 43:95–106. doi:10.1016/j.trd.2015.12.010
12. Kwon S, Park Y, Park J, Kim J, Choi KH, Cha JS (2017) Characteristics of on-road NO_x emissions from Euro 6 light-duty diesel vehicles using a portable emissions measurement system. Sci Total Environ 576:70–77. doi:10.1016/j.scitotenv.2016.10.101
13. Jaiprakash, Habib G, Kumar A, Sharma A, Haider M (2017). On-road emissions of CO, CO_2 and NO_X from four wheeler and emission estimates for Delhi. J Environ Sci (China) 53: 39–47

14. Savva NS, Hountalas DT (2014) Evolution and application of a pseudo-multi-zone model for the prediction of NO_x emissions from large-scale diesel engines at various operating conditions. Energy Convers Manag 9(5): 373–388. doi:10.1016/j.enconman.2014.05.103
15. Wen L, Eglese R (2016) Minimizing CO_2e emissions by setting a road toll. Transp Res Part D Transp Environ 44:1–13. doi:10.1016/j.trd.2015.12.019
16. Holman JP (2007) Experimental method for engineers, 7th edn. Tata Mcgraw hill Publishing Company, New Delhi
17. Ganesan V (2017) Internal combustion engines, 4th edn. Tata Mcgraw hill Publishing Company, New Delhi
18. National Ambient Air Quality Standards (1998) Central Pollution Control Board, Gazette of India Extraordinary (Vide No. S.O 384(E), dated 11th April, 1994 and S.O 935(E), dated 14th October 1998)
19. National Ambient Air Quality Status and Trends in India (2010) P R division, Central Pollution Control Board, NAAQMS/35/2011–2012

Gasoline Direct Injection Engines and Particulate Emissions

Nikhil Sharma and Avinash Kumar Agarwal

Abstract Gasoline direct injection (GDI) engines are increasingly used in transport sector worldwide in recent years due to the advantages they offer. These include superior fuel economy and better engine response and control due to introduction of electronic control unit (ECU) and high-pressure fuel injection system. One of the main challenges of using GDI engine vehicles is that they emit particulates, which are not an issue in case of multipoint port fuel injection (MPFI) engines. However, there is potential to further improve GDI engines for lower particulate matter (PM) emissions. Particulates from GDI engines are of different sizes such as coarse, fine and ultra-fine, and they also vary in composition and origin. The particulate of different sizes is known to cause adverse health effects. In this chapter, fundamental aspects of both homogeneous and stratified modes of combustion of GDI engines have been discussed, in addition to wall, spray and air-guided GDI engine concepts. A section of the chapter covers detailed comparison of particulate emitted by GDI and MPFI engines. Various size and concentration-based PM measurement techniques and instruments available commercially are included in this chapter. A discussion on influence of engine load, fuel type and spray characteristics on particulate emissions is elaborated towards the end of this chapter in addition to GDI soot morphological studies.

Keywords Particulate matter · Measuring techniques · Legislation
Health effects

List of abbreviations

BMEP	Brake mean effective pressure
Dp	Diameter of primary soot particle
ECU	Electronic control unit
EGR	Exhaust gas recirculation

N. Sharma · A. K. Agarwal (✉)
Engine Research Laboratory, Department of Mechanical Engineering,
Indian Institute of Technology Kanpur, Kanpur 208016, India
e-mail: akag@iitk.ac.in

N. Sharma et al. (eds.), *Air Pollution and Control*, Energy, Environment, and Sustainability, https://doi.org/10.1007/978-981-10-7185-0_6

EGT	Exhaust gas temperature
ELPI	Electrical low-pressure impactor
FESEM	Field emission scanning electron microscope
FIP	Fuel injection pressure
GDI	Gasoline direct injection
GPF	Gasoline particulate filter
HRTEM	High-resolution transmission electron microscopy
IARC	International agency for research on cancer
IC	Internal combustion
ICP-OES	Inductively coupled plasma optical emission spectrometry
MPFI	Multipoint port fuel injection
PFI	Port fuel injection
PM	Particulate matter
WHO	World health organization

1 Introduction

GDI engine is considered as one of the most promising gasoline engine technologies for future, since it offers relatively higher BSFC and specific power output compared to existing engine technology, such as MPFI. In addition, GDI offers superior knock resistance characteristics compared to MPFI engines because of the charge cooling effect inside the engine cylinder experienced due to vaporisation of the injected liquid fuel droplets. However, particulate emissions from GDI engines are a major area of concern among scientists and industries because they pose serious health risk [1–4].

Figure 1 shows different zones of human respiratory system affected by particulates of various sizes. Particulate emissions affect human health adversely [5] because they penetrate deep into the human respiratory system (nasopharynx, trachea, bronchi, bronchioles, alveoli) [6, 7] via inhalation and contaminate the respiratory system with various toxic substances, which are adsorbed onto its surface (Fig. 1). As particulate size becomes finer, the severity of health impact increases because of increased surface-to-volume ratio and can eventually cross the cellular membranes to enter the blood-stream [8]. PM_{10} is filtered out in the nose, throat and bronchial tubes. $PM_{2.5}$ can, however, penetrate into the bronchial tubes, and $PM_{0.1}$ can even reach up to the alveoli and enter the blood-stream. Few important health problems due to PM emissions, which may lead to premature death [9] include asthma, irregular heartbeats, decreased lung function, non-fatal heart attack and severe coughing. Hence, there is a need to understand reasons for particulate emissions and their effects on public health, when using alternative/conventional fuels in IC engines and ultimately control their formation in the combustion chamber.

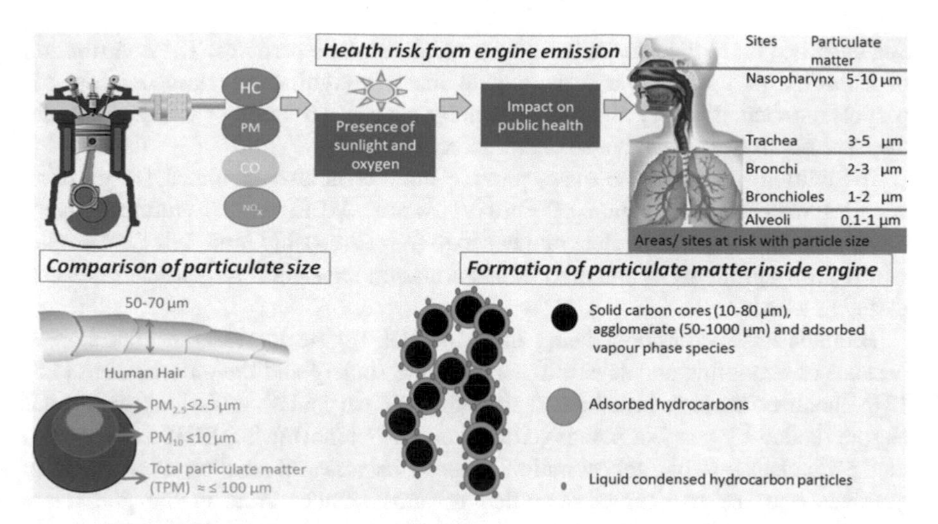

Fig. 1 Particulate emissions from IC engines [42]

Figure 2 shows the relationships between typical engine exhaust particle numbers, particle mass and particle surface area with respect to particle size. Particle size distribution in this graph can be further divided into nuclei mode particle (Dp: 5–50 nm), nanoparticles (Dp < 50 nm), ultra-fine particles (Dp < 100 nm), accumulation mode particle (Dp: 100–300 nm) and coarse mode particles (Dp < 10 μm). The particulate of different size distributions has different composition, e.g. (a) nuclei mode particles are made of hydrocarbons, which form the nucleation mode particles and (b) soot agglomerates with hydrocarbons condensed

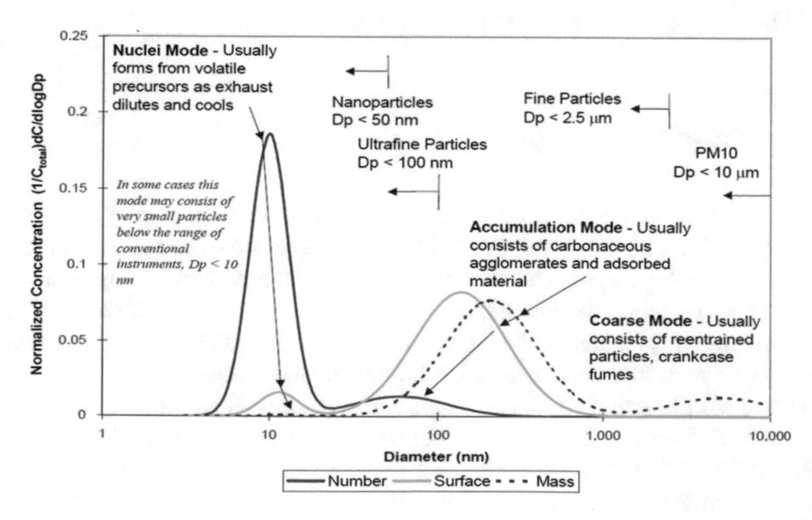

Fig. 2 Typical engine exhaust particle size distribution with particle mass and number weightings [50]

and adsorbed on to the surface form accumulation mode particles. The contribution of ultra-fine particles towards particulate mass is significantly less but towards particle numbers it is very high. Their toxicity increases because of smaller size, and they are prone to cause adverse health effects [10].

In addition to particulate mass, particle number is also regulated for gasoline vehicles with implementation of Euro-VI norms. MPFI engine emits relatively lower particle mass and particle number than the permissible limit, but GDI engine will require significant efforts to meet this emission legislation for particle mass and particle number.

Biofuels are increasingly being used in GDI engine in research laboratories, because of economic and scientific needs of the society and the environment [11–13]. Zhang et al. [14] investigated the effect of particulate emissions from GDI engine fuelled by gasohol blends (ethanol and n-butanol) using EGR and reported that particulate number concentration reduced because of addition of alcohols in gasoline, and the number peak shifted towards smaller sizes in the particulate number–size distribution curve, with increasing percentage of EGR. Higher proportions of finer particles were obtained with increasing percentage of alcohol blended in gasoline. Bai et al. [15] performed investigations using exhaust gas trap (EGT) for controlling the part-load emission characteristics of a GDI engine. They used single-stage injection to obtain homogeneous combustion mode and double injection to obtain stratified combustion mode. They concluded that cyclic variations increase with increasing EGT. They also reported that stratified mixture using the two-stage injection strategy can reduce the cyclic variations and combustion duration, thus improving the thermal efficiency. In this chapter, discussions on GDI engine combustion concept, applicable emission legislations, health effects of PM emission, comparison between MPFI and GDI engines and measurement techniques for particulate have been discussed. Effect of type of fuel, EGR, engine load and EGR on particulate morphology is also discussed.

2 Combustion Concept

GDI technology is regarded as a key technology with huge potential for the transport sector. In comparison to MPFI, GDI engine delivers superior efficiency and specific power output. This difference between the two technologies is essentially due to fuel-air mixture preparation method. In a conventional SI engine, fuel-air mixture is prepared in the intake manifold, outside the combustion chamber. There is sufficient time available to the homogeneous fuel-air mixture preparation. Moreover, MPFI technology-based engine operates on stoichiometric or near stoichiometric fuel-air mixture. On the other hand, mixture preparation in GDI engine is very different from that of MPFI. In a GDI engine, fuel-air mixture is prepared inside the combustion chamber, and the fuel is injected directly inside the engine combustion chamber. GDI engine works on two different combustion modes. Figure 3a shows homogeneous combustion mode, and Fig. 3b shows stratified

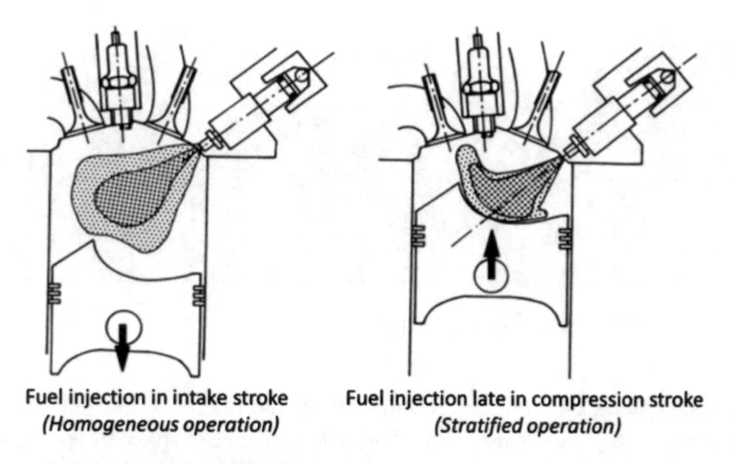

Fig. 3 Two modes of combustion in GDI engine

combustion mode. In homogeneous mode of combustion, mixture is stoichiometric, and in stratified mode of combustion, overall mixture is lean, but the charge is inhomogeneous and stratified. The local fuel-air ratio near the spark plug (at the flame) is richer.

In homogenous mode of combustion, fuel is injected early in the intake stoke so that there is sufficient time available for homogenous fuel-air mixture preparation. On the other hand, in stratified mode of combustion, fuel is injected late in the compression stoke. In both the cases, fuel-air mixture is ignited with the help of a spark plug. Due to comparatively lesser time available, the mixture is stratified, and it essentially leans out spatially with increasing distance from the fuel spray core. Stratified mode of combustion is required when the demand for power is less, and homogenous mode of combustion is required when the demand of power is comparatively higher. Lesser fuel quantity is injected in the stratified mode of combustion, and relatively higher fuel quantity is injected in the homogeneous mode of combustion.

Apart from the two modes of combustion, there are three types of engine designs for GDI engine as shown in Fig. 4 (spray-, wall- and air-guided), which are commercially available.

In the spray-guided GDI engine, the gap between the spark plug and the injector is comparatively lesser. This type of design is typically favourable to create stratified combustion mode by forming ignitable stratified mixtures near the spark plug. The moment fuel is injected, a spark instigates combustion in the richer part of the stratified charge near the spark plug. Parameters such as combustion chamber geometry, spray dynamics, FIP and fuel droplet size distribution are the factors responsible for charge stratification. This design causes spark-plug fouling and soot formation but eliminates wall-wetting. Therefore, fuel spray should be precisely controlled. Minute variation in fuel quantity may result in large cyclic variations and misfiring.

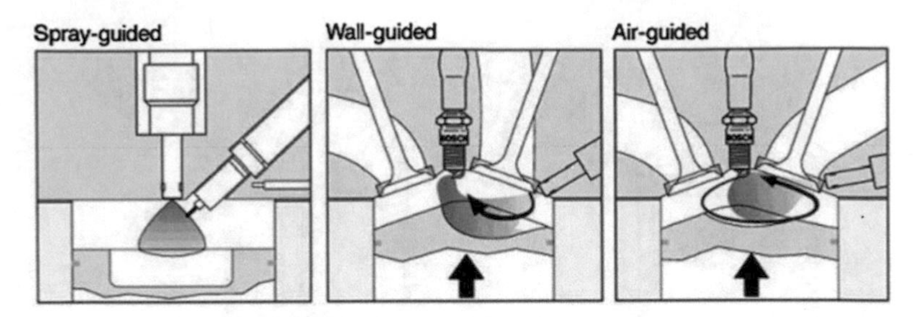

Fig. 4 Different DISI combustion systems [51]

In a wall-guided GDI engine, the injector is mounted on one side, and curved nose shape of the piston guides the charge nearer to the spark plug. There exists sufficient gap between the spark plug and injector. This type of engine is typically used to operate in homogeneous combustion mode. This is because there exist significant cyclic variations in stratified combustion mode for this design. Costa et al. [16] concluded that piston head geometry played an important role in the mixture formation in wall-guided GDI engines. Since fuel directly impinges on the top surface of the piston, this design of engines generates relatively higher soot emissions than spray-guided GDI engines.

In the air-guided GDI engine, interaction between the fuel spray and the motion of air charge inducted in the cylinder is responsible for charge stratification. The fuel is essentially injected into the air, which is directed towards the spark plug. In such designs, shape of inlet port plays an important role to direct the fuel near the spark plug. In this technique, the butterfly valve/throttle controls the mass flow rate of air, and fuel does not wet the piston top or the cylinder.

In a GDI engine, one combustion system and both combustion modes are present. Most recent engines use a combination of wall- and air-guided designs.

A typical torque vs. BMEP curve is presented in Fig. 5. Stratified combustion mode is desirable at low engine speeds and at lower engine torque. Homogeneous combustion mode is required, when the driver demands higher torque and higher engine speed. The amount of throttle pressed by drivers gives a message to the ECU to operate the engine in a specific mode. There is no separate button to change the mode of engine operation.

3 Legislation

Emission legislations play a significant role in improving the urban air quality; therefore, they tend to become increasingly more stringent with time. Although there is no scientifically proven and defined threshold where one can say that PM will not cause adverse health effects. In recent years, strict restrictions have been applied by governments world over to limit emissions from transport sector to improve air quality, especially in urban settings.

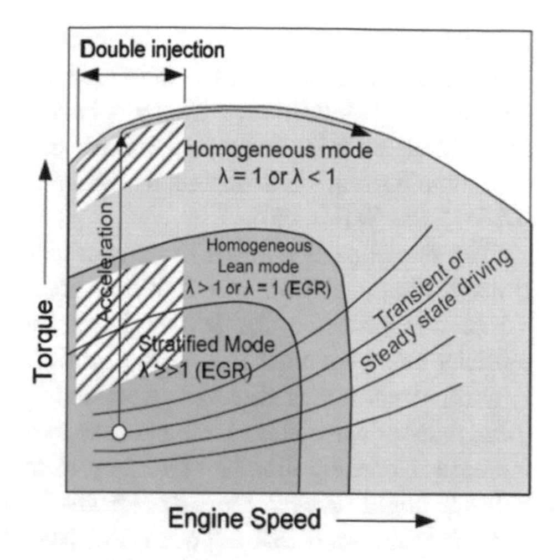

Fig. 5 Speed versus BMEP for different combustion modes

Indian emission legislations closely follow European Union legislations. In case of four-wheelers, Indian emission legislations are behind European legislations by ~ 5 years in major cities of India and ~ 10 years behind throughout the nation. However, two- and three-wheeler emission legislations are developed independently by India, depending upon country-specific requirements. At present, many Indian metro cities are struggling with severe air-quality issues. In this context, India has decided to directly move from BS-IV to BS-VI emission legislations by April 2020 in the entire nation. With this courageous decision, India has set a benchmark for other developing nation to also provide a clean and healthy environment for sustainable development of their citizens. There is a PM mass emission limit applicable to all gasoline engines in EU-V emission standards, which limits it to 0.005 g/km. To meet even more stricter emission legislations of EU-VI, GDI engines must be equipped with gasoline particulate filter (GPF).

4 Health Effects of Particulates

Rapid industrialization and urbanization for creating economic growth and job opportunities are the reason for increased PM levels in the metro cities. PM contains solids/liquid droplets which are considered as harmful to the human health because they have long atmospheric retention time and they also travel longer distances. Government implements emission legislations and introduces policies from time to time to reduce PM; however, current levels of PM emissions still cause prominent risk to human health. People with pre-existing heart and lung diseases and children and elderly people are more prone to adverse health impact of PM emissions. Air pollutants in combination with particulates accumulate and increase the severity of adverse health effects. Particulates change the blood circulation in human body and increase the mortality rate [17]. Mortality rate increases with continuous exposure to particulates. It is roughly estimated that exposure to diesel particulates could decrease projected life expectancy between one year [18] to five and half years [19]. In addition to this, the exhaust from diesel engines has been classified as carcinogenic (Group 1) to humans by the International Agency for Research on Cancer (IARC) and WHO [20].

The size of particulates is directly associated with the severity of health problems caused. $PM_{2.5}$ is about 1/28th in diameter to that of a human hair or even smaller. These fine particles can invade lungs and alveolar regions [21], and ultra-fine particles can even enter the blood vessels [22]. Exposure time to these fine particles is another parameter directly associated with health hazard. Longer the exposure time, more adverse effects it may cause to the human health. This is primarily due to increase in cardiopulmonary and lung cancer mortality. Exposure of particulates to children could be even more dangerous. It directly affects lungs and its growth rate [23]. It is reported that exposure to $PM_{2.5}$ is directly linked with the risk of cardiopulmonary mortality by 6–13% per 10 $\mu g/m^3$ of $PM_{2.5}$ [24–26]. In 2005, 130,000–320,000 premature deaths took place in the USA due to high exposure to $PM_{2.5}$. In another study, reduction in long-term exposure to $PM_{2.5}$ in the USA resulted in increased average life expectancy [27]. Various adverse health effects caused by particulates are summarized below in Table 1.

Table 1 Various health effects caused by particulates to humans and animals

Diseases	Causes/remarks
Asthma	Asthma symptoms can be worsened by increase in PM levels in a city [52, 53]
Lung cancer/decreased lung function/lung irritation	Fine particles directly affect the bronchi, which effects health of lungs and results in cancer. It causes lung irritation, leading to increased permeability in lung tissues [54, 55]
Cardiovascular diseases	Fine particulate affects heart and its functions [56]
Premature delivery and birth defects	PM may pass from mother to the child, resulting in a wide range of birth defects. Lowers birth rate [57–59]
Premature death	Higher in regions with high level of particulate emissions [60, 61]
Vascular inflammation	This is a result of plaque in arteries [62]
Atherosclerosis	Arteries become hard, reduce elasticity, plaque builds up in arteries. This leads to heart problems [63, 64]
Inflammation of lung tissues	Releases chemicals, which directly attack heart and reduce its functions [65, 66]
Blood chemistry changes	Results in clots that may lead to heart attack [62, 67]
Sensitivity to viral and bacterial germs	This leads to pneumonia in vulnerable person [68]
Others diseases	Increased blood pressure (BP), increased stress, autonomic imbalance and arrhythmias, prothrombotic and coagulant changes [67]

5 Reasons for Particulate Formation in GDI Engines

Particulates from diesel engines are being studied by researchers for last many decades. Particulate emissions were not a major concern in gasoline engines until GDI came into reality. Figure 6 shows the steps of soot formation mechanism and reasons for soot formation in GDI engines.

PFI of gasoline results in significantly lower PM emissions compared to GDI engines. This is because PFI results in strongly premixed and almost homogeneous charge resulting in stoichiometric mixture inside the combustion chamber. Hence, there are no fuel-air rich mixture zones in the combustion chamber; therefore, PFI results in negligible soot formation. On the other hand, homogeneity of fuel-air charge in GDI engine is relatively lower. The reason for relatively lower fuel-air mixture homogeneity is due to limited time available for fuel droplets to evaporate in the combustion chamber. This also results in high-temperature zones where there is relatively lower amount of oxygen present and higher concentration of

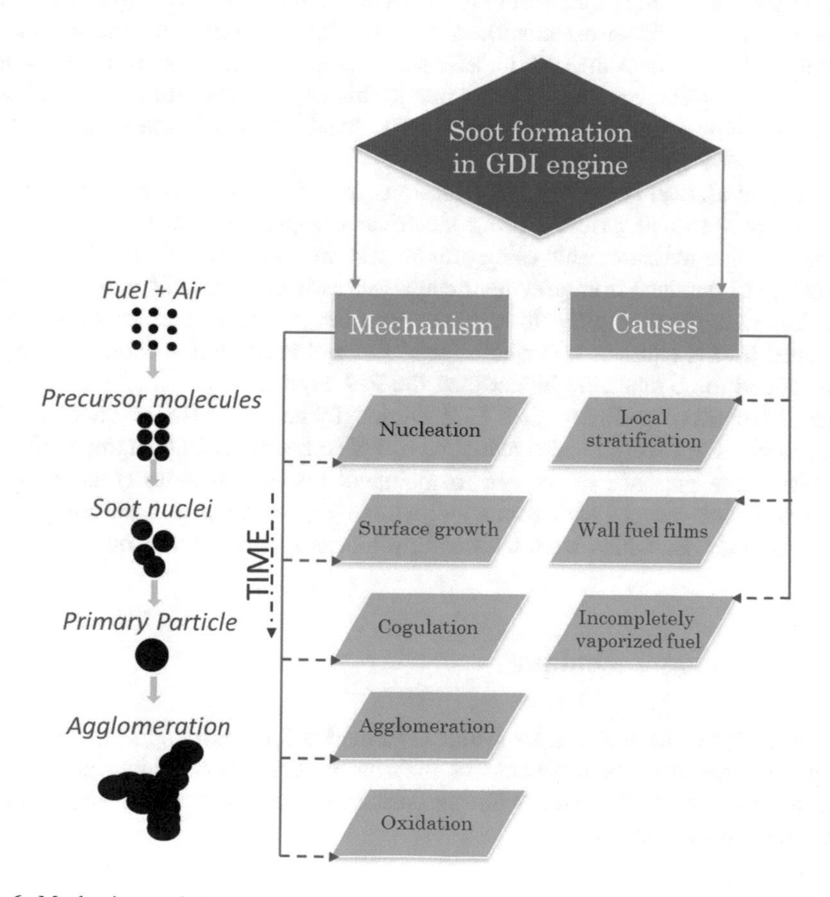

Fig. 6 Mechanism and steps responsible for soot formation in GDI engines

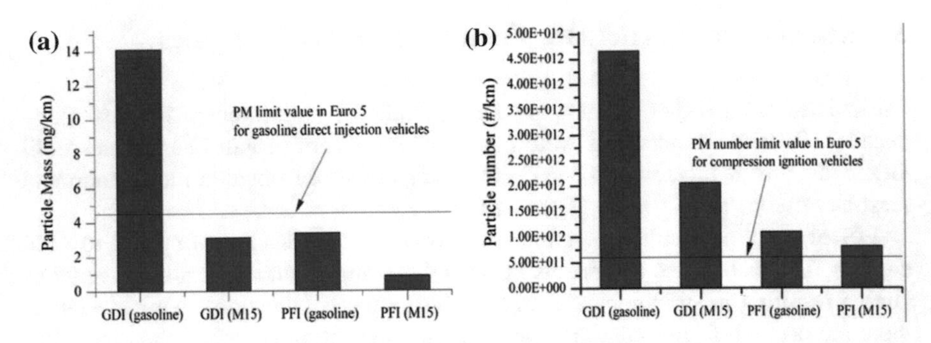

Fig. 7 **a** PM mass emissions using gravimetric measurement method. **b** total PN emission using ELPI [28]

hydrocarbon fuels. This may result in endothermic pyrolysis reactions, which in turn lead to soot formation. Another aspect, which is responsible for relatively higher particulate emissions from GDI engines, is fuel spray impingement on the piston surface (Leidenfrost effect). A fuel vapour film forms on the piston top surface, which reduces the heat transfer rate. Therefore, evaporation of subsequent fuel spray droplets is relatively lesser due to this vapour film formation, leading to diffusion combustion. This may also lead to locally fuel-rich combustion, which results in particulate formation.

Liang et al. [28] compared PM emissions from GDI vehicles and PFI vehicles fuelled by M15 and gasoline using electrical low-pressure impactor (ELPI) and mass emission measurements using gravimetric method. It was found that for each vehicle, PM obtained from gravimetric analysis estimated that PM mass from M15 fuelling was comparatively lower than gasoline fuelling. Similar trends were obtained for PN emissions. For both fuels, PM and PN emission from GDI engine were found to be relatively higher than the PFI engine.

It is reflected by Fig. 7 that GDI engine fuelled by M15 resulted in 78% reduction in PM mass and 56% reduction in PN compared to GDI engine fuelled by gasoline. The presence of oxygen in methanol resulted in more complete combustion, which reduced soot formation. Moreover, methanol did not contain aromatic components which were the building blocks for soot formation.

6 Measuring Techniques

There are numerous instruments available in market for measuring size distribution of particulates. Important instruments measure particle concentration and particle size distributions [29]. Figure 8 shows various methods and instruments used for measurements of particulates.

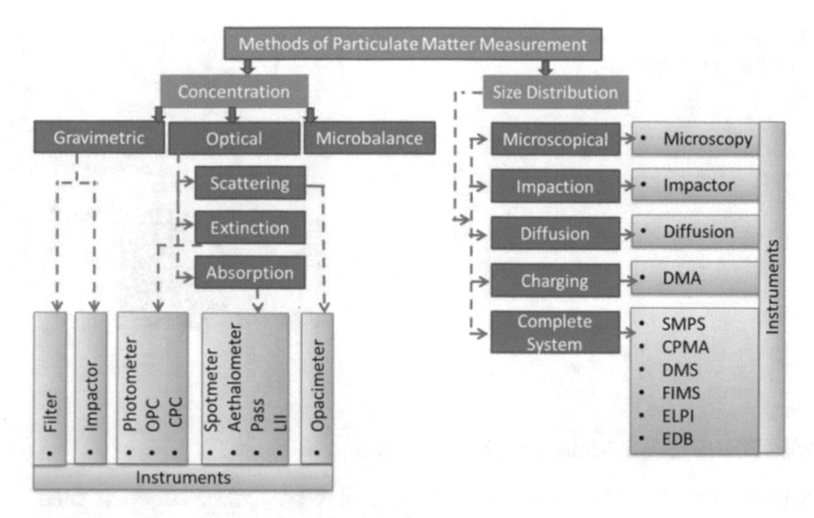

Fig. 8 Various methods for measurement of particulates

a. Concentration Measurement Methods

The particulate concentration can be measured in terms of particulate mass (*m*), particulate number (*N*) and particulate surface area (*S*) distributions with particulate size. There are several correlations used in such instruments. These measuring instruments work on different principles.

i. Gravimetric Method: In this method, particulate sampling is done on a quartz filter paper in a dilution tunnel for a fixed time. Particle mass concentration is found by weighing the quartz filter paper before and after the sampling. This offline measurement technique is time-consuming and costly.
ii. Optical Methods: In these methods, light (visible, infrared or laser) is made to pass through exhaust gas, where each single particle scatters the incident light in all directions. Each scattering signal is detected by a photo-diode. Optical instruments generally work on the principles of light extinction, or scattering, or absorption. These methods have an advantage of measuring particles of several sizes simultaneously. The estimation of particle size is based on many assumptions related to characteristics of particles. Particle diameter obtained is equivalent diameter for particles with the same refractive index as that of the calibration particle of the instrument employed. Details of such instruments are given by Giechaskiel et al. [30].
iii. Microbalance Method: Such methods use a vibrating substrate to collect soot particles. The frequency of oscillations of vibrating substances changes with changing amount of particulate collected on the substrate. This change in frequency is used to calculate the mass concentration. Giechaskiel et al. [30] described this measurement technique in detail.

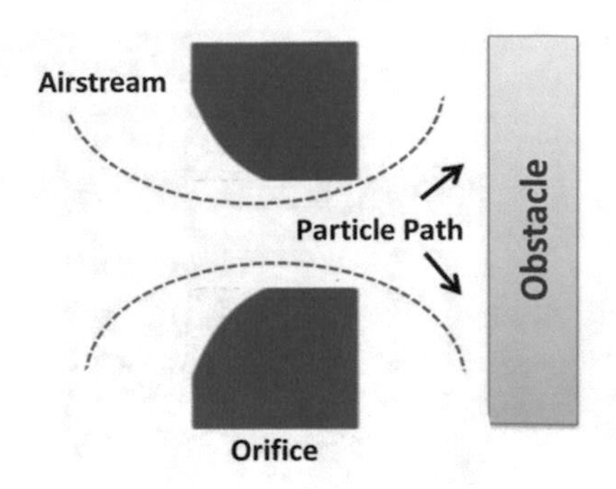

Fig. 9 Typical design of an impactor-based segregation scheme

b. **Size Distribution Measurement Methods**

i. Microscopy: In this technique, soot samples are collected on a filter paper through a dilution tunnel. Further, samples are prepared for SEM and TEM analysis. The time spent for microscopic analysis is significantly higher than other measuring techniques.
Impactor: An impactor consists of a series of orifices or circular slots and an obstacle designed for the collection of size-segregated particulates. A typical impactor is shown in Fig. 9. Particles larger than specific aerodynamic size strike the obstacle surface, while smaller particles proceed through the sampler [31]. Smaller particles are collected in the next stage with smaller orifice sizes. This process continues until the smallest particles get filtered out through the impactor. The particles get separated based on their inertia [32].

The details of few other size distribution measurement methods with their measuring principles and schematic are given in open literature [33–36].

7 Impact of Fuel Composition/Properties on PM Emissions

There are many fuel properties/composition which can effect PM emissions. The following fuel properties are the prominent ones:

(1) Aromatic content
(2) Olefin content
(3) Sulphur content
(4) Volatility
(5) Oxygen content

(1) Aromatic content: Aromatic hydrocarbons are those hydrocarbons, which have sigma bonds and delocalized pi electrons between carbon atoms forming a circle. Higher aromatic hydrocarbons have very high octane number, and they

reflect poor quality of mineral diesel. Increasing the aromatic content in the fuel exacerbates fuel pyrolysis, yielding more soot [37]. Hence, particulate emissions increase with increasing aromatic content of the fuel. Combustion characteristics such as heat release rate and the cylinder pressure are not affected significantly by the aromatic content of the fuel [38].

(2) Olefin content: The class of hydrocarbon compounds that has at least one carbon–carbon double bond in the linear chain is called olefins. Due to the presence of double bond, these olefins are more reactive than paraffinic or aromatic compounds present in gasoline. Olefins have higher formation potentials, and their quantity in gasoline is legislated by the specifications/ government. Moreover, olefins have a lower octane number, and they are known to cause carbon deposits in engines. High olefin content in the fuel can lead to plugging of the nozzle holes in the fuel injector and valves deposits due to higher particulate emissions.

(3) Sulphur content: Sulphur exists naturally in small quantities in petroleum products. However, oil companies have to remove fuel sulphur because of two reasons. First reason is that it combines with oxygen to produce sulphur dioxide, which is a catalyst poison and ruins engine's catalytic converter system. The other reason is that it causes acid-rain. SO_2 emissions contribute to formation of secondary inorganic aerosols, and fine particles, which are harmful to the human health. The sulphur compounds are primarily converted by the engine to SO_2, which is not really a problem to be resolved by the automotive manufacturers. This needs to be resolved by the petroleum refiners, who are responsible for removing fuel sulphur. SO_2 gets convert into sulphuric acid under atmospheric conditions in the presence of moisture and sunlight, which encourages gas-to-particle conversions and ultimately soot formation. Hence, SO_2 becomes important for the automotive manufacturers because the diesel oxidation catalysts (DOCs) convert SO_2 into sulphuric acid, which encourages particulate formation that the dilution tunnel filter captures.

(4) Volatility: Volatility is a measure of the tendency of a fuel to vaporize and change from liquid to vapour. Different fuels have different volatility, and this further depends on in-cylinder pressure and temperature. PM emissions are highly dependent on fuel volatility. Lesser the volatility of fuel such as biodiesel, higher will be particulate emission generated by it under identical engine operating conditions. PM emissions are related to both, FIP and fuel volatility. If the fuel volatility is higher, lower FIP can produce lesser particulate emissions from the engine. This is because higher volatility enhances fuel-air mixture, resulting in relatively lesser particulate formation. Fuel film formation on the piston of wall-guided GDI engine is also reduced in case of higher volatility fuels. They also result in reduction in PM emission from a GDI engine.

(5) Oxygen content: PM is generally formed when fuel does not completely oxidize due to insufficient oxygen availability because of some reason [39]. If a fuel contains inherent fuel oxygen content, then more complete combustion can take

place. Oxygenated fuels such as ethanol, methanol, butanol and propanol are usually blended with gasoline to increase fuel oxygen content. These alcohols can be added in various proportions to the gasoline. Inherent oxygen present in ethanol helps oxidize particles and ultimately reduce PM emissions [40].

8 Impact of Engine Load on PM Emissions

In-cylinder temperature and pressure change with changes in engine load. At lower engine loads, in-cylinder temperature is low as well because lower amount of fuel is injected into the combustion chamber. With increasing engine load, amount of fuel injected in the cylinder also increases, which increases the peak in-cylinder temperature. Cylinder pressure also increases with increasing engine load. Both the peak temperature and in-cylinder pressure of the engine combustion chamber have a significant influence on particulate formation and emissions. Higher engine loads shorten the ignition delay, which advances the start of combustion in the expansion stroke, resulting in higher peak in-cylinder pressure. Xing et al. [41] reported that with increasing engine load, the relative percentages of soot emission increased, while the relative percentages of organic particles decreased. Sharma et al. [42] reported that particulate emissions were higher at no load and decreased at intermediate loads to again increase at high loads.

9 Effect of EGR on PM Emissions

Influence of EGR on the PM emissions is a complex phenomenon [43–45]. EGR is one of the most common techniques employed to control NO_x emissions. With addition of EGR to the combustion chamber, amount of oxygen available in the engine cylinder decreases, which increase the PM emissions. This also results in reduction in peak power output. Therefore, EGR valve should not be opened, when peak power output is required. EGR can be employed in the engine as hot EGR or cold EGR. Alger et al. [46] reported a reduction in PM and PN emissions by employing cooled EGR from a turbocharged PFI engine. Zhao [47] reported that EGR increased nucleation mode particle emission and decreased accumulation mode particle emission. Lattimore [48] reported opposite trend to the one reported by Zhao [47] for nucleation and accumulation mode particles. PM emissions change with changing EGR rate and engine operating condition.

10 Influence of Spray Characteristics on PM Emissions

In order to improve the combustion characteristics and combustion efficiency of a GDI engine, it is important to understand the influence of fuel spray characteristics on the particulate emissions. Parameters such as in-cylinder pressure, temperature, fuel-air mixture flow-field and CAD position of piston play a critical role. In addition, parameters such as FIP, SoI, ST, which are dependent on engine load and speed, influence PM emissions. These parameters control the fuel injection system and affect the PM emitted from the engine. These parameters must be therefore calibrated for all engine loads and speeds.

Fuel injection pressure and spray penetration length: Higher FIP enhances spray atomization and decreases fuel droplet size. This also enhances fuel-air mixture by increasing the surface area of fuel droplets in contact with the ambient air. Spray penetration length is dependent on FIP.

In a wall-guided GDI engine, fuel impingement on the piston is intentional which results in relatively higher PM. In a spray-guided GDI engine, there is comparatively less impingement on the piston hence lesser PM is produced. PM emission in spray-guided GDI engine is a result of fuel impingement on the valves and liner. Fuel penetration length should be optimized, which depends on FIP and spray break-up length at different engine operating conditions. Atomization and spray break-up are related to turbulence kinetic energy of injected fuel quantity and shear force between fuel spray and the ambient air.

11 Particulate Morphology

The term particulate morphology refers to the structural features of the soot particles such as diameter, area, inter-lamellar distance, periodicity, as well as length and tortuosity of the soot particles. The first step towards finding this is to collect the soot particles on a quartz filter paper through a dilution tunnel. The loaded filter paper can further be analysed using various analytical instruments. The term "analytical instruments" here refers to a broad category of equipment used to evaluate the composition of soot particles such as trace metals in particulate, structure (amorphous or crystalline) and morphology of soot particles for various test fuels at different engine operating conditions. FE-SEM, HR-TEM, ICP-OES are some of the analytical instruments.

HR-TEM is the most advanced technique available for characterizing nanoparticles as of now. Liati et al. [49] studied the characteristics of the soot particles collected from a GDI engine using HR-TEM images. Figure 10 shows the presence of nearly concentric multi-core particles, graphene lamellae interrupted by amorphous material (disordered atomic arrangements). Amorphous material appears due to discontinuous incorporation of organic compounds, which originates during growth of particles. Core in the above figure consists of randomly arranged graphene lamellae.

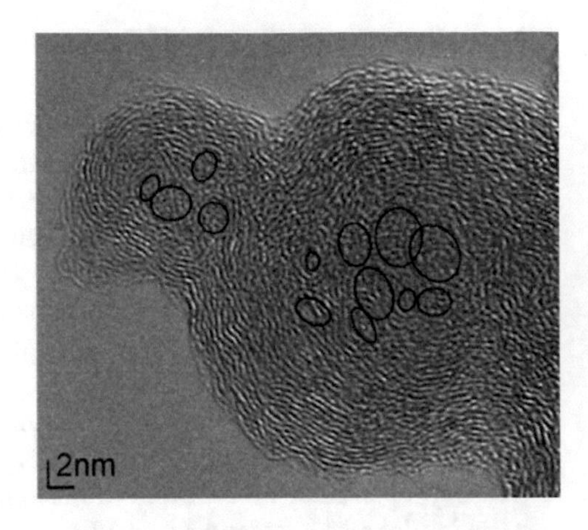

Fig. 10 HRTEM image showing: **a** multi-core particles (cores outlined for clarity); **b** crystallites in the particles [49]

12 Conclusions

Scientists from different fields are constantly working on assessing and reducing the adverse health effects caused by particulates from GDI engines. Investigations such as type of fuel, engine operating conditions, EGR give insights into methods as to how the particulate emissions from GDI engine can be reduced. Based on the research findings of various scientific studies, recommendations are made to limit the particulate emissions for the next set of emission legislations. Asthma, lung cancer/decreased lung function/lung irritation, cardiovascular diseases, premature delivery and birth defects, premature deaths, vascular inflammation, atherosclerosis, inflammation of lung tissues, blood chemistry changes, sensitivity to viral and bacterial germs are some of the important adverse health effect caused by engine particulates. A comparative analysis GDI and PFI engine technologies gave more insights of fundamental differences between gasoline-origin particulate formation processes. A discussion on the advantages of different types of GDI engine designs is also included in this chapter. Different types of particulate measuring technique/principles (mass based and concentration based) have been discussed, and instruments available commercially for such measurements are also touched upon briefly.

References

1. Zhan R, Eakle ST, Weber P (2010) Simultaneous reduction of PM, HC, CO and NOx emissions from a GDI engine. SAE technical paper
2. Samuel S, Hassaneen A, Morrey D (2010) Particulate matter emissions and the role of catalytic converter during cold start of GDI engine. SAE technical paper
3. Qin J, Li X, Pei Y (2014) Effects of combustion parameters and lubricating oil on particulate matter emissions from a turbo-charged GDI engine fueled with methanol/gasoline blends. SAE technical paper

4. Whelan I, Samuel S, Hassaneen A (2010) Investigation into the role of catalytic converters on tailpipe-out nano-scale particulate matter from gasoline direct injection engine. SAE technical paper
5. Stuart BO (1984) Deposition and clearance of inhaled particles. Environ Health Perspect 55:369
6. Chakraborty A, Gupta T (2010) Chemical characterization and source apportionment of submicron (PM1) aerosol in Kanpur region India. Aerosol Air Qual Res 10:433–445
7. Gupta T, Mandariya A (2013) Sources of submicron aerosol during fog-dominated wintertime at Kanpur. Environ Sci Pollut Res Int 20:5615
8. Oberdörster G, Sharp Z, Atudorei V, Elder A, Gelein R, Kreyling W et al (2004) Translocation of inhaled ultrafine particles to the brain. Inhalation Toxicol 16:437–445
9. Geiger A, Cooper J (2010) Overview of airborne metals regulations, exposure limits, health effects, and contemporary research. US Environ Prot Agency 25:2015 Accessed on August
10. Organization WWH (2006) Air quality guidelines for particulate matter, ozone, nitrogen dioxide and sulfur dioxide: global update 2005. WHO/SDE/PHE/OEH
11. An Y-z, Teng S-p, Pei Y-q, Qin J, Li X, Zhao H (2016) An experimental study of polycyclic aromatic hydrocarbons and soot emissions from a GDI engine fueled with commercial gasoline. Fuel 164:160–171
12. Cucchi M, Samuel S (2015) Influence of the exhaust gas turbocharger on nano-scale particulate matter emissions from a GDI spark ignition engine. Appl Therm Eng 76:167–174
13. Y-z An, Y-q Pei, Qin J, Zhao H, S-p Teng, Li B et al (2016) Development of a PAH (polycyclic aromatic hydrocarbon) formation model for gasoline surrogates and its application for GDI (gasoline direct injection) engine CFD (computational fluid dynamics) simulation. Energy 94:367–379
14. Zhang Z, Wang T, Jia M, Wei Q, Meng X, Shu G (2014) Combustion and particle number emissions of a direct injection spark ignition engine operating on ethanol/gasoline and n-butanol/gasoline blends with exhaust gas recirculation. Fuel 130:177–188
15. Bai Y-l, Wang Z, Wang J-x (2010) Part-load characteristics of direct injection spark ignition engine using exhaust gas trap. Appl Energy 87:2640–2646
16. Costa M, Marchitto L, Merola S, Sorge U (2014) Study of mixture formation and early flame development in a research GDI (gasoline direct injection) engine through numerical simulation and UV-digital imaging. Energy 77:88–96
17. Fischer PH, Marra M, Ameling CB, Hoek G, Beelen R, de Hoogh K et al (2015) Air pollution and mortality in seven million adults: the dutch environmental longitudinal study (DUELS). Environ Health Perspect 123:697
18. Organization WH (2003) Health aspects of air pollution with particulate matter, ozone and nitrogen dioxide: report on a WHO working group, Bonn, Germany 13–15 Jan 2003
19. Chen Y, Ebenstein A, Greenstone M, Li H (2013) Evidence on the impact of sustained exposure to air pollution on life expectancy from China's Huai river policy. Proc Natl Acad Sci 110:12936–12941
20. Mohapatra K, Dash C, Dash B (2016) A case study on the impact of particulate matter on health. Carbon 3(1)
21. Kampa M, Castanas E (2008) Human health effects of air pollution. Environ Pollut 151:362–367
22. Nemmar A, Hoet PM, Vanquickenborne B, Dinsdale D, Thomeer M, Hoylaerts M et al (2002) Passage of inhaled particles into the blood circulation in humans. Circulation 105:411–414
23. Organization WH (2011) Exposure to air pollution (particulate matter) in outdoor air. Copenhagen, WHO Regional Office for Europe, (ENHIS Factsheet 3.3)
24. Beelen R, Hoek G, van Den Brandt PA, Goldbohm RA, Fischer P, Schouten LJ et al (2008) Long-term effects of traffic-related air pollution on mortality in a Dutch cohort (NLCS-AIR study). Environ Health Perspect 116:196
25. Krewski D, Jerrett M, Burnett RT, Ma R, Hughes E, Shi Y et al (2009) Extended follow-up and spatial analysis of the American Cancer Society study linking particulate air pollution and mortality. Res Rep Health Eff Inst 140:5–114

26. Pope CA III, Burnett RT, Thun MJ, Calle EE, Krewski D, Ito K et al (2002) Lung cancer, cardiopulmonary mortality, and long-term exposure to fine particulate air pollution. JAMA 287:1132–1141
27. Pope CA III, Ezzati M, Dockery DW (2013) Fine particulate air pollution and life expectancies in the United States: the role of influential observations. J Air Waste Manag Assoc 63:129–132
28. Liang B, Ge Y, Tan J, Han X, Gao L, Hao L et al (2013) Comparison of PM emissions from a gasoline direct injected (GDI) vehicle and a port fuel injected (PFI) vehicle measured by electrical low pressure impactor (ELPI) with two fuels: gasoline and M15 methanol gasoline. J Aerosol Sci 57:22–31
29. Amaral SS, de Carvalho JA, Costa MAM, Pinheiro C (2015) An overview of particulate matter measurement instruments. Atmosphere 6:1327–1345
30. Giechaskiel B, Maricq M, Ntziachristos L, Dardiotis C, Wang X, Axmann H et al (2014) Review of motor vehicle particulate emissions sampling and measurement: From smoke and filter mass to particle number. J Aerosol Sci 67:48–86
31. William CH (1982) Aerosol technology. Prop, Behav Measur Airborne Part
32. Hinds WC (2012) Aerosol technology: properties, behavior, and measurement of airborne particles. Wiley
33. Nussbaumer1a T, Czasch C, Klippel N, Johansson L, Tullin C (2008) Particulate emissions from biomass combustion in IEA countries
34. Jiang R, Bell ML (2008) A comparison of particulate matter from biomass-burning rural and non-biomass-burning urban households in northeastern China. Environ Health Perspect 116:907
35. Jayne JT, Leard DC, Zhang X, Davidovits P, Smith KA, Kolb CE et al (2000) Development of an aerosol mass spectrometer for size and composition analysis of submicron particles. Aerosol Sci Technol 33:49–70
36. Elsasser M, Crippa M, Orasche J, DeCarlo P, Oster M, Pitz M et al (2012) Organic molecular markers and signature from wood combustion particles in winter ambient aerosols: aerosol mass spectrometer (AMS) and high time-resolved GC-MS measurements in Augsburg Germany. Atmos Chem Phys 12:6113–6128
37. Kidoguchi Y, Yang C, Miwa K (2000) Effects of fuel properties on combustion and emission characteristics of a direct-injection diesel engine. SAE technical paper
38. Kidoguchi Y, Yang C, Kato R, Miwa K (2000) Effects of fuel cetane number and aromatics on combustion process and emissions of a direct-injection diesel engine. JSAE Rev 21:469–475
39. Kim TY, Lee S, Kang K (2015) Performance and emission characteristics of a high-compression-ratio diesel engine fueled with wood pyrolysis oil-butanol blended fuels. Energy 93:2241–2250
40. Chen L, Stone R, Richardson D (2012) Effect of the valve timing and the coolant temperature on particulate emissions from a gasoline direct-injection engine fuelled with gasoline and with a gasoline–ethanol blend. Proc Inst Mech Eng, Part D: J Automobile Eng 226:1419–1430
41. Xing J, Shao L, Zheng R, Peng J, Wang W, Guo Q et al (2017) Individual particles emitted from gasoline engines: impact of engine types, engine loads and fuel components. J Clean Prod 149:461–471
42. Sharma N, Agarwal AK (2017) Effect of the fuel injection pressure on particulate emissions from a gasohol (E15 and M15)-fueled gasoline direct injection engine. Energy Fuels 31:4155–4164
43. Huang H, Liu Q, Wang Q, Zhou C, Mo C, Wang X (2016) Experimental investigation of particle emissions under different EGR ratios on a diesel engine fueled by blends of diesel/gasoline/n-butanol. Energy Convers Manag 121:212–223
44. Kumar BR, Saravanan S, Rana D, Anish V, Nagendran A (2016) Effect of a sustainable biofuel–n-octanol–on the combustion, performance and emissions of a DI diesel engine under naturally aspirated and exhaust gas recirculation (EGR) modes. Energy Convers Manag 118:275–286

45. Su J, Lin W, Sterniak J, Xu M, Bohac SV (2014) Particulate matter emission comparison of spark ignition direct injection (SIDI) and port fuel injection (PFI) operation of a boosted gasoline engine. J Eng Gas Turbines Power 136:091513
46. Alger T, Gingrich J, Roberts C, Mangold B (2011) Cooled exhaust-gas recirculation for fuel economy and emissions improvement in gasoline engines. Int J Engine Res 12:252–264
47. Zhao L, Yu X, Qian D, Dong W, Sun P, He L et al (2013) The effects of EGR and ignition timing on emissions of GDI engine. Sci China Technol Sci 56:3144–3150
48. Bozza F, De Bellis V, Teodosio L (2016) Potentials of cooled EGR and water injection for knock resistance and fuel consumption improvements of gasoline engines. Appl Energy 169:112–125
49. Liati A, Schreiber D, Eggenschwiler PD, Dasilva YAR, Spiteri AC (2016) Electron microscopic characterization of soot particulate matter emitted by modern direct injection gasoline engines. Combust Flame 166:307–315
50. Kittelson DB (1998) Engines and nanoparticles: a review. J Aerosol Sci 29:575–588
51. Preussner C, Döring C, Fehler S, Kampmann S (1998) GDI: interaction between mixture preparation, combustion system and injector performance. SAE technical paper
52. Baldacci S, Maio S, Cerrai S, Sarno G, Baïz N, Simoni M et al (2015) Allergy and asthma: effects of the exposure to particulate matter and biological allergens. Respir Med 109:1089–1104
53. Donaldson K, Gilmour M, MacNee W (2000) Asthma and PM 10. Respir Res 1:12
54. Ulrich MM, Alink GM, Kumarathasan P, Vincent R, Boere AJF, Cassee FR (2002) Health effects and time course of particulate matter on the cardiopulmonary system in rats with lung inflammation. J Toxicol Environ Health Part A 65:1571–1595
55. Künzli N, Tager I (2005) Air pollution: from lung to heart. Swiss Med Wkly 135:697–702
56. Brook RD, Rajagopalan S, Pope CA, Brook JR, Bhatnagar A, Diez-Roux AV et al (2010) Particulate matter air pollution and cardiovascular disease. Circulation 121:2331–2378
57. DeFranco E, Moravec W, Xu F, Hall E, Hossain M, Haynes EN et al (2016) Exposure to airborne particulate matter during pregnancy is associated with preterm birth: a population-based cohort study. Environ Health 15:6
58. Rappazzo KM, Daniels JL, Messer LC, Poole C, Lobdell DT (2014) Exposure to fine particulate matter during pregnancy and risk of preterm birth among women in New Jersey, Ohio, and Pennsylvania, 2000–2005. Environ Health Perspect 122:992–997
59. Sapkota A, Chelikowsky AP, Nachman KE, Cohen AJ, Ritz B (2012) Exposure to particulate matter and adverse birth outcomes: a comprehensive review and meta-analysis. Air Qual Atmos Health 5:369–381
60. Organization WH (2014) Ambient (outdoor) air quality and health. Fact Sheet 313
61. Chowdhury S, Dey S (2016) Cause-specific premature death from ambient PM 2.5 exposure in India: estimate adjusted for baseline mortality. Environ Int 91:283–290
62. Du Y, Xu X, Chu M, Guo Y, Wang J (2016) Air particulate matter and cardiovascular disease: the epidemiological, biomedical and clinical evidence. J Thorac Dis 8:E8
63. Pope CA, Burnett RT, Thurston GD, Thun MJ, Calle EE, Krewski D et al (2004) Cardiovascular mortality and long-term exposure to particulate air pollution. Circulation 109:71–77
64. Araujo JA, Nel AE (2009) Particulate matter and atherosclerosis: role of particle size, composition and oxidative stress. Part Fibre Toxicol 6:24
65. Sun Q, Hong X, Wold LE (2010) Cardiovascular effects of ambient particulate air pollution exposure. Circulation 121:2755–2765
66. Xing Y-F, Xu Y-H, Shi M-H, Lian Y-X (2016) The impact of PM2. 5 on the human respiratory system. J Thorac Dis 8:E69
67. Franchini M, Mannucci PM (2012) Air pollution and cardiovascular disease. Thromb Res 129:230–234
68. Medina-Ramón M, Zanobetti A, Schwartz J (2006) The effect of ozone and PM10 on hospital admissions for pneumonia and chronic obstructive pulmonary disease: a national multicity study. Am J Epidemiol 163:579–588

Part III
Particulate and Aerosols

Primary Organic Aerosols

Deepika Bhattu

Abstract Primary organic aerosol (POA) constitutes the emissions from both natural (vegetation and micro-organisms) and anthropogenic sources such as combustion of fossil fuels and biofuels, and open biomass burning (forest fire). Semi-volatile nature of POA emissions leads to overestimation in the traditional emission inventories and chemical transport models. Another class of primarily emitted volatile species, i.e., intermediate volatile organic compounds (IVOCs), present around 0.28–2.5 times of POA, potential secondary organic aerosols (SOAs) precursors, also goes unnoticed. Phase partitioning mechanisms depending on their source, dilution, and volatility distribution make the contribution of POA to overall organic aerosols (OA) budget controversial. Further, the complex and higher particle emission rates and the gas-phase chemical transformation processes lead to the conceptual ambiguity between primary and secondary organic aerosol, thus rendering physico-chemical and optical properties to be least understood. Researchers have overcome the need of complete molecular identification of gaseous species to simulate the gas-particle partitioning by developing a two-dimensional volatility basis scheme (2-D-VBS) that employs the vapor pressure and degree of oxygenation. Here, we also illustrate the chemical composition-dependent volatility distributions for different sources used to ascertain the correct POA emission factors. This suggest that the policymakers and environmental regulating authorities need to take into account the SVOCs and IVOCs causing positive and negative sampling artifacts in order to correctly account for POA source contributions.

Abbreviations

BC	Black carbon
BB	Biomass burning
C_{OA}	Total organic aerosol mass concentration
EC	Elemental carbon

D. Bhattu (✉)
Paul Scherrer Institute, OFLA/005, Villigen 5408, Switzerland
e-mail: deepika.bhattu@psi.ch

N. Sharma et al. (eds.), *Air Pollution and Control*, Energy, Environment, and Sustainability, https://doi.org/10.1007/978-981-10-7185-0_7

eBC	Equivalent black carbon
HR-ToF-AMS	High-resolution time-of-flight aerosol mass spectrometer
IVOC	Intermediate volatile organic compound
OA	Organic aerosol
OC	Organic carbon
OM:OC	Organic matter to organic carbon ratio
PM	Particulate matter
POA	Primary organic aerosol
rBC	Refractory black carbon
SOA	Secondary organic aerosol
VOC	Volatile organic compound
2D-VBS	2 dimensional—Volatile basis set

1 Introduction

Carbonaceous aerosols are the dominating components of PM_1 (particulate matter of size less than 1 μm) mass in an anthropogenically polluted site. They are made of polymerized organic material with traces of hydrogen and oxygen and are mainly categorized as elemental carbon (EC), organic carbon (OC), and carbonate minerals (present only in super-micron size range) depending on their source of origin. Due to the lack of widely accepted standard calibration materials and their complex chemical and optical properties, separation of EC and OC is challenging [13, 22]. Depending on the direct and indirect measurement techniques of physical, chemical, and optical properties, carbonaceous aerosols are sectioned into different classes.

(a) Black carbon (BC): It is defined on the basis of characteristic aggregate morphology (graphitic sp^2-bonded carbon) of primary combustion particles. BC aerosols have the characteristic property of being refractory at 4000 K, insoluble in water and common organic solvents, acids and bases, and light absorption with mass absorption coefficient (MAC) of ~5–15 $m^2\ g^{-1}$ at 550 nm [13].
(b) Elemental carbon (EC): It is measured using evolved gas analyzer (EGA) method by combusting the EC in an oxygenated helium atmosphere, soot particle aerosol mass spectroscopy, and Raman spectroscopy. It is thermally stable at ~4000 K and only gets oxidized above 340 °C.
(c) Equivalent Black carbon (eBC): It is defined on the basis of absorption coefficient which is converted to mass concentration using a constant value of MAC. However, with the known contribution of brown carbon (absorption in visible wavelength: 300–600 nm), the bias in estimating the equivalent BC due to constant MAC value can be reduced. Although, there are different filter-based and in-situ particle absorption measurement methods to derive eBC,

the filter-based methods suffer from different biases such as particle and multiple light scattering, filter loading, and RH effect. However, the in-situ particle absorption measurements include photoacoustic techniques (e.g., photoacoustic soot spectrometer: PASS), interferometric techniques, and remote sensing techniques, which are more reliable.

(d) Refractory black carbon (rBC): It is the carbon mass which is stable at $\sim$4300 K and measured by thermal emission of carbon component of the particle absorbing laser energy. It is measured by laser-induced incandescence (LII).

(e) Light-absorbing carbon (LAC): It includes both eBC and brown carbon. It absorbs light at visible wavelength.

(f) Organic carbon (OC): OC is made up of the molecules that are combined with hydrogen, oxygen and/or nitrogen, sulfur, etc. It can also be measured by EGA method by volatilizing the OC via ramping the temperature in steps in inert pure helium atmosphere and converting to CO_2 using a catalyst (e.g., MnO_2) and can be measured by either CO_2 analyzer or flame ionization detector (FID) after converting to CH_4. This method is also known as thermal analytical method. Further, the thermal optical method involves the correction for charring of OC on filter by laser transmittance or reflectance signal which otherwise gets incorrectly accounted as EC. OC and EC are measured sequentially by two widely accepted protocols: National Institute for Occupational Safety and Health (NIOSH; [3]) and Interagency Monitoring of Protected Visual Environment (IMPROVE; [8]. However, these two methods differ by a factor of 2–10 in EC measurements due to different temperature ramping regime, charring correction, presence of metal oxides (e.g., Fe_2O_3), and also dependence on the sample source and its aging [7].

These measurement techniques quantify OC which is then multiplied with organic matter to organic carbon ratio (OM:OC) elemental ratios to obtain the bulk organic aerosol mass concentration. OM:OC ratio largely depends on the organic oxygen content which corresponds to their water-soluble nature and density [1]. To overcome the constraint of large sample requirement and low time resolution of these thermal EC–OC techniques, organic aerosols (OA) elemental composition approximation from unit mass resolution dataset of high-resolution time-of-flight aerosol mass spectrometer (HR-ToF-AMS) was introduced [1]. This approach proved to be a better trade-off for the traditional offline filter analysis that suffers from various sampling artifacts like adsorption, absorption, volatilization, and chemical transformation reactions.

POA being the crucial part of carbonaceous aerosols requires further targeted and coherent observations for accurate emission estimates. This needs standardization of measurement methods for both BC and OC. Currently, large uncertainties in the quantification of emission factors depend on the source, type of combustion and environmental conditions. Overall, particle and gas-phase precursors uncertainties range between a factor of 2 and 5 [12]. In order to have consistency between observed concentrations and modeled estimates, more in-situ field and laboratory measurements are required to understand the direct particle-phase emissions and reactivity, and aerosol formation potential of gas-phase molecules.

2 Organic Aerosols (OAs)

OA makes up to ~50% of the total fine aerosol mass at continental mid-latitudes [19] and up to ~95% in tropical-forested areas [2]. POA alone contributes 35 TgC/yr to the total carbon budget where open biomass burning (BB) contributes (~70%) a major fraction [4]. EC is produced by the incomplete combustion of fossil/non-fossil fuel combustion and forest fires. However, the organic carbon has complex sources and their physico-chemical and optical properties vary depending on their origin or formation processes. Organic compounds directly emitted in the particulate form or condensed into the particle phase without undergoing chemical reactions are known as POA, whereas the ones with sufficiently low vapor pressure, formed by gas-phase oxidation of precursor volatile organic compounds (VOCs), either of biogenic or anthropogenic origin or through nucleation are known as SOA.

(a) **Sources**

Fossil fuel combustion (vehicles and energy production), domestic wood burning (heating and cooking), and forest fires (deforestation and agricultural waste) are the major sources of carbonaceous aerosols [12]. Some biogenic sources like bacteria, fungal spores, and plant debris also contribute significantly to the aerosol mass but because of their coarse size range and shorter lifetime in the atmosphere, they are considered to be of lower climatic relevance [17]. Till date, wood burning is the dominant source of POA existing globally, whereas vehicular emission is more prevalent in urban environments. According to [4], the largest identified source of global BC emissions is vegetation fires (3.3 Tg $year^{-1}$) followed by fossil fuel (3.0 Tg $year^{-1}$) and biofuel (1.6 Tg $year^{-1}$), respectively. However, for POA, vegetation fires (34.6 Tg $year^{-1}$) and biofuel (9.1 Tg $year^{-1}$) are the dominant sources followed by fossil fuel (3.2 Tg $year^{-1}$). Other studies have suggested an increase in the BC emissions in last 50 years and expect similar emission trends in POA [12]. However, there are large uncertainties associated with the measurement of POA because of their meta-stable intermediate behavior and variability within different combustion fuels and type of combustion operations [18]. In this chapter, we will focus on POA emissions and their gas-to-particle phase partitioning, an important aspect to update the current emission inventories and chemical transport models.

(b) **Gas-to-particle partitioning**

Primary PM can be either volatile (e.g., organics, mainly aromatics and inorganics like SO_4^{2-} and NO_3^-, depending on their partial pressure in the gas phase) or non-volatile (e.g., fly ash, trace metals or EC) in nature depending on their formation process [11]. The measurement methods for these dynamic primary emissions include hot conditioned ambient air dilution filter sampling close to the source with or without impingers or cryogenic traps to capture condensable vapors. If done otherwise, cooling and dilution can affect their phase distributions making the measurements challenging and under-estimating the total organic PM [18]. Several considerations need to be taken care of while designing the primary aerosol mass samplers like the sample flow rate, tunnel diameter and material, mixing length, Reynolds number, residence time, and dilution ratio. More information on sampler design consideration can be found in [11]. The major chemical constituents of primary PM measured from different sources using the dilution sampler are shown in Fig. 1.

Highest volatile POA is emitted by gasoline-powered vehicles and diesel engines at lower loads, biomass burning, natural gas-fired devices, and aircrafts, whereas sulfur dominates in high sulfur-fueled sources (e.g., ship industry). However, higher non-volatile emissions are contributed from diesel engines (mainly EC) and coal combustion (mainly fly ash). Past studies have shown the relationship between number of carbon atoms and volatility of most prevalent emission sources such as wood burning, gasoline, and diesel vehicle exhausts. The emission factors of compounds $C > 30$ are high molecular weight hydrocarbons with low vapor pressure similar to those found in lubricating oils, whereas the ones with $C < 20$ has some contribution from intermediate volatile organic compounds (IVOCs) [18].

The ambient gas-to-particle partitioning is a sorptive process where vapors can absorb/adsorb into/onto organic solution/soot, mineral dust, and other solid

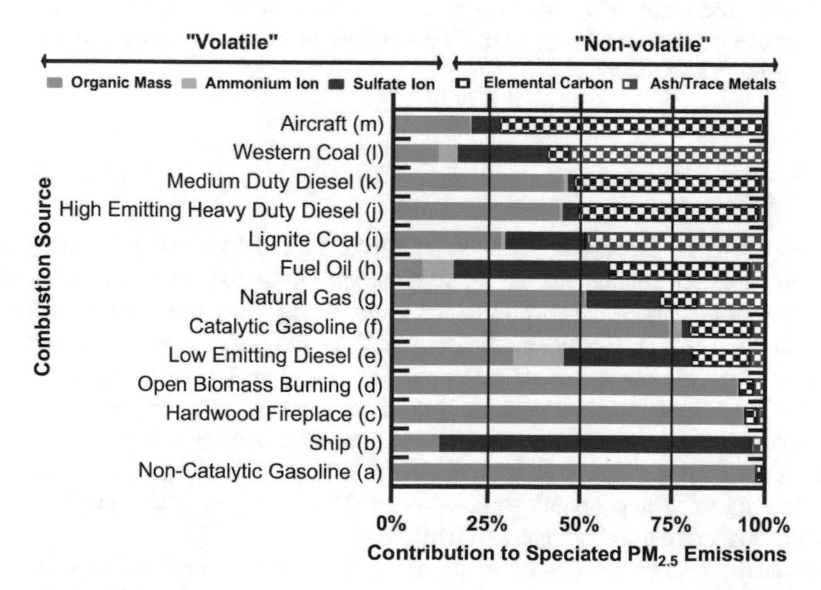

Fig. 1 Fractional contribution of primarily emitted species in various anthropogenic combustion sources. The sampling technique involves hot air dilution to capture both volatiles (solid colors) and non-volatiles (hashed bars). Trace metals are assumed to be existing in metal oxides state, and organic matter is estimated to be 1.4 times the OC fraction [18]

substrates surface. However, in source-dominated scenarios, adsorption is also equally important. Aqueous processing and chemisorption affect the gas-to-particle partitioning process by enhancing the uptake of volatile-oxygenated species, e.g., glyoxal. To understand the volatility profiles of different sources (biomass burning and vehicular emissions), studies employing thermal denuder used evaporation kinetics model to derive volatility distributions and other volatility parameters, e.g., enthalpy of vaporization [5, 10, 14]. These volatility distributions and calculated emission factors are used to correctly update the emission inventories and simulating source-specific POA emissions in chemical transport models after considering the negative (e.g., evaporation due to dilution) and positive (e.g., adsorption of vapors) sampling artifacts [14]. These derived volatility distributions from thermal denuder measurements should be used carefully as they are biased by the operational parameters, such as evaporation and re-condensation in the thermal denuder [6].

A volatility distribution of the aerosols present in phase equilibrium can be obtained using the partitioning coefficient "X_p." The partitioning of a particular species "i" between gas (G) and particle (P) phase depends on its volatility and total organic aerosol mass concentration (C_{OA}) and can be described by equilibrium constant "Kp" [15]. This theory is known as partitioning theory.

$$K_{\mathrm{p}} = \frac{P}{G * C_{\mathrm{OA}}} \tag{1}$$

Finally, the partitioning coefficient (X_{p}) of a semi-volatile species "i" in particulate phase depends on C_{OA} (μg m^{-3}) and effective saturation mass concentration (C_i^*, μg m^{-3}) as follows:

$$X_p = \frac{P}{G+P} = \frac{1}{1+\left(\frac{C_i^*}{C_{\mathrm{OA}}}\right)} \tag{2}$$

The Clausius–Clapeyron equation suggests that decreasing the temperature reduces C_i^* exponentially, and hence, higher contribution to particle phase [9]. Both C_i^* and X_{p} follow the Hinshelwood saturation curve depending on total aerosol mass concentration (C_{OA}). It suggests that when $C_i^* \gg C_{\mathrm{OA}}$, X_{p} increases linearly with increased organic aerosol mass concentration (C_{OA}) and reaches maximum and levels off at 1. Figure 2 shows partitioning of organic species as air parcel evolves in the ambient atmosphere. For example, a given organic species with $C_i^* = 1$ μg m^{-3} and $C_{\mathrm{OA}} = 1$ μg m^{-3} can equally be present in both particle and gas phase. Also, for $C_{\mathrm{OA}} = 1$ μg m^{-3}, any organic species with $0.01 \leq C_i^* \leq 100$ μg m^{-3} will be prominently present in both gas and particle phase.

Because of large complexity of organics, 90% of the composition still exists without complete molecular identification with present analytical techniques [20]. [9] presented a scheme where seven surrogate compounds were used to represent bulk POA using 1-D volatility basis set distributing the low volatile organics

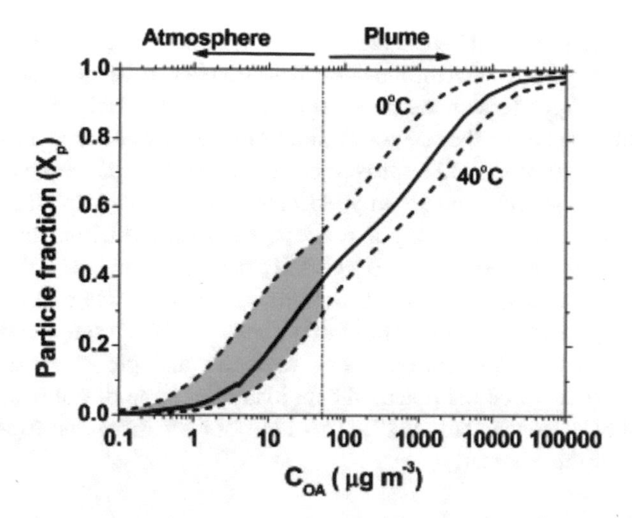

Fig. 2 Predicted particle fraction (X_p) as a function of total organic aerosol mass concentration (C_{OA}, μg m^{-3}) from wood smoke at three different temperature conditions (0, 27 and 40 °C). Here, values at 27 °C (solid black line) are the experimental data, whereas values at 0 and 40 °C are the model predictions which are presented as dashed lines. The shaded area represents ambient conditions ranging from remote to heavily polluted location in summer and winter season [21]

($C_i \leq 3 \times 10^6$ μg m^{-3}) over a logarithmically spaced bins of C_i^* at 300 K. This scheme can be easily applied to the ambient data to recover the reliable organic aerosol volatility distributions. As this scheme doesn't require molecular level information, incorporation of more identified emission sources to a chemical transport model can be easily done.

Further, a freshly emitted air parcel experiences four stages during its atmospheric processing and changes C_{OA}: (a) emission from the source; (b) mixing with another air mass, i.e., dilution (a compound of certain C_i^* can be in particle phase when it is close to the source and gas phase when close to the background concentration); (c) temperature change [increasing temperature lowers the enthalpy of vaporization ΔH_{vap} and subsequently increases C_i^*; increasing the ambient temperature from 0 to 40 °C reduces the fraction in particle phase suggesting a general trend of lower POA emission factors in summer (Fig. 2)]; (d) Chemical transformation affecting C_i^* to either low or high volatility bins depending on their production via different oxidation pathways and their order of generation [9].

The simultaneous reduction in exhaust temperature and concentration has competing effect on the partitioning of hot mixed emissions [18]. The effect of decreasing temperature is dominant at lower dilution ratios favoring increase in POA mass concentration, whereas effect of concentration is more prominent at higher dilution.

On the other hand, if aerosols are not in phase equilibrium (i.e., when ratio of residence time to the equilibration timescale <5), an evaporation kinetic model using both gas and particle phase in the 1D-VBS scheme is used [16]. The approximated equilibration time τ, i.e., the inverse of condensation sink (C_s), the mass accommodation coefficient (decides mass transfer resistance), and enthalpy of vaporization (ΔH_{vap}) decides the phase equilibrium of aerosols [5].

Further, to account for the complex oxidation pathways of organic species, their lifetime in the atmosphere and improvement in thermodynamic representation such as organic species mixing and polarity, species with similar properties are grouped together in 2-D space of volatility and average carbon oxidation state (O:C ratio) and this framework is known as 2-D volatility basis set scheme. This scheme tells us the fate of organic species present in both, gas and particle phase, and responsible for mainly SOA formation. These important classes of compounds depending should be included in the current local, regional, and global aerosol modeling schemes as they could be the potential candidates for large deviations in measured and modeled SOA concentrations [9]. This has not been discussed further as this is out of scope of this chapter.

3 Conclusions and General Atmospheric Implications

This chapter highlights the importance of correct measurement of POA and proper interpretation of the results from such measurements. Researchers and emission regulating agencies should focus on these two major aspects (i.e., total organic aerosol mass concentration and sampling/exhaust temperature) mentioned here while formulating the emission testing protocols. These can assist in reducing the potential bias due to recommended low dilution ratios and lack of measured volatility distributions. Eventually, it will improve the level of understanding the impact of tailpipe emissions in the well-mixed atmospheric conditions. In addition, these dynamic changes in gas-to-particle conversion and reconversion processes need to be updated in chemical transport models and health impact assessment studies.

References

1. Aiken AC, DeCarlo PF, Kroll JH, Worsnop DR, Huffman JA, Docherty KS, Ulbrich IM, Mohr C, Kimmel JR, Sueper D, Sun Y, Zhang Q, Trimborn A, Northway M, Ziemann PJ, Canagaratna MR, Onasch TB, Alfarra MR, Prevot ASH, Dommen J, Duplissy J, Metzger A, Baltensperger U, Jimenez JL (2008) O/C and Om/Oc ratios of primary, secondary, and ambient organic aerosols with high-resolution time-of-flight aerosol mass spectrometry. Environ Sci Technol 42:4478–4485
2. Andreae MO, Crutzen PJ (1997) Atmospheric aerosols: biogeochemical sources and role in atmospheric chemistry. Science 276:1052–1058
3. Birch ME, Cary RA (1996) Elemental carbon-based method for monitoring occupational exposures to particulate diesel exhaust. Aerosol Sci Technol 25:221–241
4. Bond TC, Streets DG, Yarber KF, Nelson SM, Woo J.-H, Klimont Z (2004) A technology-based global inventory of black and organic carbon emissions from combustion. J Geophys Res: Atmos 109: n/a-n/a

5. Cappa CD, Jimenez JL (2010) Quantitative estimates of the volatility of ambient organic aerosol. Atmos Chem Phys 10:5409–5424
6. Cappa CD, Wilson KR (2011) Evolution of organic aerosol mass spectra upon heating: implications for Oa phase and partitioning behavior. Atmos Chem Phys 11:1895–1911
7. Chow JC, Watson JG, Crow D, Lowenthal DH, Merrifield T (2001) Comparison of improve and niosh carbon measurements. Aerosol Sci Technol 34:23–34
8. Chow JC, Watson JG, Pritchett LC, Pierson WR, Frazier CA, Purcell RG (1993) The dri thermal/optical reflectance carbon analysis system: description, evaluation and applications in u.s. air quality studies. Atmos Environ Part A. Gen Top 27:1185–1201
9. Donahue NM, Robinson AL, Stanier CO, Pandis SN (2006) Coupled partitioning, dilution, and chemical aging of semivolatile organics. Environ Sci Technol 40:2635–2643
10. Grieshop AP, Miracolo MA, Donahue NM, Robinson AL (2009) Constraining the volatility distribution and gas-particle partitioning of combustion aerosols using isothermal dilution and thermo denuder measurements. Environ Sci Technol 43:4750–4756
11. Hildemann LM, Cass GR, Markowski GR (1989) A dilution stack sampler for collection of organic aerosol emissions: design, characterization and field tests. Aerosol Sci Technol 10:193–204
12. Kanakidou M, Seinfeld JH, Pandis SN, Barnes I, Dentener FJ, Facchini MC, Van Dingenen R, Ervens B, Nenes A, Nielsen CJ, Swietlicki E, Putaud JP, Balkanski Y, Fuzzi S, Horth J, Moortgat GK, Winterhalter R, Myhre CEL, Tsigaridis K, Vignati E, Stephanou EG, Wilson J (2005) Organic aerosol and global climate modelling: a review. Atmos Chem Phys 5:1053–1123
13. Lack DA, Moosmüller H, McMeeking GR, Chakrabarty RK, Baumgardner D (2014) Characterizing elemental, equivalent black, and refractory black carbon aerosol particles: a review of techniques, their limitations and uncertainties. Anal Bioanal Chem 406:99–122
14. May AA, Levin EJT, Hennigan CJ, Riipinen I, Lee T, Collett JL, Jimenez JL, Kreidenweis SM, Robinson AL (2013) Gas-particle partitioning of primary organic aerosol emissions: 3. Biomass burning. J Geophys Res: Atmos 118:1127–311338
15. Pankow JF (1994) An absorption model of gas/particle partitioning of organic compounds in the atmosphere. Atmos Environ 28:185–188
16. Riipinen I, Pierce JR, Donahue NM, Pandis SN (2010) Equilibration time scales of organic aerosol inside thermodenuders: evaporation kinetics versus thermodynamics. Atmos Environ 44:597–607
17. Robinson AL, Donahue NM, Shrivastava MK, Weitkamp EA, Sage AM, Grieshop AP, Lane TE, Pierce JR, Pandis SN (2007) Rethinking organic aerosols: semivolatile emissions and photochemical aging. Science 315:1259–1262
18. Robinson AL, Grieshop AP, Donahue NM, Hunt SW (2010) Updating the conceptual model for fine particle mass emissions from combustion systems Allen L. Robinson. J Air Waste Manag Assoc 60:1204–1222
19. Saxena P, Hildemann LM (1996) Water-soluble organics in atmospheric particles: a critical review of the literature and application of thermodynamics to identify candidate compounds. J Atmos Chem 24:57–109
20. Schauer JJ, Kleeman MJ, Cass GR, Simoneit BRT (2002) Measurement of emissions from air pollution sources. 5. C1–C32 organic compounds from gasoline-powered motor vehicles. Environ Sci Technol 36:1169–1180
21. Shrivastava MK, Lipsky EM, Stanier CO, Robinson AL (2006) Modeling semivolatile organic aerosol mass emissions from combustion systems. Environ Sci Technol 40:2671–2677
22. Wu C, Huang XHH, Ng WM, Griffith SM, Yu JZ (2016) Inter-comparison of Niosh and improve protocols for Oc and Ec determination: implications for inter-protocol data conversion. Atmos Meas Tech 9:4547–4560

Effects of VOCs on Human Health

Vipin Soni, Paramvir Singh, Venu Shree and Varun Goel

Abstract With broad commutability of pollutants, air pollution is a complicated issue that intimidates directly to human health and our environment. Various indoor and outdoor air pollution comprises of regulated and unregulated emissions. Volatile organic compounds (VOCs) are one of them which are carcinogenic and lead to photochemical reactions. Emissions of VOCs are directly associated with large number of industrial processes, emission through transportation, and various indoor and outdoor sources. Due to deleterious effect of different VOCs emitted from transportation, chemical industrial plants and from indoor on the environment make their eviction mandatory or at least degrade them under the limit set by environmental norms. So, it has turned a burning topic to diminish air pollutants and set a norm for volatile emissions. This study focuses on the major sources of carbonyl and aromatic compounds in indoor and outdoor environment. Some remedial processes like photocatalytic oxidation, plasma decomposition, chemisorption, and catalytic oxidation have been described in this study through which decomposition of these contaminants can be achieved. This chapter contains a deep study on health effects from the carbonyl and aromatic compounds.

Keywords Aromatic compounds · Carbonyl compounds · Pollutants
Indoor air · Air pollution

V. Soni · P. Singh (✉) · V. Goel
Department of Mechanical Engineering, National Institute of Technology Hamirpur,
Hamirpur 177005, Himachal Pradesh, India
e-mail: param016@gmail.com

V. Shree
Department of Architecture, National Institute of Technology Hamirpur,
Hamirpur 177005, Himachal Pradesh, India

N. Sharma et al. (eds.), *Air Pollution and Control*, Energy, Environment,
and Sustainability, https://doi.org/10.1007/978-981-10-7185-0_8

1 Introduction

Mostly, people spend their time in home, office, shopping center, car, etc. The common air pollutants include nitrogen oxides (NO_x), carbon oxides (CO and CO_2), carbonyl compounds, aromatic compounds, and particulate matter (PM). Among these pollutants, carbonyl and aromatic compounds have severe impacts on environment. The possible sources of carbonyl and aromatic compound emissions are transportation, chemical industrial plants, painting, electroplating, etc. [1]. Carbonyl and aromatic compounds, e.g., formaldehyde (HCHO), benzene, toluene, and xylene are emitted in both indoor and outdoor environments through different processes. Longtime exposure to polluted air even containing a few parts per million (ppm) of volatile organic compounds (VOCs) may cause long-term diseases [2]. Carbonyl and aromatic compounds which are generated through naturally and anthropogenic processes are having group of air pollutant in case of indoor as well as outdoor due to long-term health problems in human (WHO 2004). According to the health agencies, VOCs are considered as human carcinogen and longtime exposure of VOCs leads to disease like reduced pulmonary function, asthma [3]. Also, nasopharyngeal cancer can be caused by long exposure to formaldehyde which also comes into the category of carbonyl compounds. In addition, discharge of VOCs in the environment leads to urban smog and ozone, ozone layer depletion in stratosphere, and the greenhouse effect [4]. The increasing research to overcome detrimental effect by VOCs on human health and environment has caused fast recognition of toxic aspect of VOCs during the last two decades [5]. The different VOCs pollutants are categorized as follows: carbonyl compound (aldehydes and ketones), polycyclic aromatic hydrocarbons (PAHs), mono-aromatic hydrocarbons (MAHs), nitro-PAHs, oxy-PAHs, etc. 1,3 butadiene, aldehydes (formaldehyde, acetaldehyde), acrolein, aromatics (benzene, toluene, ethyl benzene, xylene, styrene) n-hexane, naphthalene, etc., are classified as toxic pollutant to environment and human health as per environmental protection agency (EPA) [6]. It is important to convert harmful VOCs into the environment-friendly species. In case of outdoor environmental pollution, the major origin of pollution is the transportation area. The compounds emitted through the incomplete combustion of fuel leads to various air pollutants. The pollutants from engine exhaust sometimes considered as the primary contaminants which affect the quality of air [7]. Though the formation of contaminants from compression-ignited engines is quantity wise low as compared to spark ignited engines, the formation of larger size particulate is the major problem which finally combines with the poisonous VOCs [8]. Urban air pollution and indoor air pollution come in worst pollution category. Air is polluted by various substances like through exhaust gas, liquid droplets, or solid particles. These air pollutants can be further categorized into primary, secondary, and tertiary pollutants [9]. Primary pollutants emit directly from the various sources like volcanic eruption, engine exhaust, and from different factories or industries smoke [10]. The primary pollutants include molecules like emission of carbon monoxide (CO), sulfur dioxide (SO_2), emission of VOCs, and NO_x emission from different industries and motor vehicles [11]. Apart from the primary pollutants, the pollutants which are not emitted by it are called secondary pollutants. The example of secondary air pollution is formation of ozone at surface level. The possible air pollutants are carbon monoxide (CO), carbon dioxide (CO_2), NOx, sulfur oxides (SO_2), chlorofluorocarbon (CFC), aldehydes (formaldehyde, acetaldehyde, etc.), PAHs, particulates, ammonia, radioactive particles, free radicals, etc. [6]. These pollutants produce many ill effects on both environment and human beings.

2 Sources of VOCs

2.1 Sources of Carbonyl Compounds

The organic molecule containing functional group >C=O is classified as the carbonyl compounds. These carbonyl compounds come in the category of unregulated emission which is toxic in nature, and they have various ill effects on the environment and human health [12]. The common carbonyl group found in the air pollution is grouped as follows: formaldehyde (HCHO), acetaldehyde (CH_3CHO), acetone, acrolein, benaldehyde, butyraldehyde, crotonaldehyde, 2,5-dimethyl benzaldehyde, hexadehyde, methacrolein, propionaldehyde, tolualdehyde, valeraldehyde, etc. [6]. These carbonyl compounds are also established as a part of ozone cycle. The most commonly found carbonyls are HCHO, CH_3CHO, and acetone which are the product of combustion during intermediate stage.

2.1.1 Indoor Sources

Indoor air pollutants caused by use of different paints, various plywoods, cooking, cigarette smoke, etc., which are hazardous to the human health. For this reason currently indoor air contamination has become major problem for the world. Other indoor air pollutant sources are construction work, furniture, textile, clothes, carpeting, and wood processing household appliances, particle board, painting, plywood, etc. [13]. The long-term effects of these indoor air pollutants are death threatening disease. The indoor air pollutants come in dominant position when decorative material used in buildings [14]. As the technology progression takes place, to save energy and for the enhanced thermal efficiency human usually lives in air-constrained buildings. Ultimately, they utilize the thermal insulation for improvement in the thermal efficiency of buildings. This finally results in increment of indoor air pollution. There is a lot of compounds present like cigarette smoke, CO_2, floating dust, VOCs, radon, NOx, PM, HCHO, asbestos [15]. In these pollutants, some are very hazardous to human health. These VOCs have also various sources of emission for indoor air pollution like they emit from thermal insulation, deodorant, cosmetic products, furniture, and some time through the combustion of something like woods. A new term 'sick building syndrome' has been coined for buildings with contaminations like VOCs [14]. The term SBS has covered those buildings which are recently constructed. SBS generally contains the pollutants emitted from the paints, furniture, building materials, product used for the decoration material, etc. Without any reason if people living in the buildings feel symptoms of illness, medical terminology uses SBS for that building. Features like throat irritation, headache, dizziness, fatigue, etc., are the main noticeable complaints from the people living in these types of buildings. These decorative substances contribute a lot in case of new buildings or apartments. In the recent years, several policies have passed related to diminishment and control the indoor air pollution.

2.1.2 Outdoor Sources

There are several outdoor sources like transportation sector, industrial sector, various combustion processes, and different secondary processes like smog formation. In urban regions as stated by Guo et al., the most dominant sources of carbonyl emission are the exhaust from the tail pipe of vehicles [16]. After the transportation sector, another dominating source of carbonyl emission is petrochemical plants followed by petroleum reservoirs. The primary sources of carbonyl compound include those compounds which cause direct emission, while the secondary sources include the formation of carbonyl in the atmosphere [17]. The primary mode of carbonyl compounds is motor vehicle, industries, plants, combustion process, etc. Apart from vehicle emission, secondary chemical reactions of alkanes and alkenes also produces the aldehyde in the atmosphere [18]. Photochemical process which is secondary process in the ambient is also a source of HCHO and CH_3CHO. Different industrial stages also emit carbonyl compounds. Altshuller had reported that there is substantial production of HCHO and CH_3CHO from the alkanes and alkenes due to the photo-oxidation process [19]. Various biofuels have been tested as the alternate to conventional fuel. There are various alcohols which have been used to improve the efficiency of combustion in vehicles. Since alcohol possesses high heat of vaporization, it substantially reduces the high temperature during combustion process. But different carbonyl emissions from the alcohol-operated engine go significantly high. Stump et al. had studied the emissions through ethanol-convened motor vehicles [20]. They have found that there is a significant decrease in aromatic compound but a substantial increase in HCHO and CH_3CHO due to blending of ethanol. In case of vehicle, there are several modes of emission of carbonyl compound like through tail pipe emission, tires and brakes, catchment area of fuel. The main origin of contamination in air is automobile's emissions as suggested by Faiz [21]. In carbonyl compounds, CH_3CHO and acetone are the main participants for smog promotion due to photochemical process in environment. The use of alcoholic fuels has limitation just because it is a possible source of CH_3CHO emission in environment. Poulopoulos et al. have tested the emission from the ethanol fuels [22]. The result shows that at intermediate load there is enhancement in the

CH_3CHO emission. It is clear that CH_3CHO is known for toxic element. Caplain et al. have investigated the nonregulated pollutants from diesel-fueled cars [7]. They have suggested that for diesel-operated vehicles there is 2–6.4% of carbonyl compound emission. In the possible carbonyl compound, ketone, HCHO, and CH_3CHO are the main molecules which approximately contain 73% of all carbonyl compounds. The high content of carbonyl compound emission is possible due to the cold or warm starting. The current technology advancement toward the use of alternate fuels like biodiesel has attracted researcher's attention for the emission investigation with these fuels. Graboski and McCormic have found that there is inconsistent or increased emission from biodiesel as compared to diesel in terms of ketone emission [23]. Apart from the exhaust emission, the running time period of vehicle/engine also has an effective impact on carbonyl emission. After a longtime period of engine, these can strongly participate in the formation of carbonyl emission and smog formation. The emission of HCHO sometime generated conflict between different results as increased HCHO was observed at higher load condition. Contrary to this result, higher HCHO was observed at lower loads by Takada et al. [24]. Hence, it is clear that for the emission of carbonyl emission through engines, there are several parameters to be considered as dominating factor like oxygenated fuels, engine, cycle and procedure used for running, and the temperature during the combustion process. For the increased oxygen percentage in fuel blending, alcohol and biodiesel have been used. A study has been conducted by Turrio-Baldassarri et al. on biodiesel-blended diesel oil to investigate the carbonyl compound emission [25]. The results show that CH_3CHO, HCHO, propionaldehyde, and acrolein were the common abundant compound during the emission. These are significant compounds which have the 90% of carbonyl emission. It was observed that the use of biodiesel increases the carbonyl emission percentage up to 19% as compared to mineral diesel. Graboski and McCormic have suggested that biodiesel having glycerides enhances the acrolein emission [23]. Peng et al. showed the increase in CH_3CHO emission due to ethanol mixed with biodiesel [26]. Some results show that there is evidence of clean and proper blending of biodiesel. Whether these oxygen-rich fuels provide more oxygen for helping the combustion process and reducing emissions or oxygen in fuels promotes the formation of aldehydes is still unclear. Karavalakis et al. have explained the properties of biodiesel which effects the carbonyl emission [11]. They have found that probably purity level, iodine number, and the fatty acid profile play a key role in carbonyl emission, which is due to disassociation of biodiesel into various other organic compounds. Decarboxylation process enhances the emission of carbonyl compound. This process could dissociate esters which may be processed of generation of intermediates during combustion process. Some air pollutants, namely CO, CO_2, and NO_x, can be regulated by the natural cycle, e.g., carbon cycle and nitrogen cycle but VOCs concentration which is carcinogen has not regulated till now.

2.2 *Aromatic Compounds*

As carbonyl emissions, aromatic emissions have the similar source like transportation sector, industrial sector, indoor air pollutants and various outdoor combustion processes. After the transportation sector, another dominating source of aromatic emission is petrochemical plants followed by petroleum reservoirs.

2.2.1 Indoor Sources

The organic molecules having ring structure are termed as the aromatic compounds. Due to the formation of isomers, aromatic compounds possess irregular behavior. The human-originated aromatic compounds in indoor have various sources like cooking, smoking, heating, computers, printing machines, electronic equipment, printers, and heating [27]. In general, aromatic compounds are classified in two parts; one is MAHs and other one is PAHs. Usually due to the partial combustion of organic compounds, PAHs are produced [28]. Due to their cancerous and/or ability to change genetics, PAHs are considered as the dominant compounds for the indoor air pollution [29]. These compounds are always existed in the indoors and outdoor environments. Some sources like combustion in indoors are termed as the sources of aromatics compounds in indoor as per WHO. Apart from straightforward sources of emission, other sources like individual sources also contribute to emission sufficiently. Sometimes, personal contribution in emissions may become greater rather than emission from the equipment. Sometimes, the slow emission from the polymeric equipment can affect the indoor air quality and it can produce the sick building syndrome problem. Aromatic compounds include broad range of organic molecules having distinct functional groups. Benzene, ethylbenzene, toluene, isobutyl benzene, styrene, tetramethyl benzene, xylene, naphthalene, and anthracene are the some common aromatic compounds found in the indoor environment. Benzene, toluene, and xylene are the most common indoor air pollutants which are termed as BTX commonly. The organic molecules which have more than two rings are called as the PAHs. These PAHs can be grouped in three different classes of low molecular weight (LMW), medium molecular weight (MMW), and high molecular weight (HMW) having 2–3, 4, and 5–6 rings, respectively [6]. So due to advancement in technology, a major change has been observed in buildings and working conditions. In different buildings and offices, several modifications that have been done for the use of printers, different electronic equipment, fax machines, desktop computer, etc. have strong influence on the indoor air quality. Destaillats et al. has comprehensively investigated the emission from the electronic equipment and observed that cathode ray tube directed computers and monitors of transistor having thin film are also the source of aromatic emission [30]. There are some evidence of aromatic emission from laser printers, fax machine, scanner, color printers, and ink jet printers. Other than it, in buildings, daily use products and decorative material are the important source of aromatic compounds in indoor air.

These are those emissions which are long lasting until the equipment deterioration. Coal tar having used for liquid resistive material in buildings can produce emission like naphthalene, biphenyl, and acenaphthene [31]. Yu and Crump had experimentally investigated on the emission from the building [14]. They suggested that a fresh painted building has the higher emission rate. The observation shows that evaporation is primary form of emission accompanied with diffusion process in indoor air. For the lighting and ornamentation purpose of buildings, candles have exploited as a source from very early times. Candles are also used sometime for the religious means. Due to the incomplete combustion process, inflaming of candles can originate continuous smoke which contributes majorly in the formation of benzene, xylene, toluene, etc [32]. Pagel et al. have described the wax burning as the source of polycyclic aromatic hydrocarbons [33]. Baek et al. have investigated the emission quality in Korea for the domestic buildings, restaurant, and offices [15]. The results show that benzene, ethyl benzene, toluene, m + p-xylene, o-xylene, and styrene were the major pollutants in the air during investigation. The seasonal variation also persists for the aromatic compounds which follow the trend of decreasing to increasing order from summer to winter. The cause of this type of trend is different for other activities used during the winters like different combustion process in winter, use of heat pump, and utilization of room air heaters, so a wide range of aromatic compound always exists in indoor air due to the volatile nature of these molecules. Hence solvents, cleaning product, varnishes, waxes, etc., have also ability to produce aromatic compound emission in indoor environment.

2.2.2 Outdoor Source

Movement toward urbanization and industrialization in twenty-first century broadly changes the current situation for the outdoor air pollution. Outdoor air pollution has increased due to manufacturing facilities and enhancement in motor vehicles. In all the aromatic compounds, the main concern is over PAHs emission. These PAHs are emitted from the partial combustion of organic compounds. The different combustion processes like burning of wood, coal, diesel-operated automobiles, and sometime grilled meats have the sufficient involvement in PAHs emission. The outdoor air pollution is totally hanged on emission from motor vehicle and industries. PAHs are the major component come in the unregulated emission from the engines. Many authors have given their points on the polycyclic emission from the engines [6, 9]. The use of diesel fuel in motor vehicle promotes the organic and inorganic emission for the tail pipe. These unregulated emissions from diesel exhaust are highly volatile and easily become hazardous to the environment and human beings. These emissions contain lots of organic molecules like mono-aromatic hydrocarbons and polycyclic aromatic hydrocarbons. As per the various agencies like International Agency for Research on Cancer (IARC), the exhaust from the burning of diesel fuel is considered as the carcinogen for the human health. The reason for the emission of PAHs from the diesel-fueled vehicles may be the separation of fuel during the combustion. This process leads to the pyrolysis and pyrosynthesis of aromatic compounds [34]. Lubricating oil which has been used extensively in diesel engine for improving the efficiency also contributes in the formation of PAHs [35]. Other factors which have small impact in generation of PAHs are like kind of engine used, fuel quality, load and speed of engine [36]. There are other unregulated emissions rather than PAHs like nitro- and oxy-PAHs. Singh et al. have given the trend of emission among LMWs, MMWs, and HMWs [6]. The result shows that highest content of PAHs in engine exhaust is those having LMWs along with MMWs and HMWs. Tancell et al. have resulted that about 80% benzo(a) pyrene in tail pipe emission is due to those molecules which exist in fuel [37]. Schauer et al. have studied the emission from the motor vehicle to investigate the compounds in tail pipe exhaust [38]. Result shows that benzene, toluene, and xylene are the main contaminants in air which have many sources of emission within motor vehicles. Caplain et al. have comprehensively analyzed the unregulated emission from the gasoline- and diesel-convened cars [7]. The result shows that the unregulated emission contains 54–75% aromatic compounds for a diesel-convened car. This data shows that diesel-convened engine emits large amount of aromatic compound during combustion. As the demand of diesel fuel increasing, the reservoirs of diesel fuel currently tend to deplete and cost increases simultaneously. This makes the restriction over the rapid use fossil fuels in various industries and transportation sectors. Several researches have done to investigate the optional fuel which has the similarity with diesel fuel. Vegetable oil and its derived form like biodiesel is the current area of focus. Hence due to the resemblance to diesel fuel, biodiesel has been currently used in many countries as an optional source of conventional diesel fuel [39]. There is limited information about the aromatic emission from the biodiesel fuels. Baldassarri et al. have tested the unregulated emission from the biodiesel and diesel mixture [25]. The results show that PAHs, nitro-PAHs, and low molecular weight aromatic hydrocarbons are the major source of unregulated emission through the mixture of diesel/biodiesel. There has been some evidence of low aromatic compound emission from the biodiesel fuels [6]. The cause of less aromatic compound emission from the biodiesel is that it has a sufficient amount of oxygen for the combustion process. But some result is contrary to this assumption like Baldassarri et al. have shown an increase of aromatic compound emission to 18% through blending of biodiesel with diesel [25]. They have found the compounds like xylene, benzene, toluene, benzo[j] fluoranthene, indeno[c,d] pyrene, dibenz[ah] anthracene, benzo[a] anthracene, benzo[a] pyrene, chrysene, 1-nitropyrene, and 2-nitropyrene in the exhaust of tail pipe. As biodiesel was used as the fuel, there was an increase of 16% for B20 as compared to diesel. Karavalakis et al. have used palm-based methyl ester with rapeseed methyl ester for evaluating the emission characteristic of a passenger vehicle [11]. They have found that proportion of benzene in the exhaust is very low. Various alcohols have also used for fulfilling the excess oxygen requirement of combustion process. Cheung et al. have used ethanol as the source of oxygen and dodecanol as the diluent in very low sulfur-contained diesel fuels to investigate the regulated and unregulated components in emission [40]. The result shows that at higher loads benzene molecules tend to decrease. Similar results have also depicted by Takada et al. that lower benzene contents have shown at higher loads [24]. They have explained the process of being low benzene content at higher temperature. Due to the easy oxidation of benzene at elevated temperature, the content of benzene found low at higher temperature. A similar trend has also found for the emission of xylene and toluene. At low temperature, higher BTX emission has found due to insufficient oxidation. But as compared to the diesel, biodiesel as an optional source of fuel has less aromatic compound emission in tail pipe exhaust because of rich oxygen content. As an outdoor source of emission, combustion of straw, nonroad petroleum combustion and firewood, and aluminum production are also found to be dominant factor.

3 Health Effect of VOCs

Among all the air pollutants, emission of carbonyl and aromatic compounds in air leads to various short-term and long-term diseases based on the concentration level in the air. Some carbonyl and aromatic have long-term health problem also. Some carbonyl and aromatic compounds like HCHO, CH_3CHO, benzene, toluene, and xylene have a tendency to produce cancer in human body. Hence, these compounds are known as the human carcinogen [41]. They also cause irritation in nose, eyes, skin, etc. As stated above, these carbonyl compounds and aromatic compounds play a part in the formation of smog due to photochemical process. Some carbonyl compounds like CH_3CHO, HCHO, and acrolein are considered as hazardous to human health [22]. Several PAHs and aromatic compound in the air as stated above have been taken as the greatest concern for the human beings. Due to long term and high concentration inhalation of VOCs, the lung functions becomes slow because of irritation in nose, throat etc. Hence, the primary motive is to biologically investigate these compounds and their effect on human health. As per the international agency for research on cancer, many of air pollutants have been found as the source of producing cancer to humans [42]. Direct breathing of these carbonyl and aromatic compounds can lead to lung cancer in human beings. The possible disease from these compounds due to direct exposure can proceed to some very fatal health problems like carcinogenicity, teratogenicity, and mutagenicity, so evaluating the effect of these compounds on human health is a worth point for the researchers.

3.1 Short-Term Health Effect

Common health problems reported from newly constructed buildings are headaches, sickness, atopic dermatitis, dizziness, sleepiness, irritation in skin and eyes, sick building syndrome, etc. There are otherproblems related to the buildings is multiple chemical sensitivity and sick house syndrome. These are the short-term effects of the contaminated indoor air. Other short-term effects of these contaminants include attack on the mucous layer of nose and irritation of eyes. The main problem for person who lives in indoor environment is asthma (a major lung infection) due to the contamination in indoor air [43]. There are a lot of reports of increased sickness and death rate from disease like asthma in the world. Hence, it has been cleared that these problems are arisen due to the contamination of indoor air pollutants in modern era. The sources of asthma type disease in indoor are combustion process in indoors, dust particles, waste collection, etc., which due to highly volatility of some carbonyl and aromatic compounds can easily travel through air. Various health problems have been observed with the fumes of kitchen. These kitchen fumes have aromatic compounds in excessive extent. Singh et al. [44] have tested the air sample from the kitchen. The result showed that the presence of PAHs in the kitchen air has a strong impact on the respiratory system of workers. These organic emissions are extremely lipid soluble, and ultimately it will be easily absorbed into gastrointestinal tract of mammals. These compounds instantaneously affect the tissues and simultaneously have adverse effect on the body fat. Metabolism of PAHs occurs via the cytochrome P450-mediated mixed function oxidase system with oxidation or hydroxylation as the first step. The adverse effect of aromatic and carbonyl compounds on human health is mainly depended on time duration, way of exposure, quantity of compounds exposed to human, the amount of toxic compounds, etc. There are several other factors on which the health factor depends like preexisting health of exposed human and his age [44]. Unwin et al. have listed some short-term effects on the human health due to the high exposure of PAHs which are eye irritation, confusion, nausea, diarrhea, and vomiting [45]. In spite of this, it is not clearly mentioned that which compound is responsible for these symptoms. IPCS (International Programme on Chemical Safety) has dictated effects of the specific PAHs compound like benzo(a) pyrene, anthracene, and naphthalene [46]. These mixtures of PAHs are termed as the inflammatory and irritants. Benzo(a) pyrene, anthracene, and naphthalene are termed as the direct skin irritants. These compounds are known for the skin sensitizers which mean that skin allergic effect on human and animals also.

3.2 Long-Term Health Effect

The long-term health effect of these pollutants may be life-threatening diseases like nasal tumors, leukemia, asthma, nasopharyngeal cancer, and reduced pulmonary function. The emission of dust in buildings and emission of nitrogen dioxide during the cooking process through gas increase the chances of asthma in human beings. In these indoor air contaminants, VOCs have the deep effect on human health. There is clear relation between the air pollutants and disease. The increasing research is required to overcome hazard by VOCs to human health, and environment has caused fast recognition of toxic aspect of VOCs during the last two decades. Cancer of the lung was one of the leading sites of cancer among population living in several

Indian states [44]. Tracheal, bronchus, and lung cancer rank second globally by number of incident cases in both sexes, in 50 most populous countries as per Global burden of cancer report—2013. Singh et al. have termed aromatic compounds and their subsequent as the carcinogenic, toxic, and mutagenic [6]. They have specified that there is tropospheric ozone formation due to benzene, xylene, and toluene emission. This tropospheric ozone creation is based on the concentration of unregulated emission and their structure. Carter had evaluated that some oxygenated compounds like aromatic and carbonyl are important as per the tropospheric chemistry [47]. These compounds have a sufficient contribution in the photochemical oxidants. They have found that ozone and free radicals are the important pioneers of oxygenated compounds. Correa and Arbilla have studied the ambient condition of Rio de Janeiro city and found that pioneer element for the ozone is HCHO [48]. Diesel exhaust emissions have several components which pose severe health problems. All diesel exhaust has taken as the human carcinogen as per the IARC [49]. Emmelin et al. had observed that diesel emission is a major topic in epidemiological studies due to the possibility of lung cancer [50]. Reduced immune system, damaging of kidney and liver such as jaundice, severe respiratory problems, and irregularities in functioning of lung are the long-term or long-standing health effects due to divestment of PAHs. Repeated exposure of carbonyl and aromatic compounds may also possess the intense skin problem like redness and inflammation. Common PAHs like naphthalene have intense impact on our blood system. It involves in the process of breaking of red blood cells if high concentration of this component inhaled. Apart from this, the severe health problem in humans due to the carbonyl and aromatic compounds is based on the route of divestment [51]. Although un-metabolized PAHs can have toxic effects, a major concern is the ability of the reactive metabolites, such as epoxides and dihydrodiols, of some PAHs to bind to cellular proteins and DNA. These carbonyl and aromatic compounds are responsible for the disorganization of biochemical and damaging of cell. Due to this catastrophe in the human beings, mutation, generation of tumors and cancer, and decreased rate of development in body are the grievous impacts [52]. Some other intense disease from the exposure of these compounds are gastrointestinal and bladder cancer.

4 Possible Ways to Control VOCs Emissions

There are several methods which have been used for controlling the carbonyl and aromatic compounds in indoor and outdoor. It is important to convert harmful VOCs into the environment-friendly species. Multiple modes have been introduced till now for abatement of VOCs in indoor like physical adsorption [53, 54], plasma technology [55], plant absorption [56, 57], photocatalysis [58], and catalytic oxidation [59]. Catalytic oxidation for indoor plays a crucial role for the diminishment of VOCs as this mode oxidizes it into CO_2 at room temperature condition [60]. Unlike many other pollutants in indoors which can sufficiently diminish by many process used, HCHO cannot be properly controlled just because of its higher vapor pressure (3883 mm Hg (2078 in H_2O)) at 25 °C and relatively low boiling point (−19.3 °C (−2.74 °F)) [61]. Since lower temperatures are desired for the abatement of HCHO by catalytic oxidation in indoor, the recent development is based on a catalyst for total diminishment of HCHO at surrounding temperature condition. Probably, main sources of carbonyl and aromatic emissions are building materials, industries, and transportation (mainly diesel engine running vehicles) [6]. Traditional methods like using filters to isolate the pollutants or using sorption materials to absorb different VOCs are the techniques which can only reduce the intensity of the pollutants in environment. These techniques can only change the phase of the pollutants rather than eliminating them completely. For degradation, many techniques have been used including (1) plasma decomposition with or without using catalyst, (2) biological/botanical filtration, (3) chemisorptions, (4) photocatalytic oxidation (PCO), and (5) thermal and nonthermal catalytic oxidations. VOCs can be degraded with plasma decomposition process using air ionizers developed plasma cluster ions. Plasma can be produced with the help of electronic and photo-ionization. Ion density, treatment duration, qualities, and structure of chemicals are the influencing factor for degradation of VOC with plasma techniques [62]. Plasma technology is limited with the low concentration and harmful intermediates generation. VOCs diminishment in present situation has been experimented with botanical filtration also [63]. Some questions which arise with the botanical filtration are: (1) In case of using this method, kinetics related to degradation of VOCs should be known, (2) for the efficient decomposition of VOCs through microbes in plant roots, conversion rate should be high over the time, (3) the capacity of decomposition rate should be variable to degrade varying VOCs concentration level. But many authors have given limitation of this process as technical composition needs more advancement for carrying the more pollutant to biological filter. Some passive technologies have also discussed which are used as the degradation process such as the third method, i.e., chemisorption technique for the diminishment is the adsorption method at sorbent led through chemical reaction. The sorbent used in case of this method is generally some chemical compound treated with impregnation method or other special treated. But this process is incomprehensive due to intromission of forced air. Modified PCO is also a good removal technique. The technique used under PCO is generally using semiconductor catalysts and light energy to degrade VOCs into harmless compounds like water vapor (H_2O) and carbon dioxide (CO_2). PCO process admixes oxygen, VOCs, and water on the activator surface to out-turn CO_2 and H_2O under light radiation. Release of light at TiO_2 surface originates electron and holes in the molecule. The departure of electron from conduction band to valence band originates holes in conduction band after getting the band energy. The origin of holes produces hydroxyl radicals due to the reaction of holes with soaked hydroxide ions. The formation of hydroxyl radical is due to the extreme oxidative nature of holes. Moreover, the PCO process is constrained with UV light until doped with some metals like noble metal and for this process only less than 5% solar radiation is utilized with UV photocatalyst. Visible light source was used for removal in modified PCO instead of UV light used in conventional PCO. TiO_2 photocatalysts have widely used for the abatement with different concentrations of noble metal (Pt, Au, Ag, etc.) and transition metal oxides. Pt/TiO_2 photocatalyst was investigated for diminishment of HCHO by Zhu and Wu [64]. Pt/TiO_2 photocatalyst showed 98.3% degradation efficiency for HCHO. Some limitation like formation of by-products and deactivation of photocatalyst surface is the problem with PCO. The effective degradation of HCHO is done by thermal (at higher temperature) and/or nonthermal (environmental temperature) processes using catalytic oxidation. For catalytic oxidation, noble metal (Pt, Rh, Pd, Au, Ag, etc.), transition metal, and metal oxides are some normally applied catalysts. For the outdoor emission control, there is no specific technology available which can directly convert these emissions into the harmless products. But there are some specified conditions available which can reduce the formation of emission form the engine exhaust. Many different technologies have been used for the remediation of emissions to control the harmful effect on the environment and human beings. Some authors have given many modification and conditions for the reduction of aromatic and carbonyl emissions in outdoor. Di et al. have given a condition for the reduction of benzene emission [40]. Result shows that at no-load condition the emission is high as compared to higher loads. The basics behind this reduction at higher load are due to the thorough combustion of fuels in the cylinder accompanied with exhaust system. Excess oxygen content in fuel leads to the reduction of PAHs in case of engine exhaust. But this excess oxygen in case of carbonyl emissions increases the HCHO, CH_3CHO, and acetone. For controlling the emission from the transportation, some technologies have also implied like exhaust gas recirculation system, diesel oxidation catalyst, continuous regenerating diesel particulate filter but as the name suggested these methods are not sufficient for decomposition of the carbonyl and aromatic compounds. Agarwal et al. have given a comprehensive view that some aldehyde emission shows higher trend at lower load for all tested fuels [9]. They have investigated biodiesel and methanol-blended fuels and show that at higher load there is a significant reduction of HCHO emission. Similar trend has also depicted by Sharma et al. that emission level of benzene decreased with increasing loads [65]. As from other emission, emission of organic carbon also showed decrement with increment in load. Other aromatic compounds like toluene and xylene possess different trends with the use of biofuels. These compounds have decreased concentration with biodiesel [66]. Poulopoulos et al. have observed that 10% of ethanol blending favors the complete oxidation and thus reduces the emission of acetaldehyde [22]. Increased oxygen content in the fuel during the combustion process leads to lower aldehyde emission [67]. Jo et al. have used PCO process for the vehicle exhaust cleaning [68]. They have used TiO_2 catalyst for the oxidation of VOCs using photo-oxidation technique. Result shows that approximately 100% degradation efficiency has found over the different relative humidity ranges. Bioremediation technique has also used as the promising way for the utilization of PAHs for the outdoors [44].

5 Discussion

Going through all the literature, the term involves special attention regarding carbonyl and aromatic compounds have been illustrated in this point of discussion. The main concern about the indoor and outdoor air qualities is emission of carbonyl and aromatic compounds because there are many compounds which are carcinogenic and mutagenic. These components have various paths of exposure to the human being as well as for the environment also. The severe health concern related to the carbonyl and aromatic compounds have been illustrated in Tables 1 and 2, respectively. As per the authors, the major compound related to the carbonyl and aromatic compounds are HCHO, CH_3CHO, acetone, benzene, toluene, and xylene which have intense effect on human health. The concentration of these compounds depends upon the outdoor condition, indoor condition, different agilities of human, seasonal variation, and chemical reaction. Various agencies have described these compounds into three categories like very volatile, volatile, and semi-volatile based on their boiling point. The associations of carbonyl and aromatic compounds with different severe diseases have also been broadly dictated in Tables 1 and 2. It has been observed that industrial and commercial processing, decoration work in newly made buildings, manufacturing of woods in home, transportation sector, etc., are some concerning point where the emission of carbonyl and aromatic compounds has found more as compared to other places. Common health problems reported from newly constructed buildings are headaches, sickness, atopic dermatitis, dizziness, sleepiness, irritation in skin and eyes, sick building syndrome, etc., which have been illustrated in Tables 1 and 2. Reduced immune system, damaging of kidney and liver such as jaundice, severe respiratory problems, and irregularities in functioning of lung, tumors, cancer are the long-standing health effects due to exposure of carbonyl and aromatic compounds on the human health. Severe disease like cancer, due to the long-term exposure of carbonyl and aromatic compounds, causes multi-disease that not only effects genes but also it has an intense impact on the clonal expansion and escape from apoptosis. Without doubt, there is a lot of improvement, but this should be utilized for the abatement of carbonyl and aromatic compounds. A high-quality investigation should be done for the decrement of these compounds in outdoor environment as well as indoor environment. Air quality in indoors and outdoors is the result of many toxicological and physical factors. More refined research is needed for the abatement of these compounds due to its complex nature and its intense effect on human health as well as environment. In case of indoors, various techniques have been used for the abatement for carbonyl and aromatic compounds such as plasma decomposition, biological/botanical filtration, chemisorptions, PCO, and thermal and nonthermal catalytic oxidations. Yet for the proper control of aromatic and carbonyl compound emissions from the transportation is still a gap for the researchers. Future research must be on full abatement of carbonyl and aromatic compounds from the motor vehicles. There should be a refined technique for the degradation of these compounds from the engine exhaust.

Table 1 Effect and sources of carbonyl compound

Pollutants	Short-term effect	Long-term effect	Source of emission
Formaldehyde [69]	Irritation of the eyes, nose, throat	Human carcinogen Asthma-like respiratory problems Affects reproductive system pregnancy	Power plants, automobile exhausts, and manufacturing facilities Indoor air includes consumer products building materials and tobacco smoke
Acetaldehyde [9]	Irritation of skin, eyes, mucous membrane, throat, respiratory tract Headache vomiting nausea	Carcinogen	Building materials Automobile exhausts Consumer products
Acetone [70]	Nausea, headache, drowsiness, dizziness, and confusion Mild irritation to skin	Unconsciousness adverse effect on the nervous system Red dry cracked skin (dermatitis)	Outdoor: volcanic gases, forest fires, manufacture and use, burning waste materials, exhaust of automobiles, tobacco smoke, landfills Indoor: No
Acrolein [71]	Respiratory and gastrointestinal tracts cytotoxic Histopathological Intense effects in the bronchi and/or trachea (including edema, vascular congestion, inflammation, and hemorrhagic, necrosis, exfoliation)	Effect respiratory tract	Outdoor: fermentation and ripening processes Forest fires due to improper combustion
Benzaldehyde [72]	Irritate the nose, throat coughing and shortness of breath Irritate the skin and eyes Skin rashes	Mutations seizures and passing out	Outdoor: industrial, medical evaluation
Butyraldehyde [73]	Damage to the respiratory tract	Carcinogenic	Outdoor: cigarette smoke Indoor: cigarette smoke
Crotonaldehyde [74]	Irritation of the eyes, skin affects respiratory tract	Carcinogenic	Outdoor: cigarette smoke Indoor: cigarette smoke
Hexaldehyde [75, 76]	Inhalation and skin exasperation of mucous membranes Discomfort in the eyes and discomfort in the nose	No effects on pulmonary function Nasal swelling	Outdoor: industry warehouse, ammonia Indoor: wood pellets, domestic storage room, water-based paints

(continued)

Table 1 (continued)

Pollutants	Short-term effect	Long-term effect	Source of emission
Methacrolein [77]	Exasperation in eye, nasal, and respiratory tract	Gastrointestinal mucosa in the animals Epithelial hyperplasia Ulceration hemorrhage	Outdoor: smoking, second-hand smoke Indoor: exposure to wood and plastic smoke
Propionaldehyde [78]	Exasperation to eyes, pulmonary tract, and skin Headache Consciousness problems Breathing difficulties Sore throat, spasm Edema of the larynx, and exasperation of the nose mucosa	Create problems in the lungs bronchitis to develop with cough Phlegm Shortness of breath	Outdoor: cigarette smoke Indoor: cigarette smoke
Tolualdehyde [79]	Adversely affects mucus membranes of the nasal and oral passages and the upper respiratory tract Exasperation bronchial constriction Choking and coughing exasperation on the skin of the face	–	Outdoor: exhaust from outboard motors, automobiles trucks, and emissions from fireplaces Indoor: perfumes, as a pharmaceutical and dyestuff intermediate
Valeraldehyde [80]	Acute toxicity Exasperation of skin and eyes Skin sensitization	Necrosis	Outdoor: industrial intermediate, in resin chemistry

Challenges and direction related to VOCs emissions

Carbonyl and aromatic compounds emissions have the big challenge for indoor and outdoor air condition. Approximately, all carbonyl and aromatic compounds listed above have the sources in both indoor and outdoor with their severe health effects. Vehicular, industrial, and organic compounds used in indoor are the major source for VOCs emission. As stated in above points that in outdoor, carbonyl emissions are more frequent rather than PAHs emission due to use of oxygenated fuels used for the better combustion. For outdoors, apart from primary sources of emission the formation of secondary and tertiary emission such as smog formation is also a problem to human health and for environment also. The application of new fuels like different alcoholic blends in mineral diesel and use of biodiesel are increasing the emissions of carbonyl and aromatic compounds in outdoors. Some natural process like photo-oxidation process are also becoming the concern in increasing the emission such as photo-oxidation of alkanes and alkenes also

Table 2 Effect and sources of aromatic compound

Pollutants	Short-term effect	Long-term effect	Source of emission
Benzene [81]	Drowsiness Dizziness Rapid or irregular heartbeat Headaches Tremors Confusion	Reason of not producing blood cells through bone marrow, anemia Unconsciousness Death (at very high levels) Reduces the immune system	Outdoor: volcanoes and forest fires, crude oil, gasoline, and cigarette smoke Indoor: detergents, drugs, and pesticides
Toluene [82, 83]	Headaches Dizziness Unconsciousness	Problems while pregnancy: retardation, premature delivery, congenital malformations, and postnatal developmental retardation, cognitive impairment Loss of vision and hearing	Outdoor: biodiesel, diesel fuel Indoor: paint reducer, paint thinner, or paint for their narcotic effects
m-Xylene [84, 85]	Exasperation in eyes, skin, nose, throat Dizziness Excitement Drowsiness Incoordination Staggering gait Corneal vacuolization Anorexia Nausea Vomiting Abdominal pain Dermatitis	Gastrointestinal tract blood problem Liver problem Kidneys problem	Outdoor: automobile exhaust, and during its use as a solvent, hazardous waste disposal sites
o-Xylene [86]	Dizziness Drowsiness Headache Nausea Dry skin Redness Pain Burning sensation Abdominal pain	Nervous system Hearing loss Animal tests exhibit that this component possibly causes toxicity to human reproduction or development	
Ethylbenzene [87, 88]	Exasperation in eyes, skin, mucous membrane Headache Dermatitis	Narcosis, coma	Outdoor: tobacco smoke, gasoline Indoor: household products like cleaning products or paints, pesticides, solvents, carpet glues, varnishes, paints

(continued)

Table 2 (continued)

Pollutants	Short-term effect	Long-term effect	Source of emission
Propyl benzene [89]	Exasperate the skin and eyes Irritate the nose and throat Headache Nausea Vomiting Dizziness Drowsiness and fainting	–	Outdoor: – Indoor: –
Naphthalene [90–92]	Ingestion of mothballs Hemolytic anemia Poisoning in pediatrics patients	Laryngeal or colon cancer	Outdoor: coal tar, synthetic resins, wood smoke, fuel oil and gasoline, fugitive emissions and motor vehicle exhaust, Indoor: paints, solid block deodorizer for toilets, kerosene heaters, and tobacco smoke
Anthracene [93]	Skin-related problems Blood problems Burning, itching and edema, a build-up of fluid in tissues Headaches Nausea Loss of appetite Inflammation or swelling of the stomach and intestines	Kidneys Liver Stomach and intestines	Outdoor: heavy oils, coal tar, roofing tar, or creosote Indoor: –

substantially contributed in formation of aldehyde emission. The other problems related to emission of carbonyl compounds are load conditions for the vehicle emission. As for the outdoors, combustion process in indoors is also a major concern apart from the sick building syndrome problem. These combustion processes happen in every houses, restaurants, buildings, offices, etc. Research agencies on air pollution have clearly mentioned that indoor air pollution is more as compared to outdoor air pollution. Carcinogenic and mutagenic effects of these air pollutants have given in this chapter. Due to high volatility of these compounds, these pollutants easily enter in our body through the respiration process in a continuous manner. To decompose the organic compounds in indoor air, various techniques have described in this chapter. In all these techniques, first three techniques for the removal of pollutants in air use filtration process. This implies that these compounds will only be separated not degraded. The last two processes described are sufficient for the decomposition of aromatic and carbonyl compounds. These two methods have strong capturing and abatement efficiency. In these two methods, thermal and nonthermal catalytic oxidation techniques are more powerful over the photocatalytic oxidation due to the formation of intermediate products in case of photocatalytic oxidation. In case of outdoors control of carbonyl and aromatic compounds, sufficient abatement method is still a gap.

6 Conclusion

In this article, the effects of carbonyl and aromatic compound emissions on human health have studied from different sources. So the major objective of this chapter is to study the different carbonyl and aromatic compounds and their toxicological assessment on the human health. The major conclusions which can be explored from this study are as listed here:

- Combustion products, construction equipment, paints, varnishes, calks, office materials, cooking, consumer products, etc., may be the source of emission in case of indoor air pollution.
- Emission from the vehicles and industries is the dominant sources of outdoor carbonyl and aromatic compound emissions. There is no available efficient method applied in vehicles which can convert these carbonyl and aromatic compound emissions.
- The use of biodiesel as an alternative to diesel fuels in vehicles generally reduces the aromatic compound emission but supports the aldehyde emission particularly HCHO emission.
- Carbonyl and aromatic compounds are responsible for the secondary and tertiary air pollution. Formic acid, ozone formation, and carbon monoxide are the major by-products of these emissions.
- Tropospheric ozone formations due to benzene, xylene, and toluene emissions have also adverse effect on the environment as well as human health. The secondary pollution contributes in the formation of smog.
- HCHO, CH_3CHO, benzene, toluene, and xylene are the major compound in case of carbonyl and aromatic compound emissions up to a level of 70–90%. These compounds are also responsible for the severe health disease like tumors, cancer, jaundice, decreased lung functioning.
- The technologies explained in the possible ways for the reduction of carbonyl and aromatic compound emissions are sufficient to decompose these compounds into harmless products in indoor context.

References

1. Wang S, Ang HM, Tade MO (2007) Volatile organic compounds in indoor environment and photocatalytic oxidation: state of the art. Environ Int 33:694–705
2. Tang X, Chen J, Li Y, Li Y, Xu Y, Shen W (2006) Complete oxidation of formaldehyde over Ag/MnO x–CeO 2 catalysts. Chem Eng J 118:119–125
3. Wieslander G, Norbäck D, Björnsson E, Janson C, Boman G (1996) Asthma and the indoor environment: the significance of emission of formaldehyde and volatile organic compounds from newly painted indoor surfaces. Int Arch Occup Environ Health 7:115–124
4. Liang W, Li J, Jin Y (2012) Photo-catalytic degradation of gaseous formaldehyde by TiO_2/UV, Ag/TiO_2/UV and Ce/TiO_2/UV. Build Environ 51:345–350
5. Goode JW (1985) Toxicology of formaldehyde. Advances in Chemistry 210:217–227. doi:10.1021/ba-1985-0210.ch014
6. Singh P, Chauhan SR (2016) Carbonyl and aromatic hydrocarbon emissions from diesel engine exhaust using different feedstock: a review. Renew Sustain Energy Rev 63:269–291
7. Caplain I, Cazier F, Nouali H, Mercier A, Déchaux JC, Nollet V, Joumard R, André JM, Vidon R (2006) Emissions of unregulated pollutants from European gasoline and diesel passenger cars. Atmos Environ 40:5954–5966
8. Shi X, Pang X, Mu Y, He H, Shuai S, Wang J, Chen H, Li R (2006) Emission reduction potential of using ethanol–biodiesel–diesel fuel blend on a heavy-duty diesel engine. Atmos Environ 40:2567–2574
9. Agarwal AK, Shukla PC, Patel C, Gupta JG, Sharma N, Prasad RK, Agarwal RA (2016) Unregulated emissions and health risk potential from biodiesel (KB5, KB20) and methanol blend (M5) fuelled transportation diesel engines. Renew Energy 98:283–291
10. Jia C, Batterman S, Godwin C (2008) VOCs in industrial, urban and suburban neighborhoods, Part 1: indoor and outdoor concentrations, variation, and risk drivers. Atmos Environ 42:2083–2100
11. Karavalakis G, Stournas S, Bakeas E (2009) Light vehicle regulated and unregulated emissions from different biodiesels. Sci Total Environ 407:3338–3346
12. Carlier P, Hannachi H, Mouvier G (1986) The chemistry of carbonyl compounds in the atmosphere—a review. Atmos Environ 20:2079–2099 (1967)
13. Zhang Z, Jiang Z, Shangguan W (2016) Low-temperature catalysis for VOCs removal in technology and application: a state-of-the-art review. Catal Today 264:270–278
14. Yu C, Crump D (1998) A review of the emission of VOCs from polymeric materials used in buildings. Build Environ 33:357–374
15. Baek SO, Kim YS, Perry R (1997) Indoor air quality in homes, offices and restaurants in Korean urban areas—indoor/outdoor relationships. Atmos Environ 31:529–544
16. Guo H, Lee SC, Chan LY, Li WM (2004) Risk assessment of exposure to volatile organic compounds in different indoor environments. Environ Res 94:57–66
17. Anderson LG, Lanning JA, Barrell R, Miyagishima J, Jones RH, Wolfe P (1996) Sources and sinks of formaldehyde and acetaldehyde: an analysis of Denver's ambient concentration data. Atmos Environ 30:2113–2123
18. Altshuller AP (1993) Production of aldehydes as primary emissions and from secondary atmospheric reactions of alkenes and alkanes during the night and early morning hours. Atmos Environ Part A Gen Top 27:21–32
19. Altshuller AP (1991) Chemical reactions and transport of alkanes and their products in the troposphere. J Atmos Chem 12:19–61
20. Stump FD, Knapp KT, Ray WD (1996) Influence of ethanol-blended fuels on the emissions from three pre-1985 light-duty passenger vehicles. J Air Waste Manag Assoc 46:1149–1161
21. Faiz A (1993) Automotive emissions in developing countries-relative implications for global warming, acidification and urban air quality. Transp Res Part A: Policy Pract 27:167–186

22. Poulopoulos SG, Samaras DP, Philippopoulos CJ (2001) Regulated and unregulated emissions from an internal combustion engine operating on ethanol-containing fuels. Atmos Environ 35:4399–4406
23. Graboski MS, McCormick RL (1998) Combustion of fat and vegetable oil derived fuels in diesel engines. Prog Energy Combust Sci 24:125–164
24. Takada K, Yoshimura F, Ohga Y, Kusaka J, Daisho Y (2003) Experimental study on unregulated emission characteristics of turbocharged DI diesel engine with common rail fuel injection system. SAE Technical Paper
25. Turrio-Baldassarri L, Battistelli CL, Conti L, Crebelli R, De Berardis B, Iamiceli AL, Gambino M, Iannaccone S (2004) Emission comparison of urban bus engine fueled with diesel oil and 'biodiesel' blend. Sci Total Environ 327:147–162
26. Peng CY, Yang HH, Lan CH, Chien SM (2008) Effects of the biodiesel blend fuel on aldehyde emissions from diesel engine exhaust. Atmos Environ 42(5):906–915
27. Benner BA Jr, Gordon GE, Wise SA (1989) Mobile sources of atmospheric polycyclic aromatic hydrocarbons: a roadway tunnel study. Environ Sci Technol 23:1269–1278
28. World Health oganisation (WHO) (1987) Polynuclear aromatic hydrocarbons (PAH). Air quality guidelines for Europe. World Health Organization Regional Office Europe, Copenhagen, pp 105–117
29. International Agency for Research on Cancer (IARC) (1983) Polynuclear aromatic compounds, part 1: chemical, environmental and experimental data. In: IARC monographs on the evaluation of carcinogenic risks to humans, vol 32
30. Destaillats H, Maddalena RL, Singer BC, Hodgson AT, McKone TE (2008) Indoor pollutants emitted by office equipment: a review of reported data and information needs. Atmos Environ 42:1371–1388
31. Brown VM, Cockram AH, Crump DR, Gardiner D (1990) Investigations of the volatile organic compound content of indoor air in homes with an odorous damp proof membrane. Proc Indoor Air 90:557–580
32. Lee SC, Wang B (2006) Characteristics of emissions of air pollutants from mosquito coils and candles burning in a large environmental chamber. Atmos Environ 40:2128–2138
33. Pagels J, Wierzbicka A, Nilsson E, Isaxon C, Dahl A, Gudmundsson A, Swietlicki E, Bohgard M (2009) Chemical composition and mass emission factors of candle smoke particles. J Aerosol Sci 40:193–208
34. Richter H, Howard JB (2000) Formation of polycyclic aromatic hydrocarbons and their growth to soot—a review of chemical reaction pathways. Prog Energy Combust Sci 26:565–608
35. Ravindra K, Sokhi R, Van Grieken R (2008) Atmospheric polycyclic aromatic hydrocarbons: source attribution, emission factors and regulation. Atmos Environ 42:2895–2921
36. Borrás E, Tortajada-Genaro LA, Vázquez M, Zielinska B (2009) Polycyclic aromatic hydrocarbon exhaust emissions from different reformulated diesel fuels and engine operating conditions. Atmos Environ 43:5944–5952
37. Tancell PJ, Rhead MM, Trier CJ, Bell MA, Fussey DE (1995) The sources of benzo [a] pyrene in diesel exhaust emissions. Sci Total Environ 162:179–186
38. Schauer JJ, Kleeman MJ, Cass GR, Simoneit BR (1999) Measurement of emissions from air pollution sources. 2. C1 through C30 organic compounds from medium duty diesel trucks. Environ Sci Technol 33:1578–1587
39. Yunus Khan TM, Atabani AE, Badruddin IA, Ankalgi RF, Mainuddin Khan TK, Badarudin A (2015) Ceiba pentandra, Nigella sativa and their blend as prospective feedstocks for biodiesel. Ind Crops Product 65:367–373
40. Di Y, Cheung CS, Huang Z (2009) Experimental investigation on regulated and unregulated emissions of a diesel engine fueled with ultra-low sulfur diesel fuel blended with biodiesel from waste cooking oil. Sci Total Environ 407:835–846
41. Bakeas Evangelos B, Argyris Dimitrios I, Siskos Panayotis A (2003) Carbonyl compounds in the urban environment of Athens, Greece. Chemosphere 52(5):805–813

42. IARC (International Agency for Research on Cancer) (2010) Some non-heterocyclic polycyclic aromatic hydrocarbons and some related exposures. Monogr Eval Carcinog Risks Hum 92:765–771
43. Samet J (1990) Environmental controls and lung disease. Am Rev Respir Dis 142:915–939
44. Singh A, Nair KC, Kamal R, Bihari V, Gupta MK, Mudiam MK, Satyanarayana GN, Raj A, Haq I, Shukla NK, Khan AH (2016) Assessing hazardous risks of indoor airborne polycyclic aromatic hydrocarbons in the kitchen and its association with lung functions and urinary PAH metabolites in kitchen workers. Clin Chim Acta 452:204–213
45. Unwin J, Cocker J, Scobbie E, Chambers H (2006) An assessment of occupational exposure to polycyclic aromatic hydrocarbons in the UK. Ann Occup Hyg 50:395–403
46. IPCS (International Programme On Chemical Safety) (2010) Polycyclic aromatic hydrocarbons, selected non-heterocyclic. <http://www.inchem.org/documents/ehc/ehc/ehc202.htm>
47. Carter WP (1995) Computer modeling of environmental chamber measurements of maximum incremental reactivities of volatile organic compounds. Atmos Environ 29:2513–2527
48. Correa SM, Arbilla G (2008) Carbonyl emissions in diesel and biodiesel exhaust. Atmos Environ 42:769–775
49. World Health Organization (1989) Evaluation of carcinogenic risks to humans: diesel and gasoline engine exhausts and some Nifroarines. IARC Monographs, International Agency for Research on Cancer. Lyon, France
50. Emmelin A, Nyström L, Wall S (1993) Diesel exhaust exposure and smoking: a case-referent study of lung cancer among Swedish dock workers. Epidemiology 4:237–244
51. Abdel-Shafy HI, Mansour MS (2016) A review on polycyclic aromatic hydrocarbons: source, environmental impact, effect on human health and remediation. Egypt J Pet 25:107–123
52. Bach PB, Kelley MJ, Tate RC, McCrory DC (2003) Screening for lung cancer. Chest 123:72–82
53. Domingo-Garcia M, Fernández-Morales I, Lopez-Garzon FJ, Moreno-Castilla C, Perez-Mendoza M (1999) On the adsorption of formaldehyde at high temperatures and zero surface coverage. Langmuir. 15:3226–3231
54. Matsuo Y, Nishino Y, Fukutsuka T, Sugie Y (2008) Removal of formaldehyde from gas phase by silylated graphite oxide containing amino groups. Carbon 46:1162–1163
55. Liang WJ, Li J, Li JX, Zhu T, Jin YQ (2010) Formaldehyde removal from gas streams by means of $NaNO_2$ dielectric barrier discharge plasma. J Hazard Mater 175:1090–1095
56. Xu Z, Qin N, Wang J, Tong H (2010) Formaldehyde biofiltration as affected by spider plant. Biores Technol 101:6930–6934
57. Terelak K, Trybula S, Majchrzak M, Ott M, Hasse H (2005) Pilot plant formaldehyde distillation: experiments and modelling. Chem Eng Process 44:671–676
58. Akbarzadeh R, Umbarkar SB, Sonawane RS, Takle S, Dongare MK (2010) Vanadia–titania thin films for photocatalytic degradation of formaldehyde in sunlight. Appl Catal A 374:103–109
59. Sekine Y (2002) Oxidative decomposition of formaldehyde by metal oxides at room temperature. Atmos Environ 36:5543–5547
60. Spivey JJ (1987) Complete catalytic oxidation of volatile organics. Ind Eng Chem Res 26:2165–2180
61. Pei J, Zhang JS (2011) Critical review of catalytic oxidization and chemisorption methods for indoor formaldehyde removal. Hvac R Res 17:476–503
62. Ding HX, Zhu AM, Lu FG, Xu Y, Zhang J, Yang XF (2006) Low-temperature plasma-catalytic oxidation of formaldehyde in atmospheric pressure gas streams. J Phys D Appl Phys 39:3603
63. Wu PC, Li YY, Lee CC, Chiang CM, Su HJ (2003) Risk assessment of formaldehyde in typical office buildings in Taiwan. Indoor Air 13:359–363
64. Zhu Z, Wu RJ (2015) The degradation of formaldehyde using a Pt@ TiO_2 nanoparticles in presence of visible light irradiation at room temperature. J Taiwan Inst Chem Eng 50:276–281
65. Sharma M, Agarwal AK, Bharathi KV (2005) Characterization of exhaust particulates from diesel engine. Atmos Environ 39:3023–3028

66. Tang S, Frank BP, Lanni T, Rideout G, Meyer N, Beregszaszy C (2007) Unregulated emissions from a heavy-duty diesel engine with various fuels and emission control systems. Environ Sci Technol 41:5037–5043
67. Sharp CA, Howell SA, Jobe J (2000) The effect of biodiesel fuels on transient emissions from modern diesel engines, part II unregulated emissions and chemical characterization. SAE Technical Paper
68. Jo WK, Park JH, Chun HD (2002) Photocatalytic destruction of VOCs for in-vehicle air cleaning. J Photochem Photobiol A 148:109–119
69. Agency for Toxic Substance and Disease Registry. https://www.atsdr.cdc.gov/phs/phs.asp?id=218&tid=39
70. Agency for Toxic Substances and Disease Registry (ATSDR) (1994) Toxicological profile for acetone. Atlanta, GA: U.S. Department of Health and Human Services, Public Health Service. https://www.atsdr.cdc.gov/phs/phs.asp?id=3&tid=1
71. Agency for Toxic Substances and Disease Registry (ATSDR) (2007) Toxicological profile for Acrolein. Atlanta, GA: U.S. Department of Health and Human Services, Public Health Service. https://www.atsdr.cdc.gov/phs/phs.asp?id=554&tid=102
72. New jersey Department of Health and senior Services Hazardous Substance Fact sheet. http://nj.gov/health/eoh/rtkweb/documents/fs/0196.pdf
73. New jersey Department of Health and senior Services Hazardous Substance Fact sheet. http://nj.gov/health/eoh/rtkweb/documents/fs/0299.pdf
74. The Agency for Toxic Substances and Disease Registry (ATSDR) https://www.atsdr.cdc.gov/MMG/MMG.asp?id=947&tid=197
75. Clayton GD, Clayton FE (1981) Patty's industrial hygiene and toxicology, 3rd edn. Wiley, New York. ISBN 0-471-16042-3
76. Ernstgård L, Löf A, Wieslander G, Norbäck D, Johanson G (2007) Acute effects of some volatile organic compounds emitted from water-based paints. J Occup Environ Med 49:880–889
77. Faroon O, Roney N, Taylor J, Ashizawa A, Lumpkin MH, Plewak DJ (2008) Acrolein health effects. Toxicol Ind Health 24:447–490
78. New jersey Department of Health and senior Services Hazardous Substance Fact sheet. http://nj.gov/health/eoh/rtkweb/documents/fs/1598.pdf
79. Toxicological Data Network, TOXNET. https://toxnet.nlm.nih.gov/cgi-bin/sis/search/a?dbs+hsdb:@term+@DOCNO+5361
80. http://www.inchem.org/documents/sids/sids/110623.pdf
81. IPCS (1993) Benzene. Geneva, World Health Organization, International Programme on Chemical Safety, Environmental Health Criteria 150
82. Donald JM, Hooper K, Hopenhayn-Rich C (1991) Reproductive and developmental toxicity of toluene: a review. Environ Health Perspect 94:237
83. Agency for toxic substance and disease registry. Toxic substances portal-toluene. https://www.atsdr.cdc.gov/phs/phs.asp?id=159&tid=29
84. The National Institute for Occupational Safety and Health (NIOSH). m-Xylene. Publication No. 2004-149. https://www.cdc.gov/niosh/npg/npgd0669.html
85. Agency for toxic substance and disease registry. PUBLIC HEALTH STATEMENT Xylene. https://www.atsdr.cdc.gov/phs/phs.asp?id=293&tid=53
86. The National Institute for Occupational Safety and Health (NIOSH). O-xylene, 22 July 2015. https://www.cdc.gov/niosh/ipcsneng/neng0084.html
87. National Institute for Occupational Safety and Health (NIOSH) Education and Information Division, 11 Apr 2016. https://www.cdc.gov/niosh/npg/npgd0264.html
88. Agency for Toxic Substances and Disease Registry (ATSDR) (1999) Toxicological Profile for Ethylbenzene (Update). Public Health Service, U.S. Department of Health and Human Services, Atlanta, GA
89. New Jersey department of health and senior services, hazardous substance fact sheet. http://nj.gov/health/eoh/rtkweb/documents/fs/1607.pdf

90. Hazardous Substances Data Bank (HSDB) [online database] (2010) Bethesda, MD: National Library of Medicine. http://toxnet.nlm.nih.gov/cgibin/sis/htmlgen?HSDB. [Reference list]
91. Lim HC (2006) Mothballs: bringing safety issues out from the closet. Singapore Med J 47 (11):1003
92. IARC Working Group on the Evaluation of Carcinogenic Risks to Humans (2002) International Agency for Research on Cancer, World Health Organization. Some traditional herbal medicines, some mycotoxins, naphthalene and styrene. World Health Organization
93. Agency for Toxic Substances and Disease Registry (ATSDR) (1990) Public Health Statement, Polycyclic Aromatic Hydrocarbons. Atlanta, GA: U.S. Department of Health and Human Services. https://www.atsdr.cdc.gov/phs/phs.asp?id=120&tid=25

Thermodynamics of Carbon Nanotubes and Soot Formation

R. S. Bharj, Jyoti Bharj and Vishal Vasistha

Abstract The combustion of fuel in the presence of an oxidizer is an essential requirement to produce desirable thermal effect. There exists an analogy among generation of heat energy through combustion, soot formation, synthesis of carbon nanomaterial (CNM), and producer gas production by gasification. The stoichiometric and off-stoichiometric thermodynamic chemical kinetics explain the formation of soot, production of CNM as well as generation of heat or mechanical energy. If the objective of chemical combustion process is thermal energy, then soot is generated as a by-product, and if the objective is to synthesize CNMs, then heat energy is liberated as a consequence of combustion. Gasification of combustible material is another off-stoichiometric thermodynamic chemical combustion which is used for the generation of electricity in power plants. Coal gasification produced hydrogen gas can be advantageous in many aspects such as manufacturing of ammonia and a fuel source for combustion. Additionally, coal-derived producer gas can be converted into transportation fuels such as gasoline and diesel using some appropriate treatment. A lot of literature is available on combustion of fuels in heat engines but there is a dearth of availability of the literature for off-stoichiometric combustion such as synthesis of CNMs, gasification. Synthesis of CNT has been discussed in this chapter using LPG/biogas precursor.

Keywords Gasification · Carbon nanomaterial · Off-stoichiometric combustion

---

R. S. Bharj (✉) · V. Vasistha
Department of Mechanical Engineering, Dr B R Ambedkar National Institute of Technology, GT Bye Pass Road, Jalandhar 144011, Punjab, India
e-mail: bharjrs@nitj.ac.in

V. Vasistha
e-mail: vishalnitk007@gmail.com

J. Bharj
Department of Physics, Dr B R Ambedkar National Institute of Technology, GT Bye Pass Road, Jalandhar 144011, Punjab, India
e-mail: jyoti@nitj.ac.in

N. Sharma et al. (eds.), *Air Pollution and Control*, Energy, Environment, and Sustainability, https://doi.org/10.1007/978-981-10-7185-0_9

1 Brief Introduction and Historical Perspective

As usual like all other inventions and discoveries, there has also been a slight twist to the history of discovery of carbon nanotubes (CNTs). Work on nanotubes was started in 1970s. Morinobu Endo prepared first carbon filament as carbon nanotube during his Ph.D. at the University of Orleans, France. But he was not able to measure diameter and width of carbon nanotube so his claim was rejected [1, 2]. In 1985, a new allotrope of carbon, fullerene was discovered by Kroto et al. Subsequently, CNTs were reported by Iijima in 1991 and he then got the major share for its discovery. There exist at least four well-established CNT synthesis methods namely: arc discharge, laser ablation, chemical vapor deposition, and flame synthesis; so, a single growth mechanism for all of them may not be sufficient. During the formation of CNTs, researchers have pointed out that more than one mechanism seems to be operative for a particular method. Also, the way CNTs are formed by different routes is not exactly known and that is why this subject matter is still under investigation. Herein, this need is addressed. Various interdependent physical and chemical kinetics reaction mechanisms are involved in CNT production [3, 4]. Jan Baptista Van Helmont, a Belgian physician, invented production of gas from wood or coal by off-stoichiometric combustion. In 1788, first time Robert Gardner patented off-stoichiometric gasification process. In 1800s, gasification method made available for commercial purpose such as industrial purpose, lightening. Today, it is of much importance due to stringent regulations on vehicle and industrial emissions concerned with environment [5]. The presented approach is fundamental.

2 Soot Formation in Internal Combustion Engines

In internal combustion (IC) engines, particulate matter (soot) takes form between rich fuel reaction zone and fuel spray region. Formation of soot particles during thermodynamic combustion is a highly complex conversion from gas to solid phase [6]. Poly aromatic hydrocarbons (PAHs) coagulation constructed finest soot particles. Further, PAHs adsorption enhances soot concentration particularly due to particle surface reactions. This process postulates removal of hydrogen and addition of carbon. Carbon nanomaterials produced from diesel soot show the morphology of carbon nanospheres mixed with carbon nanotubes [7].

3 Phenomenology of CNT Powder or Soot Formation

Soot formation and oxidation becomes the matter of investigation for two reasons: one being an environmental pollutants and the other being a commercial product of importance (CNTs). Complexity of particulate matter (soot) formation procedure can be understood well by proper and exact models [8–10]. In current scenario, many researchers have collected a cluster of experimental data on soot formation process and also proposed several doctrine models [11–16]. Still, there is dearth of complete and accurate knowledge of soot formation mechanism, not only in the cases of heterogeneous reactions but also in homogeneous hydrocarbon pyrolysis reactions.

There are several such thermodynamic reactions those produce novel solid carbon particles (soots) as their primary products. According to thermodynamics laws, carbon–oxygen atoms ratio comes out nearly one ($C/O \approx 1$) for soot particles in a premixed air–fuel mixture system at standardized flame temperature but it is not true in all case [17]. It is observed that soot is contained inside the combustion flame. Soot is established in oxidation environment and can be governed by the chemical dynamics.

4 Formation of Fullerenes in Combustion

In the year 1985, C_{60} and C_{70} fullerenes identification took place first time and their macroscopic quantities were produced in 1990 using graphite vaporization method [18]. A large amount of C_{60} and C_{70} fullerene ions were drawn out during 1988–1991 [19, 20]. Significant quantities of C_{60} and C_{70} were found by Howard in low-pressure premixed laminar zone of benzene–oxygen flames [21]. The highest denseness of C_{60} and C_{70} was found to be 20% of the resulting soot. Theoretical kinship between fullerenes and soot constitution was demonstrated after fullerenes invention by graphite laser vaporization method [22]. Soot and fullerene manufacturing chemical reactions were observed very alike in an arc discharge and rich fuel flames.

5 General Thermodynamic Chemical Combustion of Fuels

Thermodynamic combustion is generally defined as chemical reaction between fuel and oxidizer that results in liberation of energy. Dry air consists of 21% oxygen (approximately), 79% nitrogen (approximately), 0.9% argon (approximately), and very low quantities of some other gases such as helium, neon, hydrogen, carbon dioxide on volume basis. Hence, mole ratio of nitrogen to oxygen is established as 3.76 (0.79/0.21) for combustion requirements, as shown in Eq. (1).

$$1\,\text{kmol}\ O_2 + 3.76\,\text{kmol}\ N_2 = 4.76\,\text{kmol air} \tag{1}$$

When fuel represented as $C_\alpha H_\beta O_\gamma N_\delta$ is mixed with air represented as (O_2 + $3.76N_2$) as an oxidizer undergoes the combustion process; various species such as CO_2, H_2O, NO_2, CH_4, N_2, O_2, H_2, H, O, CO, OH, NO, N, C(s) are formed as a by-product of the combustion reaction. Thus at equilibrium, the general combustion reaction per mole of fuel can be written as:

$$\begin{aligned} & C_{\alpha}H_{\beta}O_{\gamma}N_{\delta} + a_s(O_2 + 3.76N_2) \\ & \rightarrow n_1CO_2 + n_2H_2O + n_3N_2 + n_4O_2 + n_sCO + n_6H_2 \\ & + n_7H + n_8O + n_9OH + n_{10}NO + n_{11}N + n_{12}C(s) \\ & + n_{13}NO_2 + n_{14}C_{\alpha}H_{\beta} \end{aligned} \quad (2)$$

where a_s is the stoichiometric molar air–fuel ratio.

6 Stoichiometric and Non-stoichiometric or off-Stoichiometric Combustion

When fuel's hydrocarbons react with oxygen of air and produce CO_2 and H_2O only as by-product, combustion is considered to be completed. It is called stoichiometric combustion. Conversely, if more by-products are present as seen in Eq. (2), it is often referred to as off- or non-stoichiometric combustion. Incomplete combustion occurs due to deficiency of oxygen that is required to complete the combustion, inside the combustion chamber. These are also sometimes known as lean or rich combustion. Off-stoichiometric combustion also occurs in gasification process which produces desirable products as hydrogen and methane with undesirable product as CO gas [23]. Incomplete thermodynamic combustion may also be due to the dissociation phenomenon, which dominates at higher temperature ranges. Oxygen is highly affectionate (reactive) with hydrogen than carbon. Thus, hydrogen has strong tendency to be burnt completely and convert into H_2O; even oxygen is few that is mandatory requirement for complete combustion. The mechanisms of combustion are expressed as Case I, Case II, and Case III.

Case I: Heat Transfer, IC engine

$$\text{Fuel} + \text{Oxidizer} \rightarrow \underbrace{CO_2 + H_2O}_{A} + \underbrace{\text{Heat Energy}}_{B} + \underbrace{\text{Undesirable Products}}_{C} \quad (3)$$

Maximize *A* and *B*; minimize *C*.

Undesirable products, as shown in Eq. (3), are harmful emission constituents such as CO, NO_x, HC, and PM. Higher quantities of CO_2 and H_2O indicate more complete combustion.

Case II: Synthesis of Carbon nanostructures

$$\text{Fuel} + \text{Oxidizer} \rightarrow \underbrace{CO_2 + H_2O}_{A} + \underbrace{\text{Heat Energy}}_{B} + \underbrace{\text{CNM}}_{C} \tag{4}$$

Maximize *C* w.r.t. *A* and *B*.

Some carbon, during the above chemical reaction (4), is also converted into CO and plain carbon particles (soot). These soot particles build the base for carbon nanostructures formed during the combustion process, as shown in Eq. (4).

Case III: Gasification

$$\text{Fuel} + \text{Oxidizer} \rightarrow \underbrace{CO_2 + H_2O}_{A} + \underbrace{\text{Heat Energy}}_{B} + \underbrace{CO + CH_4 + H_2}_{C} \tag{5}$$

Maximize *C* w.r.t. *A* and *B*.

Product "*C*" in Eq. (5) combined together is known as producer or syn gas that can be used as fuel, power generating source, etc.

7 CNT Powder Synthesis Chemical Reaction Mechanism

The carbon powder synthesis process can be assumed to follow the water gas equations as shown in Eqs. (6), (7), and (8):

$$CO_2 + H_2 \rightleftarrows CO + H_2O \tag{6}$$

$$2CO(g) \rightarrow C(s) + CO_2(g) \quad \Delta H = -171\,\text{kJ/mol} \tag{7}$$

$$CO + H_2 \rightarrow C(s) + H_2O \quad \Delta H = -131\,\text{kJ/mol} \tag{8}$$

The solutions for the properties of equilibrium combustion products resemble on equilibrium constant method. This route is based on minimization of Gibbs free energy of the gas. CNT powder oxidation, as shown in Eq. (9), has been modeled as a heterogeneous process where the oxidation rate is proportional to the available soot surface area.

$$CO_2 + O_2 \rightarrow C(\text{CNT Powder}) + 2O_2 \tag{9}$$

8 Producer Gas Formation Chemical Reaction Mechanism

Producer gas can be formed using gasification process [24].

$$C + O_2 \rightleftarrows CO_2 \quad \Delta H = -401.9\,\text{kJ/kmol} \tag{10}$$

$$C + H_2O \rightleftarrows CO + H_2 \quad \Delta H = +122.6\,\text{kJ/kmol} \tag{11}$$

$$CO + H_2 \rightleftarrows CO + H_2O \quad \Delta H = +42.3\,\text{kJ/kmol} \tag{12}$$

$$CO + 3H_2 \rightleftarrows CH_4 + H_2O \quad \Delta H = -205.9\,\text{kJ/kmol} \tag{13}$$

Producer gas is formed as shown in Eqs. (11), (12), and (13) using oxidation and reduction chemical reactions. Heat transfer also takes place as shown in above chemical reactions, Eqs. (10)–(13).

9 Synthesis of Carbon Nanotubes (CNTs)

Carbon nanotubes (CNTs) have the capabilities to change the properties of a fuel efficiently due to their unique chemical, mechanical, and electrical properties with nanosize advantage. Application of CNTs as nanocatalyst is among its best advantages. CNTs offer high surface area to volume ratio characteristic because of its nanosize particles which helps to overwhelm the withdrawal of emulsified fuel. CNTs are the result of hydrocarbons synthesis process which is actually thermal decomposition of fossil fuel.

Carbon nanotubes can be produced by many ways such as arc discharge, chemical vapor deposition (CVD), laser ablation out of which chemical vapor deposition method is the cheapest and highest output method. Multiwall carbon nanotubes (MWCNT) can be produced on large scale by most reliable CVD method which can grow high-purity CNTs with fewer yields and a low temperature required for synthesis [25, 26].

There are various renewable and non-renewable sources to produce CNTs such as liquefied petroleum gas (LPG), biogas. Bharj et al. manufactured cost-effective and good quality CNTs on stainless steel substrate using LPG and biogas (produced from animal waste) as a precursor with a welding torch without addition of a catalyst as shown in Fig. 1. Different lengths CNTs were observed in transmission electron microscope (TEM) analysis (100–300 nm for premixed flame and 460–600 nm for diffusion flame) using LPG. To maximize CNTs yields, three parameters have been regraded to make the decision—substrate temperature, LPG fuel flow rates, and standoff distance. The exposure time taken for the deposition of soot was 2–5 min. The yield rise of CNT can be obtained using highly precision parameters. Inexpensive domestic LPG reduces bulk production price of CNTs [27].

Soot from biogas precursor produced MWCNT of length 125 nm and diameter 7–20 nm when analyzed using TEM, scanning electron microscope (SEM), X-ray diffraction (XRD), and Raman spectroscopy. Thermogravimetric analysis showed

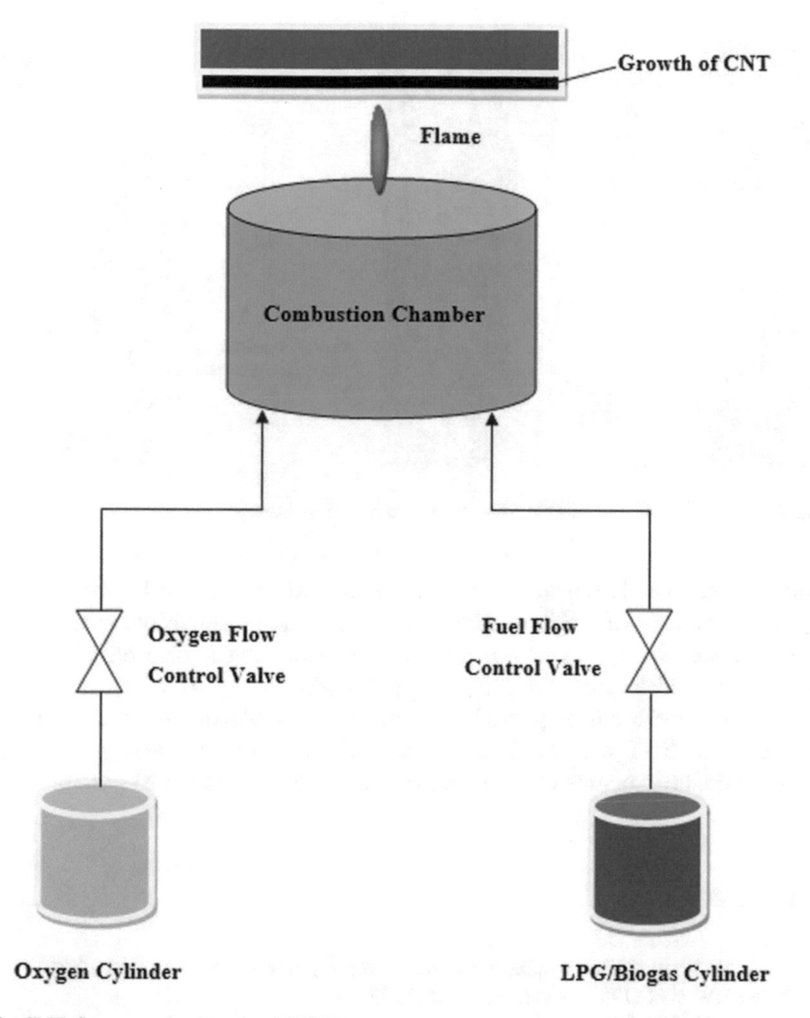

Fig. 1 CNT flame synthesis using LPG/biogas

90–95% purity of synthesized CNTs that has very close agreement with commercial CNTs [28].

10 Conclusions

Hydrocarbon fuels certainly produce desirable CNTs or undesirable soot when reacting with an oxidizer such as air. Off-stoichiometric thermodynamic combustion is also as important as stoichiometric combustion because of synthesis of CNMs that can be used as additive material to improve performance, formation of

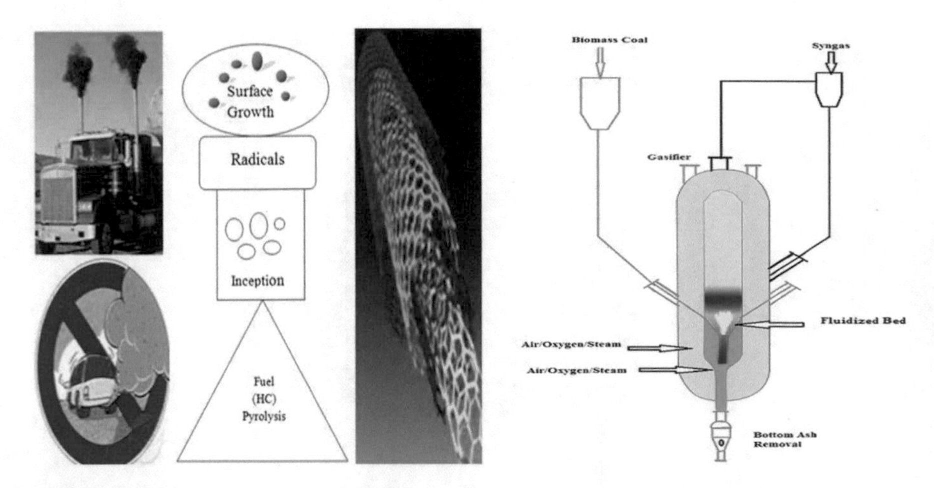

Fig. 2 Desirable and undesirable products of combustion process

producer gas that is useful for power generation, and as fuel substitute for IC engines. Synthesis of CNTs can be better understood through the phenomenon of soot formation in IC engines or pyrolysis. Gasification is also off-stoichiometric combustion which produces methane and hydrogen as useful products. Flame synthesis method has been proved cost-effective, reliable, and more accurate for the production of CNT with LPG/biogas (animal wastes) as precursor. The desirable and undesirable products of combustion process are shown in Fig. 2.

References

1. Ando Y (2010) Carbon nanotube: the inside story. J Nanosci Nanotechnol 10:3726–3738
2. Kroto HW et al (1985) Nature 318(162):1985
3. Iijima S (1991) Nature 354(56):1991
4. Iijima S, Ichihashi T (1993) Nature 363(603):1993
5. National energy technology laboratory (1999) U.S. Department of Energy. https://www.netl.doe.gov/research/coal/energysystems/gasification/gasifipedia/history-gasification
6. Haynes BS, Wagner HG (1981) Soot formation. Prog Energy Combust Sci 7:229–273
7. Manoj B et al (2012) Characterization of diesel soot from the combustion in engine by x-ray and spectroscopic techniques. Int J Electrochem Sci 7:3215–3221
8. Tesner PA (1972) Carbon formation from gas-phase hydrocarbons. Khimiya, Moscow
9. Bockhornet et al (1983) Investigation of the formation of high molecular hydrocarbons and soot in premixed hydrocarbon–oxygen flames. Ber Bunsen Ges Phys Chem 87:1067
10. Calcote HF (1989) The role of ions in soot formation. Abstracts of the III Int, Seminar on Flame Structure, Alma-Ata, p 1989
11. Tesner PA (1979) Soot formation during combustion. Combust Expl Shock Waves 15:111–119
12. Wagner HG (1979) Soot formation in combustion. In: 17th symposium (international) on combustion. Combustion Institute, Pittsburgh, pp 3–19

13. Graham SC (1977) The collisional growth of soot particles at high temperatures. In: 16th symposium (international) on combustion. Combustion Institute, Pittsburgh, pp 663–669
14. Homann KH, Wagner HG (1996) Some aspects of soot formation. In: Ray Bowen J (ed) Dynamics of exothermicity, combustion science and technology book series, vol 2. Gordon and Breach, pp 151–184
15. Glassman I (1988) Soot formation in combustion process. In: 22nd symposium (international) on combustion. Combustion Institute, Pittsburgh, pp 295–311
16. Howard JB (1991) Carbon addition and oxidation reactions in heterogeneous combustion and soot formation. In: 23rd symposium (international) on combustion. Combustion Institute, Pittsburgh, pp 1107–1127
17. Bockhorn H (1991) Soot formation in combustion. Round table discussion. Springer Verlag, Heidelberg
18. Kroto HW et al (1985) C_{60}: Buckminsterfullerene. Nature 318:162–163
19. Gerhardt P et al (1988) The formation of polyhedral carbon ions in fuel-rich acetylene and benzene flames. In: 22nd symposium (international) on combustion. Combustion Institute, Pittsburgh, pp 395–401
20. Howard JB et al (1991) Fullerenes C_{60} and C_{70} inflames. Nature 352:139–141
21. Howard JB (2004) Combustion synthesis of fullerenes and fullerenic nanomaterials for large-scale applications. In: Abstracts of International conference on Carbon, p 57
22. Zhang QL et al (1990) Reactivity of large carbon clusters: spheroidal carbon shells and their possible relevance to the formation and morphology of soot. J Phys Chem 90:525–528
23. Ahrenfeldt J (2007) Characterization of biomass producer gas as fuel for stationary gas engines in combined heat and power production. Ph.D. dissertation, Technical University of Denmark, Mar 2007
24. Beychok MR (1974) Coal gasification and the Phenosolvan process. American Chemical Society, 168th National Meeting, Atlantic City, Sept 1974
25. Bahgat et al (2011) Synthesis and modification of multi-walled carbon nano-tubes (MWCNTs) for water treatment applications. J Anal Appl Pyrol 92:307–313
26. Paradise M, Goswami T (2007) Carbon nanotubes production and industrial applications. Mater Des 28(5):1477–1489
27. Bharj et al. (2010) Flame synthesis of carbon nanotubes using domestic LPG, AIP conference proceedings, vol 1324, pp 389. Published by American institute of physics
28. Bharj et al. (2014) Biogas: a natural and renewable source for carbon nanotubes. Int J Res Advent Technol 2(3), March 2014. E-ISSN: 2321-9637

Part IV
Numerical/Simulations

A Study on Evolution and Modelling of Soot Formation in Diesel Jet Flames

M. Udayakumar and N. H. Mohamed Ibrahim

Abstract Soot emitted by diesel engines causes severe urban air pollution in the form of smog. Particularly in cities like New Delhi in India, smog presents a health risk for millions of people. To counter this problem diesel engines are to be designed with combustion systems which can minimize smoke formation, and if possible cheap and effective exhaust treatment devices are to be fitted in the exhaust of these engines. Hence, understanding of the chemistry and physical events in the soot formation is the starting point in solving this problem. Particularly, the soot formation studies on high-pressure diffusion flames burners issuing turbulent hydrocarbon fuel jets are relevant for this study. In this article, the various theories associated with the soot formation like soot inception, coagulation, agglomeration, oxidation are discussed. Also, the results of the numerical studies carried out by the authors on diesel-air flames at laboratory conditions are briefly presented.

Nomenclature

$\bar{\rho}$	Mean density
μ	Viscosity of the mixture
$\bar{\omega}_k$	Chemical production rate of species k
$\bar{\tau}_{ij}$	Viscous stress tensor
μ_k	Viscosity of species k
σ_k	Constant in the k-Ɛ turbulence model
σ_ϵ	Constant in the k-Ɛ turbulence model
$C_{\varepsilon 1}$	Constant in the k-Ɛ turbulence model
$C_{\varepsilon 2}$	Constant in the k-Ɛ turbulence model
C_μ	Constant in the k-Ɛ turbulence model
$\tilde{\varepsilon}$	Rate of dissipation of turbulence energy
$\tilde{k}$	Turbulence kinetic energy
μ_t	Eddy viscosity

M. Udayakumar (✉) · N. H. Mohamed Ibrahim
Department of Mechanical Engineering, National Institute of Technology, Tiruchirappalli, Tamil Nadu, India
e-mail: muday@nitt.edu

N. Sharma et al. (eds.), *Air Pollution and Control*, Energy, Environment, and Sustainability, https://doi.org/10.1007/978-981-10-7185-0_10

P_k Production rate of turbulence kinetic energy
$\tilde{Q}$ Production rate of thermal energy
μ_{eff} Effective viscosity
$\vec{\upsilon}$ Cartesian velocity component
P Pressure
N_A Avogadro's number
h_k Specific enthalpy of species k

Superscripts

$''$ Product, fluctuating value, density-weighted averaging
- Mean
$'$ Reactant, fluctuating value, averaged
$\sim$ Density-weighted mean

1 Introduction

Continuous exposure to particulate matter emitted from diesel engine combustion adversely affects human health. Some of the studies report that the particulate air pollution affects the mortality rates. The combustion of any hydrocarbon fuel generates soot which is the second most contributor to global warming [1]. As the very high adsorption capacity soot particles carry lots of toxic substances, soot particles having diameters ranging from 0.01 to 0.1 μm when breathed into lungs the alveolus will get damaged then the air exchange process ultimately leading to acute respiratory distress symptoms. The important soot precursor, polycyclic aromatic hydrocarbons (PAHs), is a carcinogenic agent, which affects the oxygen exchange process inside the lungs and leading to cardiovascular diseases [2].

Soot formation occurs within a millisecond which involves the following six steps: pyrolysis, nucleation, coalescence, surface growth, agglomeration and oxidation [3]. The soot formation study thus involves a detailed interaction among chemistry, transport and fluid flow, particularly at very high pressure and with high transport rates. Soot particle starts with inception, usually its initial structure being in nanoscale. The newly formed soot incipient grows through surface by a chemical reaction between the soot incipient surface and PAH molecules condensation known as the soot particle coagulation. Finally, the soot particles reduce in mass and size during oxidation and fragmentation. Surface growth is the key step by which the soot mass increases [4]. The chemical kinetics of soot studies concludes that acetylene is the primary soot precursor, and it is independent of fuel type [4, 5]. The most widely used soot formation model is Hydrogen–Abstraction–C_2H_2–Addition (HACA) for aromatic fuels [6, 7].

As highlighted by Heywood [8], most of the information on fundamentals of soot formation in combustion comes from studies on premixed and diffusion flames, stirred reactor, shock tubes and constant volume combustion bombs. Since in diesel engines a highly turbulent diffusion jet of vapourizing fuel reacts with air to form diffusion flames, the results of the soot formation studies carried out using co-flow burners or counter-flow burners at high pressures with diffusion jet flames approximate the reality [9–17]. However, the studies reported in literature were carried out using fuels like n-heptane, methane, ethane, propane. Hence, it is expected that the results obtained using diesel fuel ($C_{12}H_{23}$) will give the soot formation values more realistically.

This article reviews the formation of soot in various hydrocarbon flames and some of the soot measurement techniques available. Also, some aspects of computational modelling of soot formation are presented. This article also presents the soot formation studies carried out by the authors using CFD code (FLUENT) for diesel vapour–air mixtures using co-flow diffusion jet flames. The diesel vapour–air mixture is specifically identified for this study as in many parts of India the pollution caused by the diesel smoke needs detailed investigations of soot formation to arrive at the in-cylinder control and the suppression techniques for diesel engines.

2 Theories of Soot Formation

A combustion process is an exothermic chemical reaction between a fuel and an oxidizer. Normally, the fuel is a carbon-based solid, liquid or gas and the oxidizer is usually either oxygen or air. The combustion reaction is said to be complete or stoichiometric because the fuel/air ratio chosen could produce only carbon dioxide and water as a product. If the fuel/oxygen ratio is increased, the oxygen will be insufficient to complete the burning process in a given resident time which leads to the formation of carbon dioxide, water, carbon monoxide and hydrogen in the product gases. If the fuel/oxygen ratio is further increased, the carbon/air (C/O) ratio will become larger and larger (greater than unity) and leads to the excess amount of carbon and incomplete combustion after the available oxygen is utilized to form carbon monoxide. In many experiments it was observed that the soot is usually formed for C/O ratios of even less than one. Soot formation starts in the C/O-ratio range beginning 0.5–0.8 [8, 18].

2.1 Soot Formation Model

The incomplete combustion of hydrocarbon fuel leads to various undesirable products such as NO_X, unburned hydrocarbons (UHC) and soot. This is due to the insufficient combustion parameters like turbulence, time and temperature. In this article, we discuss some of the soot precursors and soot particle formation mechanism as the formation of soot is a very complex mechanism, which involves many chemical and physical process. The chemical kinetic analysis of soot formation usually falls in two steps,

1. Gas-phase chemistry
2. Particulate phase

2.1.1 Gas Phase

The hydrocarbon fuels are initially fragmented into smaller hydrocarbon molecules and free radicals by pyrolysis or oxidation process. The evolution of initial aromatic ring usually will be benzene or phenyl. This mechanism is valid for all parametric conditions like temperatures, pressure and mixture compositions.

First Nucleus Ring

The main attention is on the formation of the first aromatic species from the small aliphatic compound. Some of the most famous methods are the even-carbon-atom pathways as shown by Frenklach [19], which involves the addition of acetylene to n-C_4H_3 and n-C_4H_5 radicals. The reactions are given in Eqs. (2.1) and (2.4)

$$\text{n-}C_4H_5 + C_2H_2 \rightarrow C_6H_6 + H \tag{2.1}$$

They also stated that above reaction should occur predominantly and has one or more intermediate steps. Callear and Smith [20] experimentally observed that the reaction (2.1) occurs in three intermediate steps. They also showed that reaction of H with acetylene at low temperatures leads to benzene in the products. Finally, they arrive at the following reactions as a conclusion which occurs to form the reaction (2.5).

$$H + C_2H_2 \rightarrow C_3H_3 \tag{2.2}$$

$$C_2H_3 + C_2H_2 \rightarrow \text{n-}C_4H_5 \tag{2.3}$$

Frenklach et al. [21, 22] also showed that cyclization will occur primarily through the following reaction (acetylene addition to form phenyl group) (Eq. 2.4).

$$\text{n-}C_4H_3 + C_2H_2 \rightarrow C_6H_5 \tag{2.4}$$

It is also believed that the importance of resonance stabilized free radicals (RSFRs), such as propargyl (C_3H_3), are of importance in forming aromatics and PAH in flames. They proposed an odd-carbon-atom pathway through the recombination reaction of two propargyl radicals,

$$C_3H_3 + C_3H_3 \rightarrow C_6H_6 \tag{2.5}$$

The propargyl radical is a stable radical and so for a long time these species play the main role in aromatics formation [23]. The stability of the RSFRs will get reduced in reactivity with respect to O_2. The RSFRs form weaker bonds with the free radicals and are less reactive with O_2. Miller et al. [24] suggested the reaction Eq. (2.6) for the formation of cyclopentadienyl radical (c-C_5H_5).

$$C_3H_3 + C_2H_2 \rightarrow \text{c-}C_5H_5 \tag{2.6}$$

According to Melius et al. [25] the cyclopentadienyl radical (c-C_5H_5) reacts rapidly to form benzene as per reaction (2.7).

$$\text{c-}C_5H_5 + CH_3 \rightarrow C_6H_6 + H + H \tag{2.7}$$

Again, it was shown that cyclopentadienyl radical reacts to form naphthalene and hydrogen atom

$$C_5H_5 + C_5H_5 \rightarrow \text{Naphthalene} + H + H \tag{2.8}$$

This atomic hydrogen evolved during the reactions given by Eqs. (2.7) and (2.8) sustain reactions given by Eq. (2.2). It was shown by them that the above reactions shift their equilibrium condition to favour the products if the temperature varies in the range 1400–1700 K.

Aromatic Growth by HACA

The most popular mechanism of PAH growth is the HACA pathway developed by Frenklach and Wang [26]. This model reveals the repetitive reaction in two important steps,

1. Abstraction of an H atom from the hydrocarbon reactant

$$A_i + H \rightarrow A_{i-} + H_2 \tag{2.9}$$

2. Addition of a gaseous C_2H_2 to the radical being formed

$$A_{i-} + C_2H_2 \rightarrow \text{products} \tag{2.10}$$

where A_i is an aromatic molecule, and A_{i-} is its radical. Frenklach [27] described that the forward reaction of (Eq. 2.9) dominates as compared to the reverse (Eq. 2.10) increases with pressure and molecular size.

Aromatic Oxidation

Haynes and Wagner [28] and Neoh et al. [29] assumed that the hydroxyl (OH) radical is a primary oxidizing agent during soot particles formation. According to Frenklach [27] the primary oxidation of aromatic radicals is by O_2 and occurs at the very beginning of the phenyl stage growth.

2.1.2 Particulate Phase

There are numerous uncertainties which have to be identified in the soot particle formation and evolution which includes some processes like soot particle inception, surface growth and oxidation, coagulation and agglomeration.

Soot Particle Inception

Soot inception is the homogeneous process in a gas-phase environment taking place at a molecular level having mass ranges in order about 500 a.m.u. [30], 300–700 a.m.u. [31], 1600 a.m.u. and 2000 a.m.u. [32]. Above this PAH can be said to be solid particles instead of molecules. The first and the foremost soot particles formed will be spherical in shape and roughly have a C/H ratio in the order of 2. On ageing, they will fuse to form larger spherical particles, and the larger particle will undergo surface elementary reactions, dehydrogenation, oxidation and coagulation. The soot emitted from combustion devices typically has a C/H ratio in the order of 10 and forms a graphitic-like structure [33].

Soot Particle Growth

A very large part of soot (more than 95%) yield is due to the formation of surface growth rather than soot inception [34]. The soot particle growth is identical to the PAH and acetylene formation but the surface growth is a heterogeneous process, in which adsorption and desorption processes at the surface takes place. Franklach and Wang [35] framed the first-order differential equation (Eq. 2.11)

$$\frac{\mathrm{d}f_\mathrm{v}}{\mathrm{d}t} = k_\mathrm{s}\left(f_\mathrm{v}^\infty - f_\mathrm{v}\right) \tag{2.11}$$

where k_s is surface growth rate coefficient, f_v^∞ ultimate volume fraction of soot. Harris and Weiner [36] studied soot volume fraction (Eq. 2.12) on premixed acetylene—air and ethylene/air flames. They stated that C_2H_2 satisfies the requirements for soot growth, and also he proposed a very simple model (Eq. 2.11) in which soot mass growth rate is directly proportional to soot surface area and acetylene concentration [34],

$$\frac{\mathrm{d}f_\mathrm{v}}{\mathrm{d}t} = k_{\mathrm{C_2H_2}} * p_{\mathrm{C_2H_2}} * A_\mathrm{s} \tag{2.12}$$

where A_s is surface area density ($\mathrm{m^2/m^3}$) and $p_{\mathrm{C_2H_2}}$ is the partial pressure of the gas-phase acetylene and $k_{\mathrm{C_2H_2}}$ is an acetylene surface growth rate coefficient.

Soot Coagulation

Frenklach et al. [33] also explained that the coagulation process, starting form pyrene, in the free molecular regime and showed it to be irreversible. Once the PAH monomers reach a particular size they will stick each other during collisions and form PAH dimers. These dimers collide with PAH molecules forming trimers or with other dimers lead to tetramers and so on. The coagulation is a process of sticking of two particles. It takes place only for relatively small particles (up to 10 nm diameter at low-pressure premixed zone) [34]. The rate of a coagulation process can be calculated by solving Smoluchowski [37]. Coagulation rate equation (Eq. 2.13). While deriving the coagulation rate equation, the following assumptions are made.

- Compared to gas mean free path, soot particles are very small.
- Each collision between two soot particles leads to coagulation.
- All soot particles are spherical.

$$\frac{\mathrm{d}n_k}{\mathrm{d}t} = \frac{1}{2}\sum_{i+j=k} N_{ij} - \sum_{i=1}^{\infty} N_{ij} \tag{2.13}$$

where n_k is number density of the molecules in size class 'k', with mass which results in a collision between other molecules of different classes 'i' and 'j'. N_{ij} denotes the rate of collision between molecules of classes 'i' and 'j'. The collision between two molecules forms a new molecule 'k', with a mass of newly formed molecules $m_k = m_i + m_j$. The rate formation of the new molecules 'k' is given by (Eq. 2.14)

$$\frac{1}{2}\sum_{i+j=k} N_{ij} = \sum_{i+j=k} \beta(m_i, m_j) n_i n_j$$

where

$$\begin{aligned}\beta(m_i, m_j) &= \sqrt{\frac{6k_\mathrm{B}T}{\mu_{i,j}}}(r_i + r_j)^2 \\ &= 2.2\left(\frac{3}{4\pi\rho}\right)^{1/6}\sqrt{\frac{6k_\mathrm{B}T}{\rho}}\sqrt{\frac{1}{m_i} + \frac{1}{m_j}}\left(m_i^{1/3} + m_j^{1/3}\right)^2\end{aligned} \tag{2.14}$$

where $\mu_{i,\,j} = m_i m_j/(m_i + m_j)$ is the reduced mass, r_i is the radius of the molecules in class i and ρ is the density of the molecules. Graham [38] found soot coagulation (Eq. 2.15) in shock-heated hydrocarbon/argon mixtures and derived a coagulation rate expression in terms of the rate of decrease of the particle number density $[n]$,

$$-\frac{\mathrm{d}n}{\mathrm{d}t} = \frac{25}{72}\left(\frac{3}{4\pi}\right)^{1/6}\sqrt{\frac{6k_\mathrm{B}T}{\rho_\mathrm{soot}}}f_\mathrm{v}^{1/6}[n]^{11/6}G.\propto \tag{2.15}$$

where f_v is soot volume fraction, G is collision factor, $\propto$ is polydisperse system factor, ρ is the condensed particle density, k_B is the Boltzmann constant.

Soot Particle Oxidation

Soot particle oxidation process occurs parallel to that of surface growth. Some potential soot oxidants are O, O_2, OH and CO_2. Frenklach et al. [33] observed that the major oxidation process occurs at the beginning of soot particle growth. According to Neoh et al. [39] and Lucht et al. [40], the hydroxyl radical is the most abundant oxidizing species under fuel-rich conditions. Lucht et al. [40] showed that OH is the limiting oxidative reactant in a fuel-rich condition as the soot decreases with an increase in OH concentration.

Soot Agglomeration

Due to lack of surface growth soot agglomeration takes place in the later phase, coagulation is no longer possible [34] which results in the formation of openly structured aggregates, containing from 10 to 100 primary particles (spherules). A relationship between the number N of primary particles and the maximum length L of the aggregates (Eq. 2.16) can be derived as

$$N = k_f\left(\frac{L}{3d_\mathrm{p}}\right)^{D_\mathrm{f}} \tag{2.16}$$

where k_f is a constant, d_p the particle diameter and D_f a fractal dimension.

3 Soot Studies in Diffusion Flames

The effect of temperature on soot formation in a propane–air premixed flame is discussed in detail by Prado and Lahaye [9] for a given fuel-air ratio. They found that local soot volume fraction and soot agglomerate diameter decreases throughout the flame with increasing temperature. Mixture of methane-oxygen was studied in a highly pressurized cylindrical bomb at about 12.5 atm. for equivalence ratios of 2, 3 and 4, and it was observed that 20% of the carbon in the fuel gets converted into soot [10]. Similarly, no soot deposition was observed when the ethylene, propene, trimethene and n- and iso-butene mixed with oxygen at an equivalence ratio of 3 and exploded in a glass bulb between 0.5 and 1 atm. Parker and Wolfhard [11] analysed acetylene flames in a counter-flow pipe in which they have supplied the equal mass of air and the fuel at sub-atmospheric pressures. They found that at certain low-pressure carbon formation was absent, whereas presence of carbon increased with pressure. They also stated that the flame height is independent of pressure with change in gas velocity. They concluded in acetylene flame the height of the luminous carbon zone is pressure sensitive.

Milberg [12] measured the rates of soot formation, in an acetylene-air flame at sub-atmospheric pressures. They stated that the soot formation varies linearly with pressure, while it is independent of equivalence ratio. They observed the behaviour of premixed ethylene and methane with air or oxygen at pressures up to 40 atm. and found that at a given pressure; as the equivalence ratio was increased, soot formation also increased to a maximum and then decreased and finally disappeared due to reaching the rich flammability zone. Schalla and McDonald [13] observed the diffusion characteristic of 9 hydrocarbons flames at pressures up to 4 atmospheres. They found that the maximum smoke-free fuel flow is inversely proportional to pressure, and they extended their studies up to 22 atm. in an ethylene and ethane flames by measuring the flame height at smoking point. They observed that smoke point height is inversely proportional to pressure. MacFarlane et al. [14] conducted the experiments on C_5 and C_6 hydrocarbons at a pressure up to 20 atm both in laminar and turbulent jet flames. They observed that the soot yield increases with pressure and flame temperature by the order of $P^{2.5}$–P^3. Flower and Bowman [15] studied soot particle diameter, particle number density and soot concentration. At a specific height, they observed that the particle diameter increases from 60 nm at 1 atm to 90 nm at 2 atm. Again at the same height, particle number density increases and also there is an increase in particle size with pressure rise from 5 and 40 atm.

Kim et al. [16] observed the behaviour of acetylene flames from 0.125 atm. to atmospheric pressure, and they found that mean particle diameter was 10 and 20 nm for all pressures along the centreline locations. They also reported that there was a significant increase in mean particle diameter with pressure. In the study given in [16], mean particle diameters for ethylene flames were reported as a function of axial location. It is stated that the mean diameters between 10 and 20 nm along the centreline at atmospheric pressure found to increase to 30 and

40 nm at 8 atm. Milberg [17] observed the soot particles are deposited on a glass filter. Using the deposited soot, he measured the smoking rate. He found that the particle diameters were within 20 and 50 nm, and particle sizes have not yet varied with equivalence ratio.

4 Soot Measurement Techniques in Hydrocarbon-Air Flames

Improvements in soot diagnostics allow detailed numerical simulations and better validation of the results. Methods of soot diagnostics are generally divided into two categories depending on how flames are accessed

(a) Intrusive techniques
(b) Non-intrusive techniques

Mostly in all research works, non-intrusive techniques were preferred because of their versatility and non-interference. The optical diagnostics usually fall short in studying soot morphology [41]. Main parameters in soot study are soot volume fraction and soot morphology. The interpretation of these results gives the fundamental idea about flame properties such as flame temperature, reaction rate, reaction zone thickness and species pool concentration and flow rate.

4.1 Intrusive Techniques

Combustion measurements using probes to collect samples of the flow have been extensively done. Some probing techniques employed are pitot tubes for velocity measurements, thermocouples probes for temperature measurements and various types of iso-kinetic and sonic probes for species concentration measurements [42]. The advantage of probing techniques is the simplicity, ease of use, and they are cheaper compared to optical instrumentation. Under experimental condition it has been observed using probes, the main species mass fractions and velocity in a turbulent flame can be measured within 10% and temperature within 5% of their turbulence-averaged means [43]. The accuracy is worst in measuring and tracing radicals and minor species. Probe sampling methods are also used to collect soot particle inside a flame.

Probe collection techniques used in combustion research are many, each having its own strengths and shortcomings. Some researchers use multiple probing techniques, or they will select the probe that fits best to the flame configuration [44]. Same flame configuration results obtained using different probe designs will not match always. The variance is due to different residence times or different probe characteristics that affect the flame such as soot growth or different thermophoretic forces and diffusion and heat transfer rates that exist in a probe. Again soot particles measured by different probe techniques may exhibit differences in their structures, primary particle size, the arrangement of aggregates and the number of particles per aggregate. Hence appropriate sampling method is to be followed in soot morphology studies. In the method proposed by Maricq [44] a surface element (mesh) is attached to a probe, and the probe is immersed into the flame for various orientation. Thermophores are the driving force to deposit soot on the surface of the grid when exposed to a short period of time (30 ms) and are often enough for adequate soot deposition. When temperature gradients are large, the difference in the kinetic energies of the molecules at opposite sides of a particle repels the particle towards the lower temperature. Then, the collected samples are examined for their morphology under a transmission electron microscope (TEM) for soot visualization, and then this sampling method was extended for soot volume fraction measurements. This diagnostic technique has an advantage over optical techniques that soot optical properties such as refractive index are not needed.

4.2 *Non-intrusive Techniques*

Advanced optical diagnostics are developed and used in order to investigate combustion characteristic on flames. Laser spectroscopy techniques are employed for temperature measurements and species concentration determination in combustion systems. Polycyclic aromatic hydrocarbons (PAH) and soot volume fraction in a diffusion and rich premixed flames, aerosol and pollutant analysis are the key parameters to investigate strongly using this method. In the research work, the following techniques are normally used.

4.2.1 Light Extinction Technique

This technique is widely used for measuring soot concentration in a flame. If a beam of light is passed through an aerosol region some of the light intensity is partly absorbed by the aerosol particles, then the difference in the intensities is used to measure the soot particles size in a flame. It is one of the simple, effective optical methods. The total light intensity extinction by a particle can be expressed as the sum of the absorption, scattering, diffraction, reflection and refraction [45]. The scattering (S) is given by Eq. (4.1)

$$S = \frac{\pi^4}{4\lambda^4} d^6 f(m, N) \tag{4.1}$$

where λ represents the wavelength, d represents the diameter of the particle and f is the function depends on soot index (m) and soot density (N). The total intensity of light absorbed is expressed in terms of absorption coefficient K_a as Eq. (4.2)

$$K_a = \frac{\pi^2}{\lambda} N d^3 E(m) \tag{4.2}$$

where $E(m)$ is the function of soot refractive index.

4.2.2 Spectral Soot Emission Technique (SSE)

This technique depends on the emitted soot radiation along its chord. Measurement of soot volume fraction is without an illuminating light source. Since this method depends on soot radiation so it is best suited for high pressure because at high pressure the effect of soot radiation will be high so it will be very effective in measuring soot concentration, soot volume fraction without any error. Another advantage of SSE is analysing flame through a single port, whereas other diagnostics systems normally have two ports which use two illuminating light sources.

4.2.3 Laser-Induced Incandescence (LII)

Here, the soot particles in a flame are heated to their vapourization temperature along a line or plane by a short-duration high-intensity laser pulse. The energy transferred to the soot by laser light is partly released from soot within a few hundred nanoseconds following the laser pulse. The underlying principle is that the magnitude of the incandescence signal is proportional to soot concentration, and the decaying incandescence is related to the particle size distribution. For the high temporal resolution of the turbulent diffusion flames, LII is used in the measurement of soot concentration. The accuracy of the LII system depends on pressure range in which it operates. At sub-atmospheric pressure, the collision effects are very less compared to the absorption of the laser energy by soot particle and energy loss is through sublimation, conduction and radiation of heat [46]. LII is limited to low-pressure combustion analysis because particle volume increases with a particle diameter by the order of three, whereas particle sweep area is directly proportional to second power. Hence, smaller particles lose their internal energy faster relative to higher particle size per unit volume. Within a Rayleigh regime, the particle diameter is the function of radiation loss from the particle per unit volume. The use of LII is more problematic at elevated pressures. But this problem can be avoided at elevated pressure by marinating the flow as laminar in a co-flow burner.

5 Computational Modelling on Soot Formation

The computational modelling on soot particulate formation uses polyaromatic hydrocarbon model (PAH) given by Frenklach [19] along with physical model proposed by Brookes and Moss [47] and discussed in the following sections.

5.1 Soot Modelling

The soot formation in the aromatic fuels involves inception, coagulation, surface growth and oxidation rate. The rate of soot formation can be determined by the soot number density (*N*) and mass density (*M*). Brookes and Moss [47] proposed the relation for soot formation rate given by

$$\left(\frac{DN}{Dt}\right)_{\text{soot}} = \left(\frac{\mathrm{d}N}{\mathrm{d}t}\right)_{\text{inc}} + \left(\frac{\mathrm{d}N}{\mathrm{d}t}\right)_{\text{coa}} \tag{5.1}$$

$$\left(\frac{DM}{Dt}\right)_{\text{soot}} = \left(\frac{\mathrm{d}M}{\mathrm{d}t}\right)_{\text{inc}} + \left(\frac{\mathrm{d}M}{\mathrm{d}t}\right)_{\text{gro}} + \left(\frac{\mathrm{d}M}{\mathrm{d}t}\right)_{\text{oxi}} \tag{5.2}$$

5.2 Soot Inception

Solid soot particles are formed from gas-phase species. This transition from gas-phase species to solid soot particles is proposed, using charged species [48], poly acetylenes [49] and polycyclic aromatic hydrocarbons as precursors. The soot particle inception process is believed to be controlled by the formation and growth of these ring-structured molecules. The presence of aromatics in the diesel increases the soot inception rate relative to non-aromatic fuels. The rate of change of soot mass density is related to the number density by Brookes and Moss [48]

$$\left(\frac{\mathrm{d}M}{\mathrm{d}t}\right)_{\text{inc}} = \frac{M_p}{N_{\mathrm{A}}}\left(\frac{\mathrm{d}N}{\mathrm{d}t}\right)_{\text{inc}} \tag{5.3}$$

where M_p is the mass of a soot nucleus and has a value of 1200 kg/kmol (based on the assumption that the soot size corresponds to 100 carbon atoms [49]), and $N_{\mathrm{A}}\left(= 6.022045 \times 10^{26}/\text{kmol}\right)$ is Avogadro's number.

5.3 PAH Inception Model

Polyaromatic hydrocarbon soot inception model is the one in which the rate of soot formation of aromatic ring structures like $C_{10}H_7$ and $C_{14}H_{10}$ are from the acetylene (C_2H_2), Benzene (C_6H_6) and phenyl radicals (C_6H_5). Puri et al. [50] proposed the mechanism involved in the formation of two- and three-ring aromatics, and the rate of soot formation is as follows

$$2C_2H_2 + C_6H_5 \rightarrow C_{10}H_7 + H_2$$

and

$$C_2H_2 + C_6H_5 + C_6H_6 \rightarrow C_{14}H_{10} + H + H_2$$

$$\left(\frac{dN}{dt}\right)_{inc} = 8c_2 \frac{N_A}{M_P}\left[\rho^2\left(\frac{Y_{C_2H_2}}{W_{C_2H_2}} \cdot \frac{Y_{C_2H_2}}{W_{C_2H_2}}\right)\frac{Y_{C_6H_5}W_{H_2}}{W_{C_6H_5}Y_{H_2}}\right] e^{-\left(\frac{4378}{T}\right)} + 8c_3 \frac{N_A}{M_P}\left[\rho^2\left(\frac{Y_{C_2H_2}}{W_{C_2H_2}} \cdot \frac{Y_{C_6H_6}}{W_{C_6H_6}}\right)\frac{Y_{C_6H_5}W_{H_2}}{W_{C_6H_5}Y_{H_2}}\right] e^{-\left(\frac{6390}{T}\right)} \quad (5.4)$$

where $c_2 = 1.27 \times 10^{8.88}$ and $c_3 = 1.78 \times 10^{9.50}$, and when compared to the acetylene inception model, the soot particle formation rate is eight times faster in PAH inception model. Hence, this model is best suitable for aromatic liquid fuels compared to the other gaseous fuels.

5.4 *Soot Coagulation*

As the soot particle propagates along the flame sheet, each soot particle collides with the other. The formation of soot particle rate depends on the mean free path and the collision frequency. Initially, the soot particle is in condensed phase and as it collides with other particles fusion of carbon atoms will take place. The rate of fusion will determine the size of the particle. Puri et al. [50] proposed the coagulation rate of soot particle by the following expression

$$\left(\frac{dN}{dt}\right)_{Coa} = -\left(\frac{24R}{\rho_{soot} \cdot N_A}\right)^{1/2} \cdot \left(\frac{6}{\pi\rho_{soot}}\right)^{1/6} T^{1/2}M^{1/2}N^{11/6} \quad (5.5)$$

where R is universal gas constant.

5.5 *Soot Growth*

The rate of soot surface growth is a function of activation energy, the level of acetylene in the flame and number of active molecules on the soot surface. Frenklach [51] proposed the effect of surface growth on soot at the activation temperature of 2,100 K.

$$\left(\frac{\mathrm{d}M}{\mathrm{d}t}\right)_{\mathrm{gro}} = c_4\left(\frac{Y_{\mathrm{C_2H_2}}}{W_{\mathrm{C_2H_2}}}\right)\mathrm{e}^{-\left(\frac{12100}{T}\right)}\left[\left(\frac{6M}{\rho_{\mathrm{soot}}}\right)^{2/3}(\pi N)^{1/3}\right] \tag{5.6}$$

where $c_4 = 9000.6$ kg m/kmol s.

5.6 *Soot Oxidation*

Lee et al. [52] formulated the relation for the effect of OH and O_2 on the soot oxidation in the laminar diffusion flame (Eq. 5.7). According to them, OH species and O_2 are the main sources for the soot oxidation process. The collision efficiency 'η' refers to the fraction of a single carbon atom from the soot molecule that involves in the oxidation process. The soot oxidation rate is given by

$$\begin{aligned}\left(\frac{\mathrm{d}M}{\mathrm{d}t}\right)_{\mathrm{oxi}} = &-c_5\left(\rho\eta\frac{Y_{\mathrm{OH}}}{W_{\mathrm{OH}}}\right)T^{1/2}\left[\left(\frac{6M}{\rho_{\mathrm{soot}}}\right)^{2/3}(\pi N)^{1/3}\right] \\ &- c_6\left(\rho\frac{Y_{\mathrm{O_2}}}{W_{\mathrm{O_2}}}\right)T^{1/2}\mathrm{e}^{-\left(\frac{19778}{T}\right)}\left[\left(\frac{6M}{\rho_{soot}}\right)^{2/3}(\pi N)^{1/3}\right]\end{aligned} \tag{5.7}$$

The values proposed by them are $\eta = 0.3$, $c_5 = 105.81$ kg m/kmol K$^{1/2}$ s, $c_6 = 8903.51$ kg m/kmol K$^{1/2}$ s.

6 CFD Soot Modelling for Diffusion Flame

6.1 *CFD Mathematical Model*

6.1.1 Favre-Averaged Conservation Equations

The nature of the turbulent flow is irregular with rapid fluctuations in velocity, density, temperature and composition. This fluctuating nature makes turbulent flow highly diffusive resulting in enhanced transport of momentum, mass and energy. The introduction of mean quantities makes the numerical handling of turbulent flow simpler. However, it also causes new challenges, especially related to the handling of chemical kinetics [53, 54]. The source terms in the Favre-averaged transport equations for the individual species are exponential functions of temperature. Using the mean temperature to compute a mean source term will lead to an erroneous result. Accordingly, the averaged chemical production rates need special treatment and are approximated by a combustion model.

Mass conservation of species k

$$\frac{\partial}{dt}(\bar{\rho}\tilde{Y}_k) + \frac{\partial}{dx_j}(\bar{\rho}\tilde{Y}_k\tilde{v}_J) = \frac{\partial}{dx_j}\left(\frac{\mu}{\sigma}\frac{\partial\tilde{Y}_k}{dx_j} - \bar{\rho}\tilde{Y}_k''\tilde{v}_J''\right) + \bar{\rho}\tilde{\omega}_k \tag{6.1}$$

Conservation of overall mass

$$\frac{\partial\bar{\rho}}{\partial t} + \frac{\partial}{dx_j}(\bar{\rho}\tilde{v}_J) = 0 \tag{6.2}$$

Conservation of momentum

$$\frac{\partial\bar{\rho}}{dt}(\bar{\rho}\tilde{v}_J) + \frac{\partial}{dx_j}(\bar{\rho}\tilde{v}_J\tilde{v}_J'') = \frac{\partial\bar{p}}{dx_j} + \frac{\partial}{dx_j}\left(\frac{\mu}{\sigma}\frac{\partial\tilde{v}_J}{dx_J} - \bar{\rho}\tilde{v}_J\tilde{v}_J''\right) + \bar{\rho}g_i \tag{6.3}$$

Conservation of energy

$$\frac{\partial}{dt}(\bar{\rho}\tilde{h}) + \frac{\partial}{dx_j}(\bar{\rho}\tilde{h}\tilde{v}_J) = \frac{\partial\bar{p}}{dx_j} + \frac{\partial}{dx_j}\left(\frac{\mu}{\sigma}\frac{\partial\tilde{h}_J}{dx_J} - \bar{\rho}\tilde{Y}_k''\tilde{v}_J''\right) + \bar{\rho}\tilde{Q} \tag{6.4}$$

To solve the averaged conservation equations, they have to be approximated in some way. Methods approximating these unknown terms are called turbulence models. A large number of turbulence models have been developed. However, for simulating turbulent combustion, there are mainly two different types of models which are in use at present. The most commonly used model is the k-ε model, and this model is used in the present work. Brookes-Moss-Lee model is used to predict soot formation.

6.1.2 k-ε Turbulence Model

The k-ε model was presented by Jones and Launder [53] and is based on the assumption that the turbulent transport can be described by using an effective viscosity

$$\mu_{\text{eff}} = \mu + \mu_t$$

where μ_t is the eddy (or turbulence) viscosity. The turbulence stresses are now expressed by the eddy viscosity from an analogy with the expression for viscous stresses

$$-\bar{\rho}\tilde{v}_k''\tilde{v}_J'' = \mu_t\left(\frac{\partial\tilde{v}_J}{dx_J} + \frac{\partial\tilde{v}_J}{dx_k}\right) - 2/3\left(\bar{\rho}\tilde{k} + \mu_t\frac{\partial\tilde{v}_J}{dx_J}\right)\delta_{kj} \tag{6.5}$$

where k is the turbulence kinetic energy and is given by

$$\tilde{k} = \frac{1}{2}\tilde{v}_i''\tilde{v}_i$$

By the aid of the instantaneous and the Favre-averaged momentum equations, a transport equation can be obtained [53]

σ_t	σ_k	σ_ε	$C_{\varepsilon 1}$	$C_{\varepsilon 2}$	C_μ
0.7	1	1.3	1.44	1.92	0.09

The variable represents the dissipation of turbulence energy into heat. The dissipation of turbulence energy is found from a modelled transport equation [53]

$$\frac{\partial}{dt}(\bar{\rho}\tilde{\epsilon}) + \frac{\partial}{dx_j}(\bar{\rho}\tilde{\varepsilon}\tilde{v}_J) = \frac{\partial}{dx_j}\left[\left(\mu + \frac{\mu_t}{\sigma_t}\right)\frac{\partial\tilde{\varepsilon}}{dx_j}\right] + C_{\epsilon 1}\frac{\tilde{\varepsilon}}{\tilde{k}}\bar{\rho}P_k - C_{\epsilon 2}\frac{\tilde{\varepsilon}}{\tilde{k}}\bar{\rho}\tilde{\varepsilon} \tag{6.6}$$

The term present in the two transport equations is the production of turbulence kinetic energy [53]

$$\bar{\rho}P_k = \mu_t\left(\frac{\partial\tilde{v}_J}{dx_j} + \frac{\partial\tilde{v}_J}{dx_k}\right)\frac{\partial\tilde{v}_J}{dx_k} - \frac{2}{3}\left(\bar{\rho}\tilde{k} + \mu_t\frac{\partial\tilde{v}_J}{dx_j}\right)\frac{\partial\tilde{v}_J}{dx_k} \tag{6.7}$$

The eddy viscosity is expressed as

$$\mu_t = C_\mu\bar{\rho}\frac{\tilde{k}^2}{\tilde{\varepsilon}}$$

6.2 *Simulation Boundary Conditions*

The present CFD analysis is carried out for computing radial temperature, axial temperature, mean mixture fraction, soot volume fraction, soot coagulation and soot growth rate at different downstream locations in a diesel vapour–air flame burning at atmospheric pressure. A turbulent diesel-air jet diffusion flame is investigated for the combustor design as shown in Fig. 1. The nozzle diameter is 1.5 mm where the diesel vapour jet flows with the velocity of 22.28 cm/s surrounded by a coaxial annulus of 76.75 mm in which the air flows with the velocity of 0.234 m/s. For the combustion analysis, the domain of the combustor length is limited to 600 mm. The important boundary conditions are listed in Table 1. The geometry is discretised using GAMBIT version 15 into 25,957 nodes and 25,600 elements using quadri-lateral high fine and smooth mesh having standard deviation of 0.0086144 with mesh metric maximum Jacobian ratio as 1.6535. Here, X axis represents the radial direction and Y axis represents axial direction. Standard k-ε model is used for

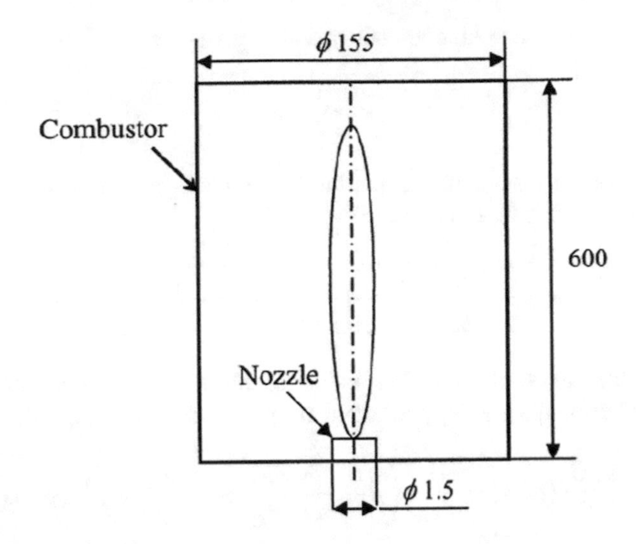

Fig. 1 Combustor model [55]

Table 1 Boundary conditions [55]

	Properties	Diesel ($C_{12}H_{23}$: 100%)	Air (O_2: 23.3%; N_2: 76.7%)
Inlet	Velocity (m/s)	22.28	0.234
	Turbulent intensity	0.03	0.03
	Eddy length scale	0.02	0.02
	Temperature (K)	598	288
	PDF option	Inlet diffusion	
Wall	Adiabatic smooth wall		
Outlet	Pressure = 1 bar		

turbulent analysis. Brooke and Moss soot mechanism [48] was used for soot formation considering polyaromatic hydrocarbon (PAH) soot inception model with C_6H_6 as the precursor with mixture fraction as the turbulent interaction mode.

The simulation is run using Newton–Raphson linearization method (using CFD solver FLUENT), and all the results were tabulated on convergence of all the residuals. Initially, the study was carried out using kerosene/air mixture with the optimum number of grids to get confirmation of trends given by Wen et al. [54]. The fuel is then changed to diesel vapour, and with the same geometry and boundary conditions, the simulation was carried out to study soot formation.

6.3 *Results and Discussion*

Temperature is the main factor in the formation of soot and Fig. 2 shows the temperature contour for the computational domain and Fig. 3 shows the flame

Fig. 2 Axial temperature

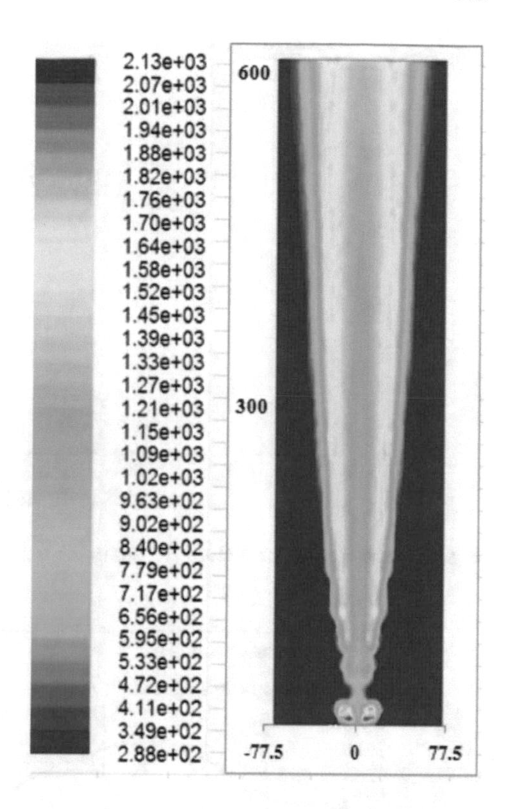

temperature along the axial directions. It is observed that the flame temperature varies from 598K to 2130 K along the axial length. It is also observed that the maximum temperature 1288 K occurred at 600 mm from the inlet along the axis. The thermal analysis is further extended to the radial position at a distance 300 and 600 mm from the inlet. It is found from Fig. 4 that at 300 mm above inlet when we move from the axis towards the wall there is a rise in temperature in radial direction up to 26 mm and there after the temperature decreases. The maximum temperature attained at 26 mm is 1440 K. At 600 mm from the inlet, the maximum temperature still increases to 1610 K. This is the place where the flame glows in yellowish colour indicating the presence of solid soot.

Figure 5 presents the mass fraction of OH variation along the axial direction. It is noticed that the maximum value of OH is found to be 0.00169 and is located in the fuel-lean mixture zone. As mentioned earlier, OH and O_2 are the main species responsible for the soot oxidation process. From Fig. 6, the variation of O_2 along the axial and the radial distance can be seen. It also tells us about the diffusion of the oxygen species radially for the oxidation process. Maximum volume fraction of soot occurs at a distance 600 mm and is computed as 3×10^{-9} (Fig. 7). Also, the variation of soot volume fraction along the radial distance at axial locations of 300 and 600 mm is shown in Fig. 8. It is observed that along the radial distance from centreline the soot volume fraction initially increases and then decreases. It is also

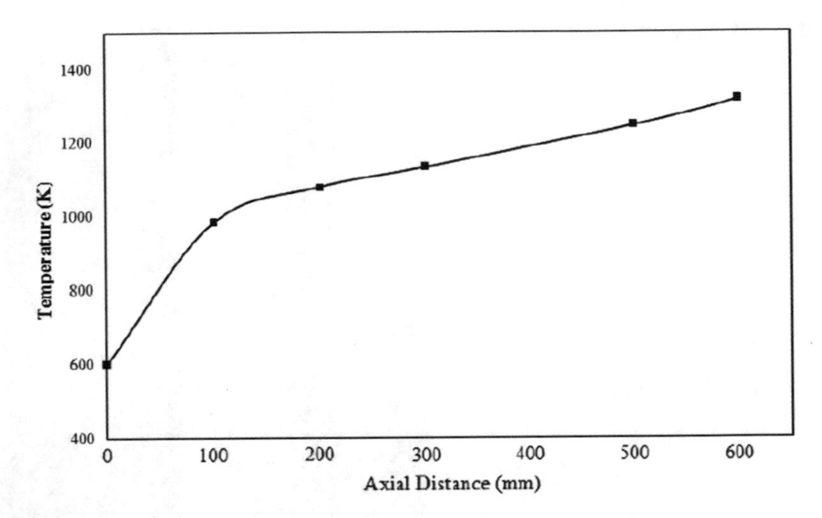

Fig. 3 Variation of temperature (K) along the axial distance (mm)

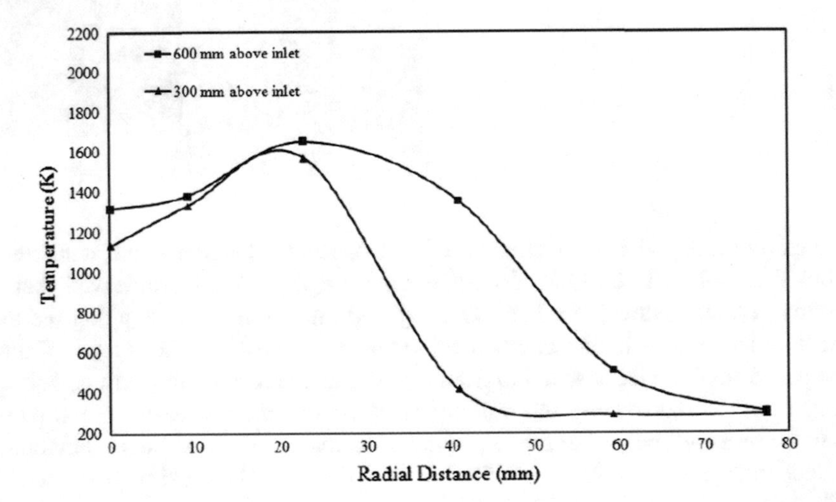

Fig. 4 Variation of temperature (K) along the radial distance (mm)

seen that as we move axially, the fraction of maximum soot volume increases by 57.6% (approximately). Figure 9 shows the contour for the mass fraction of H_2. H atom from H_2 is the main precursor for soot formation in Hydrogen–Abstraction–C_2H_2–Addition mechanism, and its maximum value is found to be 3.35×10^{-4}. As discussed earlier, this H atom concentration occurs at the maximum temperature location and is responsible for the high rate of soot formation as seen from Fig. 10. It is concluded that the soot density is maximum at the flame end, and the maximum value is found to be 7.47×10^{-6} kg/m^3, and as we move radially the soot density decreases confirming the trends shown by Fig. 8.

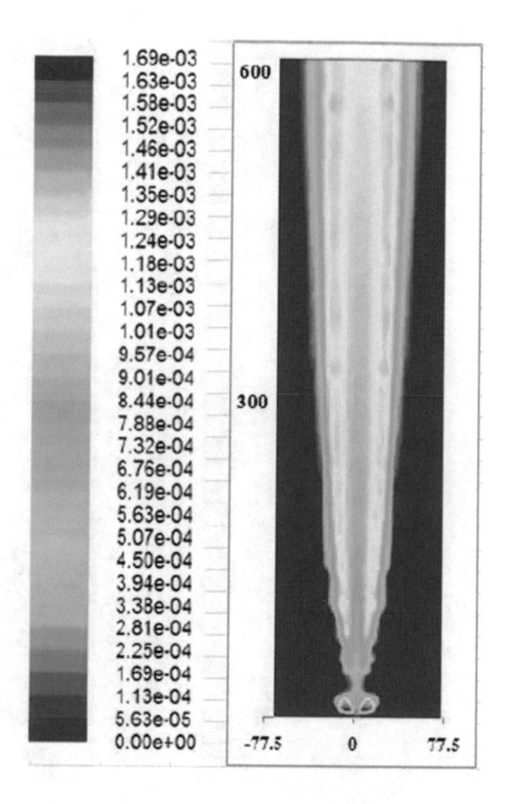

Fig. 5 Mass fraction of OH

Figure 7 shows the variation of soot volume fraction along the axial distance from the fuel inlet. It is observed from the figure that along the axial distance the soot volume fraction increases up to 600 mm. The variation of mean mixture fraction along the axial direction is given in Figs. 11 and 13 and along radial direction in Fig. 14. Along the axial direction from the inlet, the mean mixture fraction continuously decreases. The analysis is further extended to radial location at two different axial heights of 300 and 600 mm. Figure 14 shows the mixture fraction to be maximum at 300 mm from inlet when compared to 600 mm, and it decreases along the radial directions. Figure 12 shows the rate of soot growth proposed by Frenklach and Wang [6], for mono-dispersed spherical particles. It is found that the maximum rate of surface growth is 6.86×10^{-4} kg/m^3 s. Figure 16 shows the coagulation rate along the radial and axial positions. It is noted that the maximum coagulation is found to be 2.08×10^{20} particle/m^3 s, and the rate of formation of soot nuclei is presented in Fig. 15, and it says that the soot nuclei formation rate for diesel jet diffusion flame is 1.78×10^{19} particle/m^3 s.

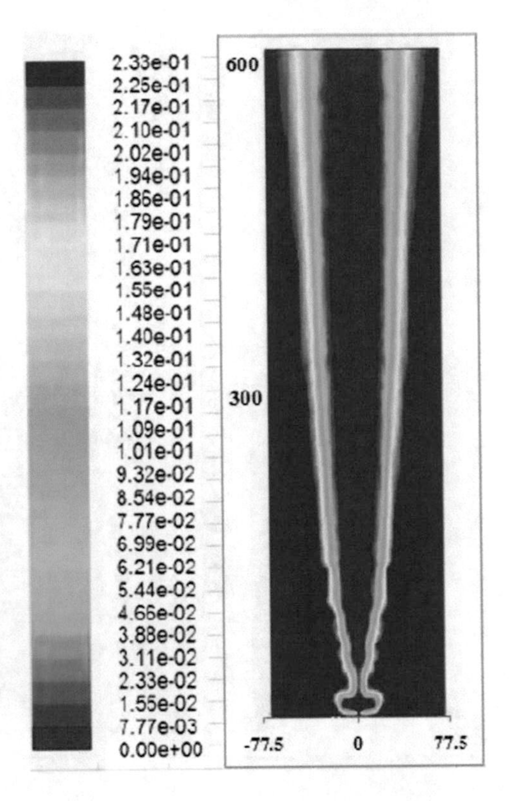

Fig. 6 Mass fraction of O_2

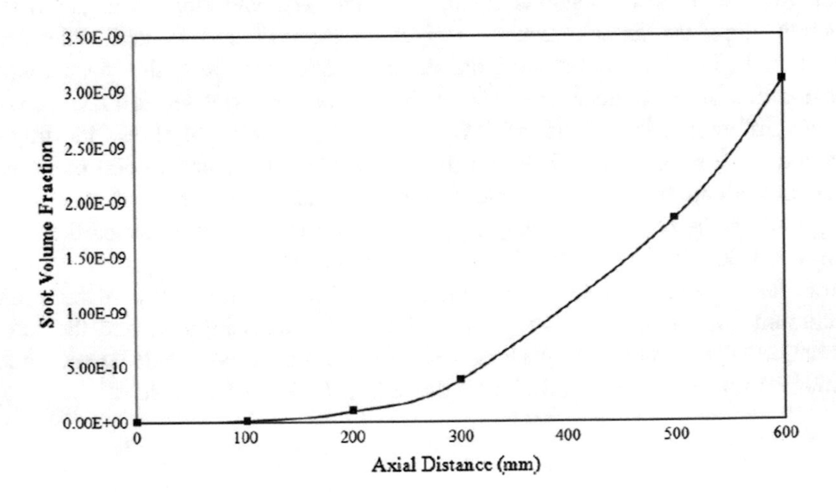

Fig. 7 Variation of soot volume fraction along the axial distance (mm)

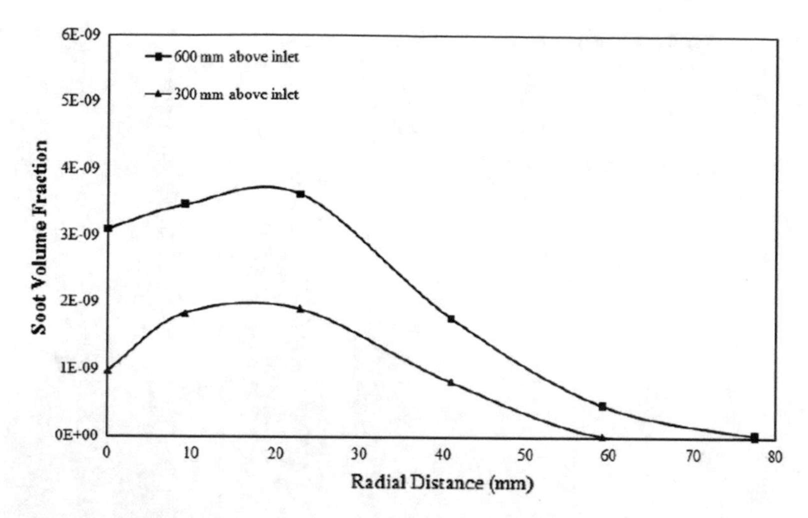

Fig. 8 Variation of soot volume fraction along the radial distance (mm)

Fig. 9 Mass fraction of H_2

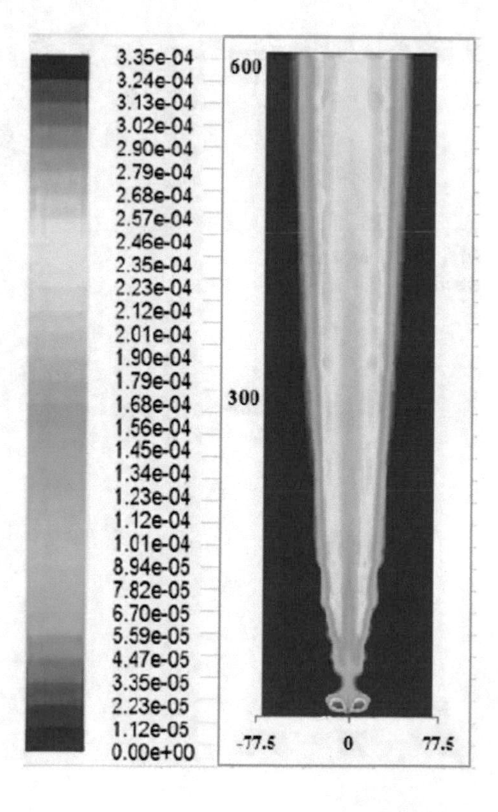

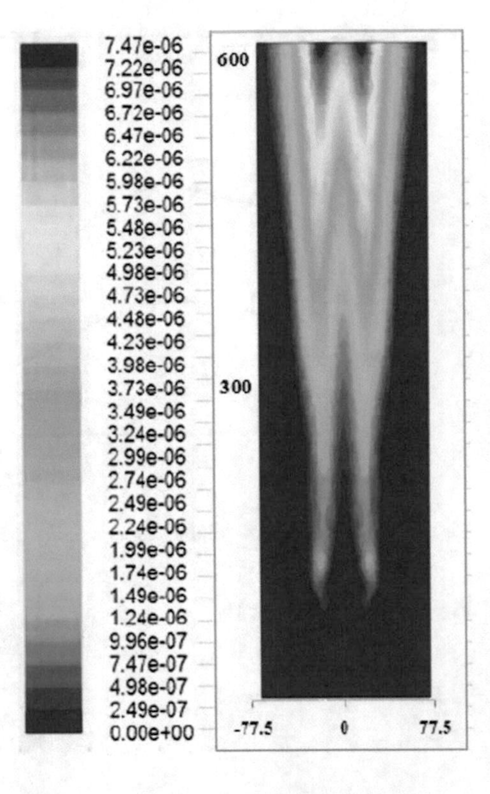

Fig. 10 Soot density (kg/m^3)

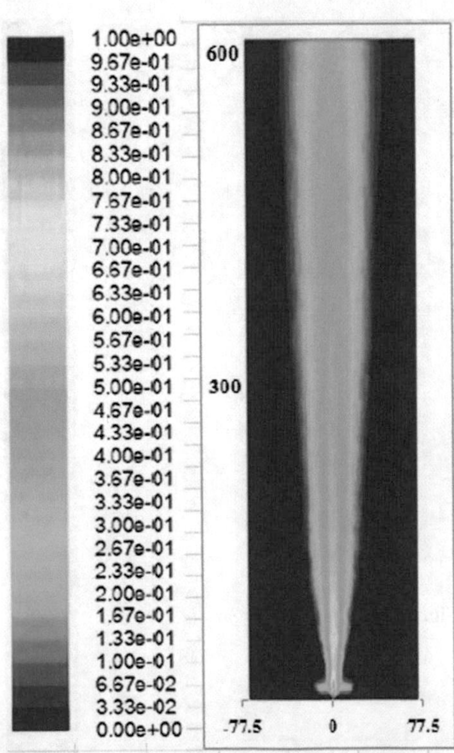

Fig. 11 Mean mixture fraction

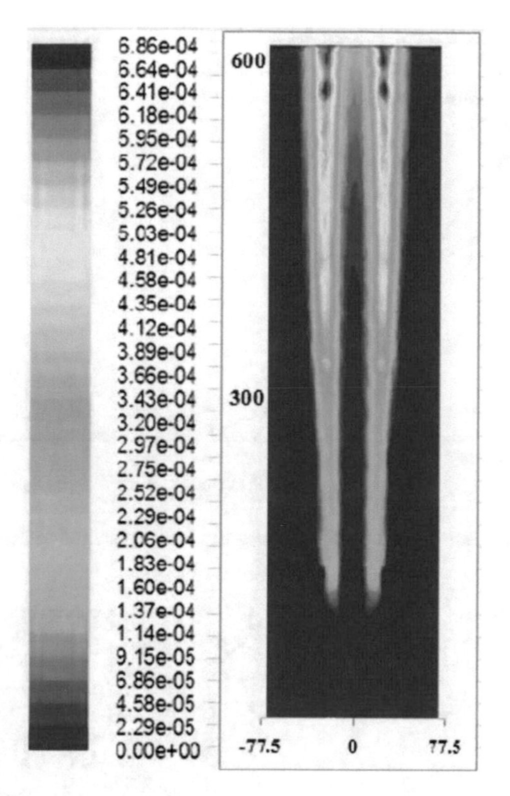

Fig. 12 Rate of surface growth (kg/m^3 s)

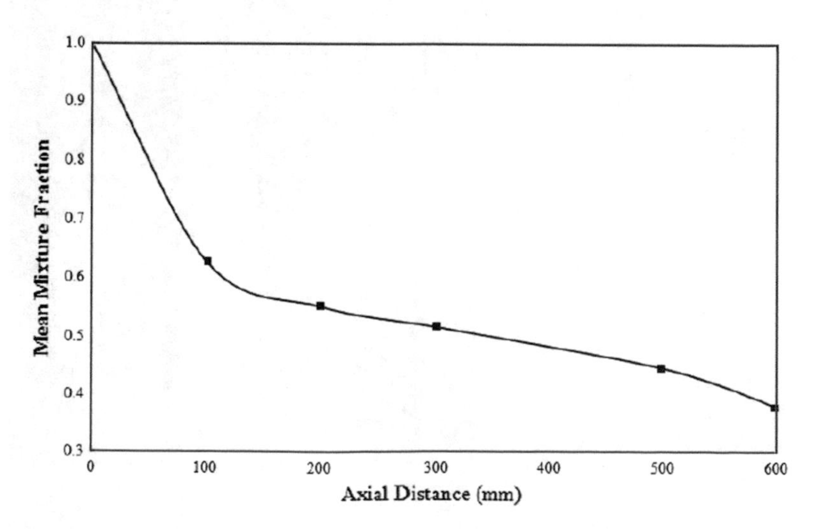

Fig. 13 Variation of mean mixture fraction along the axial distance (mm)

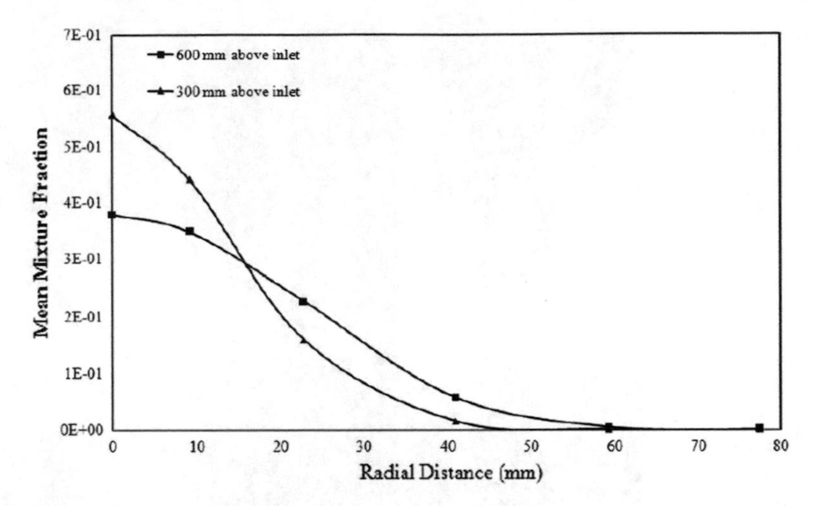

Fig. 14 Variation of mean mixture fraction along the radial distance (mm)

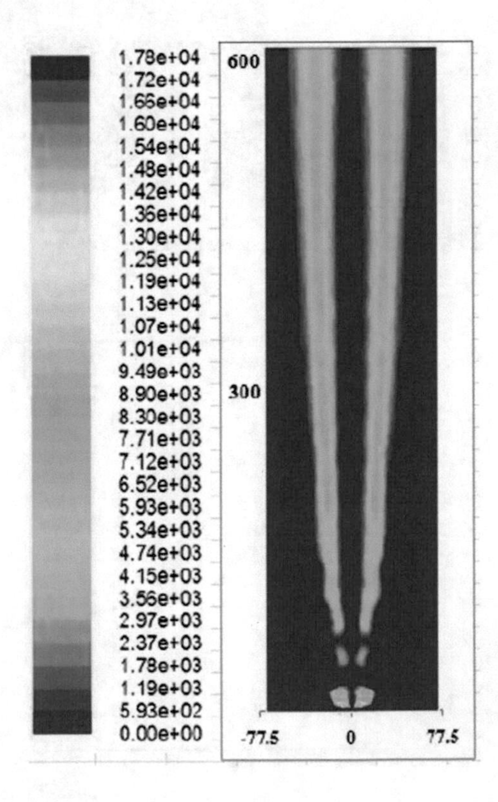

Fig. 15 Rate of nucleation (1×10^{15} Particle/m^3 s)

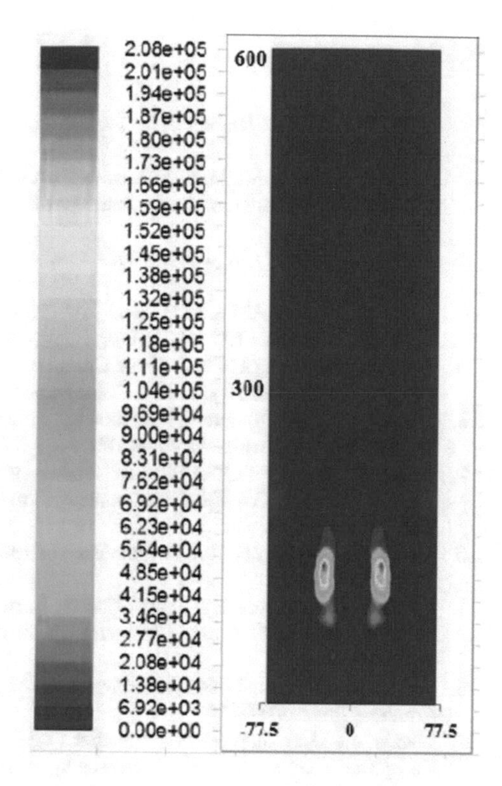

Fig. 16 Rate of coagulation (1×10^{15} Particle/m^3 s)

7 Summary

This chapter reviewed theory on soot formation in hydrocarbon flames at various stages of combustion at constant pressures. The chemistry of soot formation using acetylene model in gaseous phase and polyaromatic hydrocarbon (PAH) model in the particulate phases is reviewed.

All the reviewed theories available as models are invoked along with turbulence model to solve the Favre-averaged Navier–Stokes equation in two dimensions together with necessary boundary conditions. The soot particulate formation is evaluated using Brookes-Moss-Lee model along with PAH model for constant atmospheric pressures. The numerical analysis was performed using commercial CFD code initially for kerosene/air, and the results were found to be in agreement with experimental values published by Wen et al. [54]. The soot particulate formation study then was carried out by the authors using diesel fuel vapour ($C_{12}H_{23}$)–air mixtures, and some of the results obtained are presented. Further validation of these results using an experimental facility employing instrumentation as outlined in this chapter needs to be carried out.

References

1. Lim SS, Vos T, Flaxman AD, Danaei G, Shibuya K, Adair-Rohani H (2012) Lancet 380:2224–2260
2. Al-Omari S-AB, Kawajiri K, Yonesawa T (2001) Soot processes in a methane-fuelled furnace and their pact on radiation heat transfer to furnace walls. Int J Heat Mass Tran 44(13):2567–2581
3. Kennedy IM (1997) Models of soot formation and oxidation. Prog Energy Combust 23:95–132
4. Harris SJ, Weiner AM (1985) Annu Rev Phys Chem 36:31–52
5. Haynes BS, Wagner HG (1981) Prog Energy Combust Sci 7:229–273
6. Frenklach M, Wang H (1991) Proc Combust Inst 23:1559–1566
7. Haynes BS, Wagner HG, Phys Z (1982) Chem N F 133:201–213
8. Heywood JB (1988) Internal combustion engine fundamentals. In: McGraw-Hill series in mechanical engineering. McGraw-Hill, New York
9. Prado G, Lahaye J (1981) Physical aspects of nucleation and growth of soot particles. In: Siegla DC, Smith GW (eds) Particulate carbon formation during combustion. Plenum Press, New York, pp 143–176
10. Smyth KC, Miller JH (1987) Chemistry of molecular growth processes in flames. Science 236:1540–1546
11. Parker WG, Wolfhard HG (1950) Carbon formation in flames. J Chem Soc 2038–2049
12. Milberg ME (1959) Carbon formation in an acetylene air diffusion flame. J Phys Chem 63:578–582
13. Schalla RL, McDonald GE (1955) Mechanism of smoke formation in diffusion flames. Proc Combust Inst 5:316–324
14. Macfarlane JJ, Holderness FH, Whitcher FSE (1964) Soot formation rates in premixed C5 and C6 hydrocarbon air flames at pressures up to 20 atmospheres. Combust Flame 8:215–229
15. Flower WL, Bowman CT (1984) Measurements of the effect of elevated pressure on soot formation in laminar diffusion flames. Combust Sci Technol 37:93–97
16. Kim CH, Xu F, Faeth GM (2008) Soot surface growth and oxidation at pressures up to 8.0 atm. in laminar non-premixed and partially premixed flames. Combust Flame 152:301–316
17. Milberg ME (1959) Carbon formation in an acetylene air diffusion flame. J Phys Chem 63:578–582
18. Prado G, Lahaye J (1981) Physical aspects of nucleation and growth of soot particles. In: Siegla DC, Smith GW (eds) Particulate carbon formation during combustion. Plenum Press, New York, pp 143–176
19. Frenklach M (2002) Reaction mechanism of soot formation in flames. Phys Chem Chem Phys 4:2028–2037
20. Callear AB, Smith GB (1984) The addition of atomic hydrogen to acetylene chain reactions of the vinyl radical. Chem Phys Lett 105(1):119–122
21. Prado G, Lahaye J (1981) Physical aspects of nucleation and growth of soot particles. In: Siegla DC, Smith GW (eds) Particulate carbon formation during combustion. Plenum Press, New York, pp 143–176
22. Frenklach M, Clary DW, Gardiner WC, Stein SE (1985) Detailed kinetic modelling of soot formation in shock-tube pyrolysis of acetylene. Proc Comb Inst 20:887–901
23. Frenklach M, Yuan T, Ramachandra MK (1988) Soot formation in binary hydrocarbonmixtures. Energy Fuels 2:462–480
24. Miller JA, Pilling MJ, Troe J (2005) Unravelling combustion mechanisms through a quantitative understanding of elementary reactions. Proc Comb Inst 30:43–88
25. Melius CF, Colvin ME, Marinov NM, Pitz WJ, Senkan SM (1996) Reactionmechanisms in aromatic hydrocarbon formation involving the C5H5 cyclopentadienyl moiety. Proc Comb Inst 26:685–692

26. Frenklach M, Wang H (1991) Detailed modeling of soot particle nucleation and growth. Proc Comb Inst 23:1559–1566
27. Frenklach M (2002) Reaction mechanism of soot formation in flames. Phys Chem Chem Phys 4:2028–2037
28. Haynes BS, Wagner HG (1981) Soot formation. Prog Energy Combust Sci 7:229–273
29. Neoh KG, Howard JB, Sarofim AF (1981) Particulate carbon formation during combustion, vol 261. Plenum Press, New York
30. Frenklach M, Wang H (1991) Detailed modeling of soot particle nucleation and growth. Proc Comb Inst 23:1559–1566
31. Frenklach M, Ebert LB (1988) Comment on the proposed role of spheroidal carbon clusters in soot formation. J Phys Chem 92:561–563
32. Miller JH (1990) The kinetics of polynuclear aromatic hydrocarbon agglomeration in flames. Proc Comb Inst 23:91
33. Frenklach M, Clary DW, Gardiner WC, Stein SE (1985) Detailed kinetic modelling of soot formation in shock-tube pyrolysis of acetylene. Proc Comb Inst 20:887–901
34. Warnatz J, Maas U, Dibble RW (2006) Physical and chemical fundamentals, modelling and simulation, experiments, pollutant formation, 4th edn. Springer, Berlin
35. Frenklach M, Wang H (1990) Detailed kinetic modelling of soot particle nucleation and growth. Proc Comb Inst 23:1559–1566
36. Harris SJ, Weiner AM (1983) The surface growth of soot particles in premixed ethylene air flames. Combust Sci Technol 31:155–167
37. Smoluchowski MV (1917) Versuch einer mathematischen Theorie der Koagulationskinetik kolloider Loesungen. Z Phys Chem 92:129–168
38. Graham SC (1976) The collisional growth of soot particles at high temperatures. Proc Comb Inst 16:663–669
39. Neoh KG, Howard JB, Sarofim AF (1985) Effect of oxidation on the physical structure of soot. Proc Comb Inst 20:951–957
40. Lucht RP, Sweeney DW, Laurendeau NM (1985) Laser-saturated fluorescence measurements of hydroxyl radical in atmospheric pressure methane/oxygen/nitrogen flames under sooting and non-sooting conditions. Comb Sci Technol 42:259–281
41. Eckbreth AC (1988) Laser diagnostics for combustion temperature and species, 1st edn. Abacus Press, Cambridge
42. Fristrom RM (1976) Probe measurements in laminar combustion systems. In: Goulard R (ed) Combustion measurements: modern techniques and instrumentation. Academic Press, New York, pp 287–317
43. Clark HR, Stawicki RP, Smyth IP, Potkay E (1990) Collection and characterization of soot from an optical fiber preform torch. J Am Ceram Soc 73:2987–2991
44. Maricq MM (2009) Electrical mobility based characterization of bimodal soot size distributions in rich premixed flames. In: Bockhorn H, D'Anna A, Sarofim AF, Wang H (eds) Combustion generated fine carbonaceous particles(proceedings of an international workshop held in Villa Orlandi, Anacapri, 13–16 May 2007). KIT Scientific Publishing, pp 347–366
45. Modest MF (2003) Radiative heat transfer, 2nd edn. Academic Press, Boston
46. Santoro RJ, Shaddix CR (2002) Laser induced incandescence. In: Kohse-Höinghaus K, Jefries B (eds) Applied combustion diagnostics. Taylor & Francis, pp 252–286
47. Brookes SJ, Moss JB (1999) Combust Flame 116:486
48. Launder BE, Spalding DB (1974) The numerical computation of turbulent flows. Computer Methods in Appl Mech Eng 3:269–289
49. Leung KM, Lindstedt RP, Jones WP (1991) Combust Flame 87:289
50. Puri R, Richardson TF, Santoro RJ (1993) Combust Flame 92:320
51. Frenklach M (2002) Reaction mechanism of soot formation in flames. Physical Chemistry 407. Chem Phys 4(11):2028–2037
52. Lee KB, Thring MW, Beer JM (1962) Combust Flame 6:137–145

53. Jones WP, Launder BE (1972) The prediction 432 of laminarization with a two-equation model of turbulence. Int J Heat Mass Transf 15:301–314
54. Wen et al (2003) Modelling soot formation. Combust Flame 135:323–340
55. Puri R, Richardson TF, Santoro RJ (1993) Combust Flame 92:320

Numerical Simulation of Air Pollution Control in Hospital

Tikendra Nath Verma, Arvind Kumar Sahu and Shobha Lata Sinha

Abstract Human first created buildings to protect themselves from the adverse climatic conditions and other hazards in the natural environment. People have become more cognizant of the gist of the indoor atmosphere on health as a consequence of media publicity surrounding building-related sickness (BRS) and the sick building syndrome (SBS). Building-related sickness comprises the sensation of stuffy, stale and unacceptable indoor air, irritation of mucous membranes, headache, lethargy, and so forth. Acceptable indoor air quality (IAQ) helps to maintain healthy and productive indoor environments. This chapter deals with air pollution in healthcare place, the importance of ventilation in a hospital environment, indoor air pollutants, and transmission of contaminants and airborne particle inside the infirmary. In general, pollutants of common concern in buildings are divided into two broad classifications: *particulates* and *volatile organic compounds* (VOCs). Respirable suspended particles (RSPs) are small, easily-made-airborne particles, which can be actively measured with appropriate sensing equipment. There are many sources of airborne pollutants and odours in and around buildings. Some pollutants of particular concern to quality of indoor air are formaldehyde, VOCs, ozone, tobacco smoke, and aerosols, etc. In addition, odours, CO_2, and the moistness, which cause important effects on indoor air quality, especially in densely occupied spaces. The precise prediction of air stream within a room may improve heating, ventilation, and air conditioning (HVAC) scheme for a salubrious environment significantly. Ventilation and quality of indoor air stream are just two of the many fields which would benefit from the enhancement of room air flow. In this

T. N. Verma
Department of Mechanical Engineering, National Institute of Technology,
Imphal 795004, Manipur, India
e-mail: tikendra.me@nitmanipur.ac.in

A. K. Sahu · S. L. Sinha (✉)
Department of Mechanical Engineering, National Institute of Technology,
Raipur 492010, Chhattisgarh, India
e-mail: slsinha.mech@nitrr.ac.in

A. K. Sahu
e-mail: arvindkumarsahu25@gmail.com

N. Sharma et al. (eds.), *Air Pollution and Control*, Energy, Environment, and Sustainability, https://doi.org/10.1007/978-981-10-7185-0_11

work, simulation of airflow in a room of the ICU has carried away to examine air flow pattern using FLUENT 15 CFD software. Standard k-epsilon turbulence model is used for airflow simulation. Simulation is carried out using second-order upwind simple scheme. The study predicts room air flow information in terms of velocities, temperatures, and contaminant distributions which are beneficial for infection control, building layout investigation.

Keywords Air pollutants · Particle dispersion · CFD

Nomenclature

C_d Coefficient of drag
d Diameter (m)
F Force (N)
F_D Drag force (N)
G Acceleration due to gravity (Nm/s^2)
Re Reynolds number
S_ϕ Source term in governing equation
t Time (s)
u Velocity in x direction (m/s)
v Velocity in y direction (m/s)
w Velocity in z direction (m/s)

Greek Letters

Γ_ϕ Effective diffusivity (m^2/s)
Φ General dependent variable
k Turbulent kinetic energy (m^2/s^2)
m Dynamic viscosity (Ns/m^2)
ρ Mass density (kg/m^3)

Subscript

i Particle identifier
p Particle

1 Introduction

The main objective of ventilation system design in the hospital is to lay the patient at no peril of infection while hospitalized. Usually, hospitals are considered to be clean and detached from a pathogen which is really not true. Referable to the complex environment of the infirmary, the effective ventilation for the comfort of patients and control of infections must be given highest priority. Intensive care represents the highest stage of continuing patient care and handling. The thermal comfort depends upon velocity, temperature, relative humidity, mean radiant temperature, turbulence intensity, metabolic activity, etc.

Indoor air quality (IAQ) is characteristic of air flow which is capable to provide healthfully and well-being environment inside the room. Acceptable IAQ helps to maintain healthy and productive indoor environments. Other elements, such as aesthetics, furnishing, and personal demeanour, also influence the overall acceptability of buildings and the total is sometimes called “indoor environmental quality (IEQ)” [1]. It is usually presumed to exist if adequate ventilation air is driven home to building residents. Breathing air is defined as external air or suitably treated recirculating air. Breathing air may be supplied via the introduction of outdoor air by ventilation or infiltration.

There are many factors that define the short- and long-term success of ventilation systems in working into the objectives of designers, owners, and occupants. Many types of ventilation products are presently used in residential, commercial, institutional, and industrial buildings and many novel types of equipment will likely be brought out in the hereafter. The temperature of the air and the environment, as well as air movement, humidity, and skin moisture, affect thermal comfort. Comfort perception includes not only thermal comfort, but also satisfaction with air quality and acoustical, aesthetic, emotional, and other factors, too. Each person consciously or unconsciously determines whether he or she is comfortable in a particular time and location or not. The design must provide a ventilation system that minimizes exposure hazards to healthcare providers for a comfortable working environment. The HVAC system must also provide ventilation that minimizes the hazard exposure of visitors. The field of the ventilation air flow pattern in a confined space has become more important during the last year due to its broad popularity [2].

Gupta et al. [3] have reported ventilation conditions such as the air flow rate, air flow direction. CFD simulation is conducted out of mouth opening area of 25 human subjects coughing, taking a breathe, and speaking. Shetabivash [4] have investigated operational parameters on cross-ventilation and air flow system within a building. CFD tool FLUENT is used to solve Navier–Stokes equations. Pulat et al. [5] have proposed flow of air and temperature distributions in the well-known International Energy Agency (IEA). Annex 20 room have been forecast mathematically to investigate the effects of the inlet turbulence intensity and the length scale on the flow characteristics while considering solution multiplicity.

Prakash and Ravi Kumar [6] focused to study the occupants’ thermal comfort and indoor air flow characteristics of a room with adjacent window openings under generalized approach. Computational fluid dynamics (CFD) technique is used to study the indoor air flow for a three-dimensional room model. Romano et al. [7] have designed a ventilation scheme for operating theatres (OT), which directed to cut the patient infection risk while maintaining adequate comfort and productivity for the surgical staff. The supply air comes from a ceiling filter system composed of 23 H14 filters, which convinces a unidirectional flow with different air velocities over the protected region. The objective of this study is to examine mathematically and experimentally an OT with arrangements as per the air conditioning and ventilation standard 1946–4. Balocco and Lio [8] have investigated numerical transient simulations for the air flow designs, delivery, and air flow rate. For the particulate dispersion within an existing distinctive hospitalization room furnished with an innovative heating ventilation air conditioning (HVAC) with variable air volume (VAV). The main air scheme intended for immune-suppressed patients and 3D model of the room considers diverse, typical, locations of the patients. The resolution shows the fine weather for the high induction air inlet diffuser and the method of pressures imposed in the room to supply the efficient means of controlling flows containing virus droplets. Niu et al. [9] have examined the spread of respiratory droplets among two seated occupants equipped with one type of custom-made ventilation (PV) device using round movable panel (RMP) in an office room. The office was freshened by three diverse total volume (TV) ventilation scheme, i.e. mixing ventilation (MV), displacement ventilation (DV), and under-floor air distribution (UFAD) system.

Yau et al. [10] have the spotlight on the freshening of multiple-bed healthcare wards in the tropical environment, taking into account the scheme, indoor environment, and engineering controls. The required indoor situations such as temperature, moisture, air motion, and indoor air value in the ward places are concise based on the existing strategy and practices. Robinson et al. [11] have shown up the potential for airborne spread in the indoor location. Respirable pathogen-transport droplets give a vector for the spatial transmission of contamination with droplet transfer determined by the diffusive and convective route. The consequence of an ambient airflow, as an infection control, with droplet density depends on the communicable density at a former time. Bhamjee et al. [12] have investigated that the increase of temperature and heat gain is upper in the natural air flow case than in the forced air flow cases. The models have been experimentally validated in terms of the air speed, airflow, and an increase of temperature. The velocity field was calculated using Laser Doppler Velocimetry (LDV) and the overall flow field was captured using smoke for flow visualization. Nielsen et al. [13] have simulated in a full-scale room with two-bed hospital ward with one standing condition physician and a roof-mounted small impulse semi-circular inlet diffuser. Tracer gas is utilized to simulate gaseous contaminants, and the concentration is measured at diverse air change rates and various postures of the patients. A textile partition between the beds, which is distinctive in a healthcare ward, is utilized for the defence of the patients in some of the experiments.

Sinha et al. [14] have examined the division of velocity and temperature in the 2D room for various views of inlet and outlet and for diverse values of Reynolds and

Grashof number under heating conditions using finite volume method. Sinha [15] have studied air circulation and temperature distribution in a room for various orientations, inlet air velocities, and dimensions of a jet for a given vertical position of the air inlet and vent on the opposite wall. Navier–Stokes equations and energy equation in the two-dimensional rectangular Cartesian coordinates have been projected out by the control volume method. Thool and Sinha [16] have examined three cases of mix ventilation systems. Single-sided high supply and low exhaust (Case-1); high supply and low exhaust (Case-2) and low supply and high exhaust (Case-3) using computation fluid dynamics (CFD) technique have abided by the effective contamination control in case-3 as thermal plumes play the predominant function. Verma and Sinha [17] have examined the trajectory of a contaminated particle in the intensive care unit of hospitals using a Lagrangian particle tracking CFD model. By observation, it is suggested to keep the nurse station and healthcare staff near the inlet to keep them tidy. Verma and Sinha [18] have simulated numerical model of eight different cases for multiple beds ICU. The works have been carried out at room inlet velocities from 0.5, 0.75, and 1.0 m/s in the ICU using k–ε model.

2 Indoor Air Pollutants

In general, pollutants of common concern in buildings are divided into two broad classifications: particulates and volatile organic compounds (VOCs). Pollutants are always present in indoor air or outdoor air may be in the form of liquid or solid or in the form of suspended particles. There are many sources in the building causes of indoor air pollutants, which are occupants itself, building, furnishing, building materials, consumable liquids, and gases. Pollutants are generated inside from internal sources or come from outside, inside generated pollutants are extremely undesirable. The particulates settle onto or adhere to materials and volatile organic compounds (VOCs) can be taken up [2]. Lungs are excellent filters of many airborne pollutants but such filtration is normally undesirable.

Some pollutants of special concern which significantly affects the quality of air inside the room are listed below [19]:

- **Odour**: Odour pollution is associated with comfort rather than health effects. The human sense of smell permits perception of very low concentrations of odours but the sensitivity varies between individuals. People who are exposed to odours for a long period of time become less sensitive to them. Odour dilution to acceptable levels is usually achieved by supplying outside air to indoor air.
- **Carbon Dioxide**: By human respiration, an average sedentary adult (M = 70 Wm^{-2} and A = 1.8 m^2) produces about 0.0051 s^{-1} (181 h^{-1}) of CO_2. Expired air contains about 4.4% by volume of CO_2. Carbon dioxide concentration increases with the increase of occupants inside the room, unlike some other contaminants (e.g. Tobacco smoke), CO_2 cannot be filtered, absorbed, or adsorbed and it is, therefore, a good measure of the staleness of indoor air.

- **Tobacco Smoke**: It produces undesirable odours, particularly to non-smokers, and some of the smoke constituents can irritate the nose and eye portions. In indisposed ventilation system tobacco products like tar, CO, nicotine, actively affects the health of smoker and no smoke occupants inside the room or building. As per World Health Organization (WHO), world cancer report tobacco smoking causes lung cancer. Large air flow requires ventilating tobacco smoke, so its population is considered at the early stage of design ventilation system.
- **Formaldehyde**: Formaldehyde (HCHO) resins are used as bonding and laminating agents, as adhesives in compressed wood products, and as plastic foam insulation and packaging products. It is widely used in touching up, lavatory, comestible packaging with 1% concentration. It is also extremely used in construction materials and building appliances. Most commonly used as a binder in the production of plywood, wood chip board, plaster board, and fibre boards. Formaldehyde takes entry in the human body through inhalation. Most of that inhaled is absorbed in the upper respiratory tract. After entering inside of body, it actively affects body tissue which contains hydrogen in the form of amino acids, proteins, DNA, and others to form stable and unstable products and subsequently causes damage to the body tissues, i.e. it is genotoxic.
- **Volatile Organic Compounds**: Volatile organic compounds (VOCs) are generated from various sources. But the classes of VOCs present in the indoor air are not well defined. The formaldehyde is considered as VOCs, but generally, it is considered separately as its measuring equipments are different. Effect of volatile organic compounds in indoor air and human health is new research area and very little information is available about it.
- **Ozone**: Ozone is considered as one of the most toxic pollutants regulated in indoor air. It is commonly present in the air and its concentration in air depends upon elevation from ground and environment condition. Ozone is also produced by electrostatic equipments. It appears to cause the anatomical and curative effect to the populace.
- **Radon**: Radon (Rn) is naturally occurring radioactive gas which arises from the decay of radium (Ra) present in little amounts in the universe and also a little in building materials. Radium originates from the decay chain of Uranium (U).
- **Aerosols**: Outdoor air pollution is a complex mixture of smokes, mists, fumes, granular particles biogenic particles, and synthetic fibres. When suspended in air, these particulates are called aerosols. An aerosol is a liquid or solid particle which is in a quasi-stable suspension in air. A sample of atmospheric dust usually consists of soot, silica, clay, lint and plant fibres, metallic fragments, living organisms such as pollen, mould spores, viruses, bacteria. A large proportion of indoor aerosols originates from outdoor sources which penetrate the building envelope through cracks and opening by the action of wind attack and ventilation system. If a mechanical ventilation system is used, the concentration of indoor airborne particles can be reduced by air filtration. The efficiency of the most of air filtration systems is the highest for large or ultra-fine particles, and it is the lowest for fine particles.

Aerosol diameters below 0.01 μm are usually formed from combustion of fuels and radon; those up to 0.1 μm are produced by cocking and cigarette smokes; 0.1–10 μm is typically airborne dust, micro-organisms, and allergens; and particles of 100 μm and larger are airborne soil, pollen, and allergens.

The effect on health due to aerosols is mainly due to biogenic pollutants such as fungi, moulds, mites, bacteria, viruses, and pollens. Indoor dust, which is a mixture of human and animal skin flakes, fibrous material from textiles, organic particles, and mites, is also a potential allergen and a cause of bronchial asthma and allergic rhinitis for about 1% of the population. Pollens are the most widespread allergies which cause fever seasonally. The main source of pollen is outdoor vegetation, but their indoor concentrations can be reduced by air cleaners in the ventilation system.

3 Hospital Environment and Ventilation

Hospital-acquired infection (HAI) is an important public health issue with unacceptable levels of morbidity and mortality. Febrile respiratory illness is the most common reason for patients to admit in ICU and also for the patients to deteriorate in ICU. One of such a febrile illness is Ventilator Associated Pneumonia (VAP) which is the result of endogenous infection. These infections are mostly caused by antibiotic resistant organisms. Healthcare facility system is required to meet a variety of demands and applications, at a high standard of performance, in many ways unique to the buildings they serve. Heating ventilation and air conditioning (HVAC) systems are required to establish comfortable environmental conditions through the control of temperature, air movement, relative humidity, noise, and objectionable odours. Environmental control is important, not merely in providing personal comfort, but in facilitating the healing process as comfortable patient heals faster. In addition, healthcare facility HVAC systems are called upon to support a variety of medical functions, practices, and systems critical to health and safety, including the following:

- **Infection control**: Medical facilities are the places where relatively high levels of pathogenic (disease-causing) micro-organisms are generated and therefore require stringent practices and controls to safeguard the staff and patient population. The HVAC system is one of several tools and processes used in the control of infection.
- **Environmental control for specific medical functions**: Certain medical functions, treatments, or healing processes demand controlled environmental temperature and relative humidity conditions that exceed the requirements of mere personal comfort.
- **Hazard control**: Many medical facilities include functions where chemicals, fumes, or aerosols are generated that pose health or safety hazards. HVAC equipments are used in such applications to remove, contain, or dilute the environmental concentration of such contaminants to safe levels.

- **Life safety**: HVAC systems contribute for the detection and containment of fire and smoke and may be called upon to evacuate or exclude smoke from atria or exit enclosures. Engineered smoke control systems may be called for to provide complex zone pressurization control.
  Depending upon the type of medical facility, the characteristics of its patient population, and the nature of medical procedures performed, the range and criticality of services required in the above-listed categories will vary. Similarly, the complexity of the HVAC system design and the need for the close coordination with the design of other major building systems will vary by facility.

4 Modes of Transmission: Direct Contact and Airborne

The disease may be transmitted through two primary means: direct contact (including ingestion) and airborne. The means of transmission is determined by the nature of the infectious organism and/or how it enters or exits within the building environment.

4.1 Direct Contact Transmission

In this case, pathogens enter the body through a wound, open sore, or vulnerable body location (mouth, eyes, etc.) via contact with unwashed hands, infectious body fluids, droplets from sneezes or coughs, or other infected objects or material. An example of direct contact infection opportunities include:

- **Hand contact**, with an infection source (an ill patient, a contaminated equipment surfaces, etc.) transfers the organism by touching a vulnerable part of one's own or other's body;
- **Contact of a vulnerable body part with an infected body fluid**, such as might occur in an accidental splash of contaminated blood droplets from a laboratory specimen;
- **Needle stick**, whereby a healthcare provider accidentally sticks a contaminated syringe needle into human body;
- **Insect transmission** by bite or by direct transfer of pathogens from a contaminated substance (trash, animal droppings, etc.) to human food or food preparation surfaces;
- **Contact with infected liquid droplets produced by a sneeze, cough, or talk by a person with contagious disease**: Many of these droplets are of a size greater than 5 μm that cause them to settle out of the air quickly, limiting "infectivity" to a radius of several feet. A single sneeze can produce 100,000 aerosolized particles. Coughing can produce of the order of 10,000 particles per minute.

4.2 Airborne Transmission

It is usually distinguished as resulting from respiration of particles or aerosols of low mass and size (1.0–5.0 μm) that can remain indefinitely suspended in air. Infectious bacteria, fungi, and viruses normally are transmitted into the air in forms larger than the individual microbe, such as via attachment to organic or inorganic dust and particles, e.g., soot, skin cells, or the "droplet nuclei" that are the residual of aerosolized liquid droplets. Particles of this size are easily respirated deeply into the lungs, where in a suitably vulnerable host or in high enough concentration, they can overcome the body's immune system and cause disease. Typically means of airborne transmission include the following [20]:

- **Sneezing, coughing and talking by an infected** person produce many particles light enough to remain suspended in air. These activities can, therefore, spread infection by both the direct and airborne infection routes;
- **Resuspension** into air of in situ microbes, settled or trapped in building dust or debris, furnishing materials (including bed coverings), equipment and room finishes and released by disturbing activities such as bed making, maintenance, and construction work;
- **Aerosolization of contaminated water droplets** via shower heads, spray humidifiers, or evaporative cooling equipment (including cooling towers). Aerosolization of infectious particles or droplets also can occur via surgical and autopsy procedures, particularly those involving powered cutting or abrasion tools;
- **Carriage on human skin flakes** (squames) which the average person sheds into the environment at a rate of about 1,000 squames per hour;
- **Amplification (reproduction) of microbes within** HVAC air flow equipment, especially areas where moisture and dirt can accumulate, such as cooling coil drain pans, wet filters, and porous duct linings exposed to direct moisture.

5 Governing Equation

It is important to achieve good ventilation system design for the removal of pathogens airborne particles. Computational fluid dynamics (CFD) plays an important character to simulate and analyze the breathing system of an ICU. In this investigation, turbulent flow modelling has been applied to study air flow in ICU. The speed and temperature distributions in an ICU room have been computed by solving the underlying equations of fluid mechanics in steady, incompressible, turbulent flow with uniform thermo-physical properties and under Boussinesq's approximation. The equations describing the room airflow are conservation of mass, conservation of momentum, and conservation of energy. The general transport equation for turbulent flow is given as follows [21–23].

For two-dimensional turbulent flow:

$$\frac{\partial(\rho\phi)}{\partial t} = \frac{\partial}{\partial x_i}\left(\Gamma_f \frac{\partial\phi}{\partial x_i} - \rho u_i \phi\right) + S_\phi \tag{1}$$

For three-dimensional turbulent flow:

$$\frac{\partial(\rho\phi)}{\partial t} = \frac{\partial}{\partial x}\left(\Gamma\phi\frac{\partial\phi}{\partial t} - \rho u\phi\right) + \frac{\partial}{\partial y}\left(\Gamma\phi\frac{\partial\phi}{\partial y} - \rho v\phi\right) + \frac{\partial}{\partial z}\left(\Gamma\phi\frac{\partial\phi}{\partial z} - \rho w\phi\right) + S_\phi \tag{2}$$

6 Particle Trajectory

The motion of contaminated particles in the indoor or outdoor area is affected by various factors like airflow pattern of air flow, properties of the particle, building geometry, air flow rate and direction, diffuser locations. The physical systems of the contamination existing may involve matter in a number of states, including gases, vapours, liquids, solids (including very finely divided particles, called particulate) and or mixture of states, such as aerosols and mists (liquid in a gas), dusts (solid in gas), foams. It may comprise infectious organisms, including bacteria, their spores, fungi, viruses, yeasts.

There are two methods to calculate the trajectory of particle (a) Lagrangian method (b) Eulerian method. As per Newton's second law, the motion of an individual particle can be described by the equation below [21–23]:

$$\frac{\mathrm{d}u_p}{\mathrm{d}t} = F_i, F_D = \frac{18\mu C_\mathrm{d}\mathrm{Re}}{\rho_p d_p^2 \times 24} \tag{3}$$

$$\frac{\mathrm{d}u_p}{\mathrm{d}t} = F_\mathrm{D}(u_i - u_p) + \frac{g(\rho_p - \rho)}{\rho_p} + F_X \tag{4}$$

7 Results and Discussion

The literature survey reveals that a lot of work has been done to predict the velocity fields, turbulence levels, and temperature distributions in enclosed spaces by numerical simulations and experimental studies. Such predictions are often used in the design of isolation room ventilation and air conditioning systems to verify if the velocity, turbulence level, and temperature, etc., are within the tolerable limits as prescribed by the prevailing standards [24]. However, a systematic study of the effectiveness of different hospital ventilation system designs on the removal of contaminants from a confined space has not received much attention. In India, hospitals are normally designed for multi-patient in a room. Economically, these rooms are the most efficient in which patients can be placed along one corridor or in two rows: facilitating the supervision of patients and reducing the amount of time which nurses spend in travelling.

Lots of studies carried out to look feasible transference routes of diseases, but not too many have observed the impact of ventilation system on diseases transmission directly. The final selected studies are based on the direction and the rate of air flow indoor or outdoor area with the diffusion of certain communicable diseases. The infective diseases which are able to transmit between patient and ventilation are measles, small pox, pulmonary tuberculosis (TB), and chickenpox. All these diseases are airborne diseases.

The crucial verdicts of various researchers are given as under for airborne disease:

- The absence of breathing or poor ventilation rates is directly associated with the increase of airborne diseases.
- The risk of infection can be decreased with high ventilation rate. The ventilation rate is associated with transmission of airborne diseases, as high ventilation rate has high dilution capacity. It decreases infection inside affected room or hospital.
- The impact of overall indoor air flow and air flow direction on the transmission of droplet transmitted diseases is not too much, also according to the transmission physics of droplet transmission, basic ventilation does not affect largely to the droplet transmission.
- The rate of infection depends upon the source of infection, i.e. infected person. The rate of infection is associated with distance from infected person, it increases when the distance from infected person decreases. Low distance from infected person and low ventilation rate are not desirable for human health.

The trouble in developed ventilation guidelines for infection control is an absence of sufficient data to urge at least air flow rate to control indoor infection against droplet nuclei. Ventilation can diminish the concentration of airborne pathogens through eliminating or reducing airborne droplet nuclei. A higher ventilation rate can provide a higher dilution capability and thus potentially reduces the risk of airborne infections.

In simulation work, inlet and outlet for different values of Reynolds number have been considered with and without enclosures. In general total, 18 cases have been considered for numerical investigation. Figure 1 shows the multi-patient intensive care unit room of size 5.8 m × 3 m × 6.3 m. The inlet and outlet area is kept constant throughout the simulation as 0.6 m × 0.4 m located on the north and south wall, respectively. The air is supplied at three different velocities which have been categorized in three different cases. The numerical simulation is based on the following assumptions:

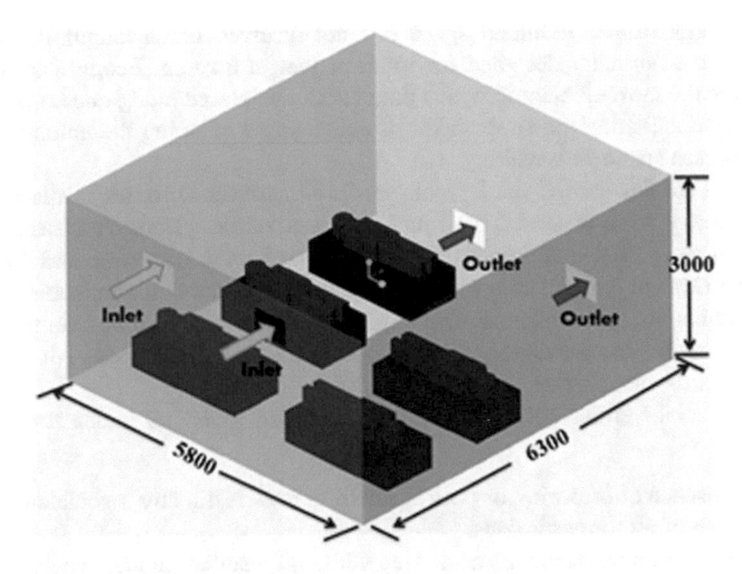

Fig. 1 Room layout of multiple patient intensive care (case 1)

- The bed, light source, and enclosure are in the shape of rectangular boxes.
- All walls, roof, and ground of the room are considered to be well insulated.
- The patient is considered to be lying on the bed in the east–west direction.
- When one patient is considered to be infected, other patients are assumed to produce no contamination;
- All contaminated particles are of spherical solid shape;
- No particle rebounds on solid surfaces;
- The movement of only one particle has been considered during the study for clarity in the figures;
- Transfer of mass and heat among contaminated particle trajectory and air are neglected;
- Contaminated particles are assumed to be of uniform diameter and mass less.

Case 1: Multiple patients with two inlets and two outlets Figure 1 shows the layout of multiple-bed (five beds) intensive care units. The layout of ICU consists of two inlets for the entry of fresh air and two outlets for removal of infected air from the ICU. For this layout of ICU, air flow is simulated for inlet air velocity 0.2–1 m/s. The temperature of inlet air is maintained 20 °C for cold flow air flow simulation and 30 °C for hot flow air flow simulation. The properties of air are taken as per the temperature of the air stream. Inlets and outlets are rectangular in cross-section and dimensions of each inlet and outlet are 600 mm × 400 mm. Position of both inlets are kept constant, i.e. 2300 mm above from the floor and 1500 mm, 3700 away from the east wall for inlet 1 and inlet 2, respectively. Same like inlets, the position of both outlets are kept at constant position, i.e. 300 mm above the floor and is located 1500 and 3700 mm in front of the east wall for outlet 1 and outlet 2, respectively. The dimensions of the beds are 2000 mm × 900 mm × 500 mm and patients are 1800 mm × 500 mm × 300 mm. According to change in inlet velocity Reynolds number and ACH (air changes per hour) also vary which is shown in Table 1. Figure 2 shows velocity vector of air flow and Fig. 3 shows temperature contours from the plane created at centre of the inlet.

It is clearly observed that clean cold air is entering through the inlet moves horizontally and bends near the south wall and falls down before leaving through the outlet. The cold main stream moves above the occupied zone of patients. One clockwise recirculation cell is formed between the first patient and north wall and another small anti-clockwise recirculation zone is formed between the patients.

Case 2: Multiple patients with two inlets and two outlets and heat source The position of the inlet, outlet, and patient is same as the first case. As shown in Fig. 4, four heat sources of 1000 mm × 100 mm × 50 mm (two tube light in each rectangular heat source) is located at 1300 mm from north and south wall and 400 mm from the roof. Numerical studies are carried out with different rating of light 28, 36 and 40 W for inlet air velocity 0.2 m/s.

Figure 5 shows velocity vector of air flow, and Fig. 6 shows temperature contours from the plane created at centre of the inlet. Same as the first case, It is clearly observed that clean cold air is entering through the inlet moves horizontally and bends near the south wall and falls down before leaving through the outlet.

The cold main stream moves above the occupied zone of patients and one clockwise recirculation cell is formed between the last two patients. Small anti-clockwise recirculation zone is formed between the first patient and north wall.

Case 3: Multiple patients with two inlets and five outlets In this case, extra vent dimension 400 mm × 300 mm for isolation (outlet) is created whose position is at the centre of two patients and 200 mm above the floor, as shown in Fig. 7. Figure 8 shows velocity vector of air flow and Fig. 9 shows temperature contours from the plane created at centre of the inlet. Same as above cases, clean cold air is entering through inlet moves horizontally and bends near the south wall and falls down. One clockwise recirculation zone is formed between the last two patients. Because of extra vent provided for flow air stream, it is now fluctuating towards the vent.

Figure 10a–e show the motion of the contaminated particle in case 1, which is coming out from the mouth of patients oriented at various positions in the ICU. It is clear that particles are after coming from patient mixes with fresh air coming from inlets and going out without affecting other patients.

Table 1 shows the time required by contaminant particle to leave ICU. As envisioned from Fig. 10a–e contaminated particle starts from the mouth of the patients and moves through a tortuous path without affecting to other patients and leaves through the outlet. In cold air flow case 1, the total time taken is 3 min for a particle which is getting out from patient 4 and patient 5 to leave the ICU for inlet

Table 1 Time required by contaminated particle to leave ICU for case (1) and case (2)

Cases	Air velocity	Reynolds No.	ACH	Time to leave the ICU (in minute)				
				Patient 1	Patient 2	Patient 3	Patient 4	Patient 5
Cold flow	0.2	12,289	2	7	11	10	7	14
	0.4	24,578	4	7	7	5	3	7
	0.6	36,867	6	10	7	10	12	14
	0.8	49,156	8	10	12	4	6	3
	1	61,445	10	4	10	5	6	2
Hot flow	0.2	12,289	2	16	11	9	6	10
	0.4	24,578	4	4	4	5	5	5
	0.6	36,867	6	9	9	8	2	7
	0.8	49,156	8	9	9	6	5	5
	1	61,445	10	6	6	5	5	5
Extra vent cold flow	0.2	12,289	2	25	27	13	23	33
	0.4	24,578	4	17	15	9	15	6
	0.6	36,867	6	2	3	3	9	7
	0.8	49,156	8	10	10	10	4	4
	1	61,445	10	10	8	9	8	10

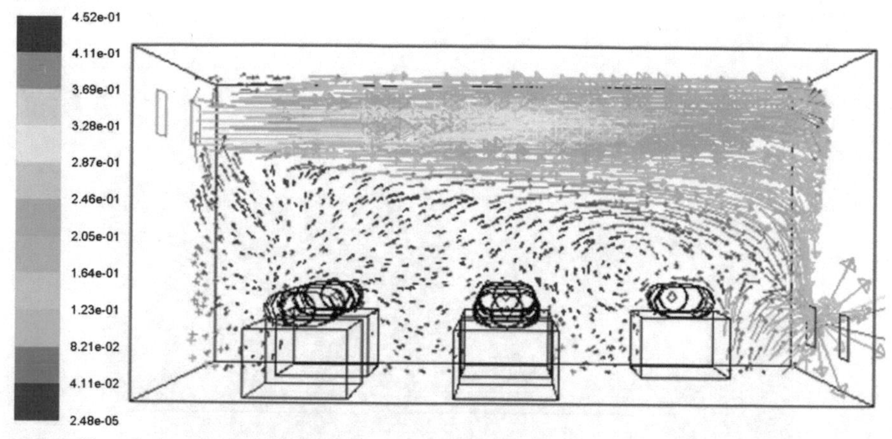

Fig. 2 Velocity vector for case 1 on the plane centre of inlet and outlet

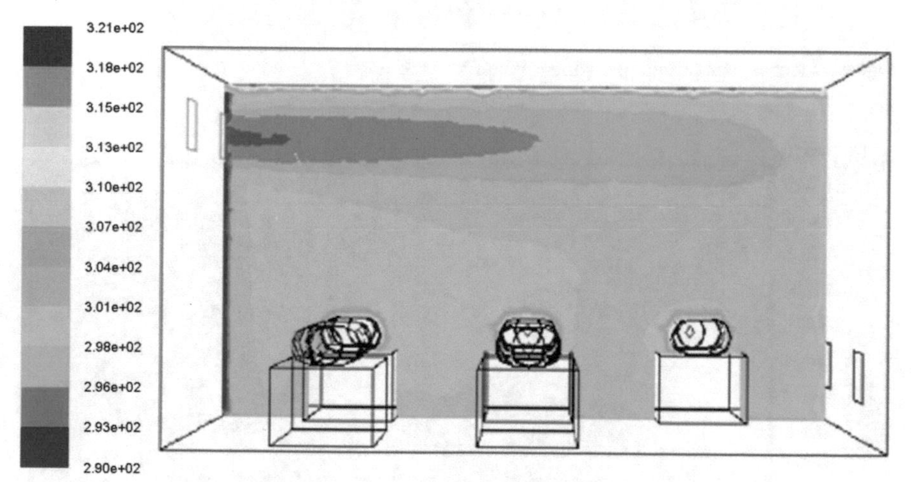

Fig. 3 Temperature contour for case 2 on plane centre of inlet and outlet

air velocity 0.4 and 0.8 m/s, respectively. The maximum time taken by particle coming out from patient 5 to leave ICU is 14 min for inlet air velocity 0.2 and 0.6 m/s. It is clear that for this arrangement, contaminant particle which is getting out from the patient 5 is taking larger time. For hot air flow case 2, a contaminant particle which is getting out from the patient 4 is taking 2 min for inlet air velocity

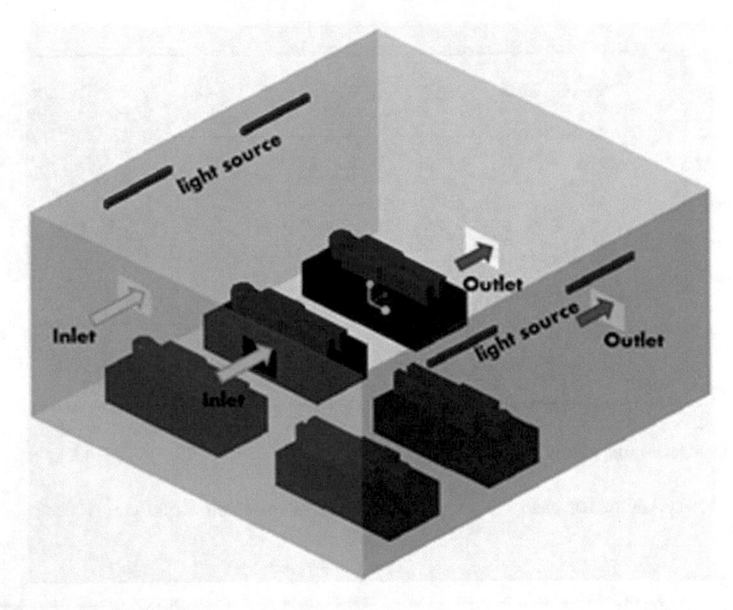

Fig. 4 Room layout of multiple patient intensive care (case 2)

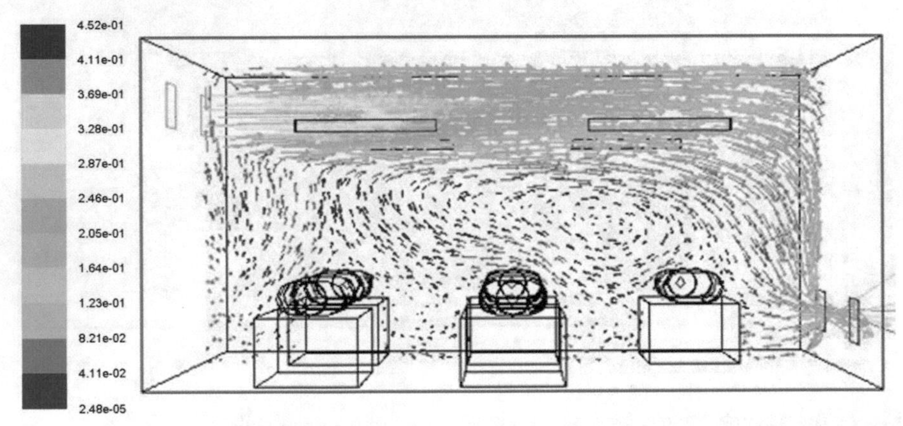

Fig. 5 Velocity vector for case 2 on plane centre of inlet and outlet

0.6 m/s. In this case, a contaminant particle which is getting out from patient 1 is taking maximum 16 min to leave the ICU for air velocity 0.2 m/s. In case 3, where the extra vent is provided for contaminant particles to leave the ICU. A contaminant particle which is getting out from patient 1 is taking minimum 2 min for inlet air velocity 0.6 m/s to leave ICU. A contaminant particle which is getting out from the patient 2 is taking maximum 27 min to leave the ICU for inlet air velocity 0.2 m/s.

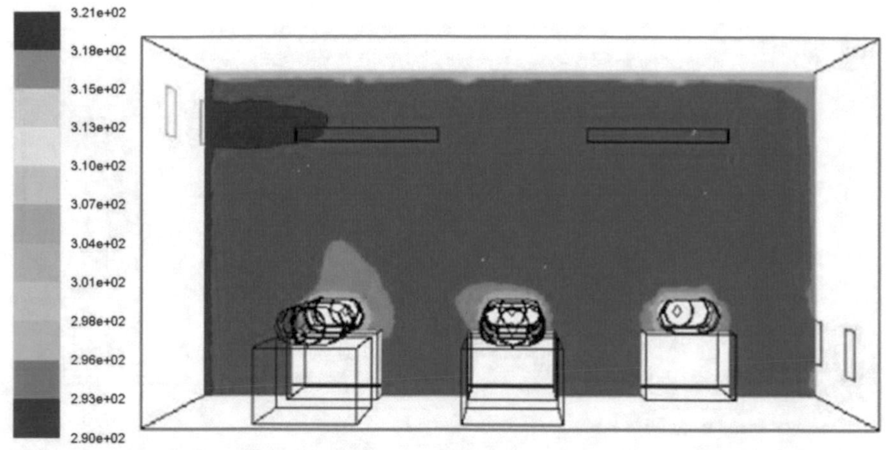

Fig. 6 Temperature contour for case 2 on plane centre of inlet and outlet

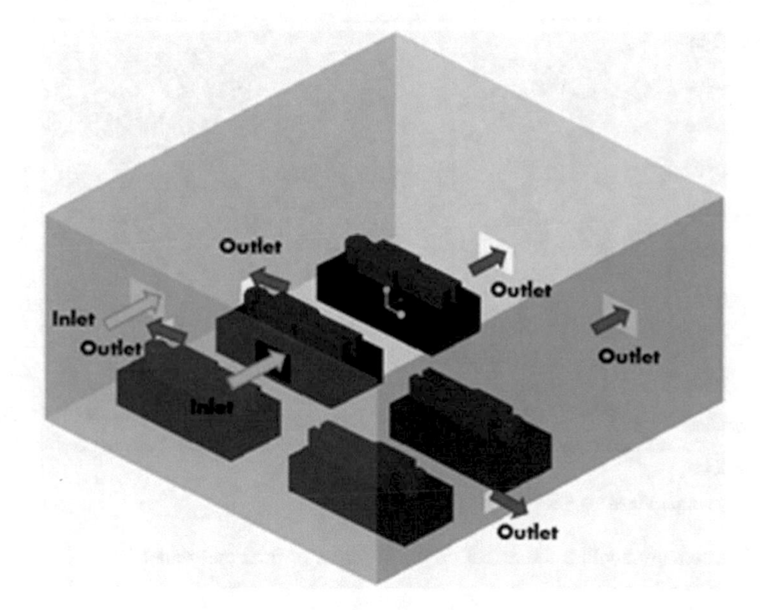

Fig. 7 Room layout of multiple patient intensive care (case 3)

It is clear from Table 1, for inlet air velocity 0.4–0.6 m/s contaminant particle is taking comparatively less time to leave ICU.

Table 2 shows the total time required by the massless contaminant particle for the different light rating to leave the room of the ICU. In this arrangement inlet, air

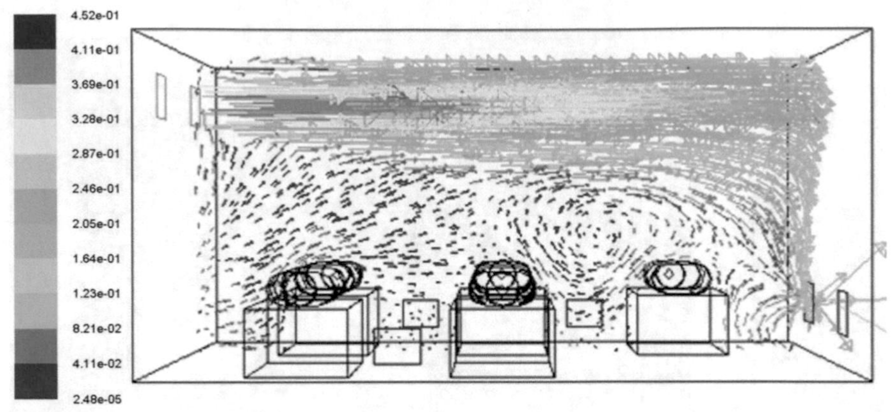

Fig. 8 Velocity vector for case 3 on plane centre of inlet and outlet

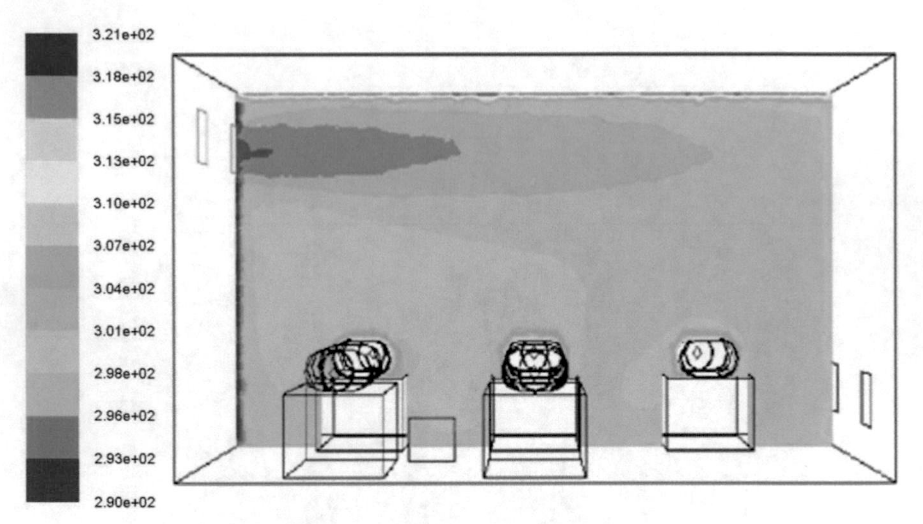

Fig. 9 Temperature contour for case 3 on plane centre of inlet and outlet

velocity is taken as 0.2 m/s. It is seen that time required for contaminant particle which is coming out from the patients is nominal and healthy for the hospital environment. It is clear, in this arrangement; time required for a contaminant particle to leave ICU is comparatively less than above cases for same air velocity also light rating does not affect much the time required by contaminant particle to leave ICU.

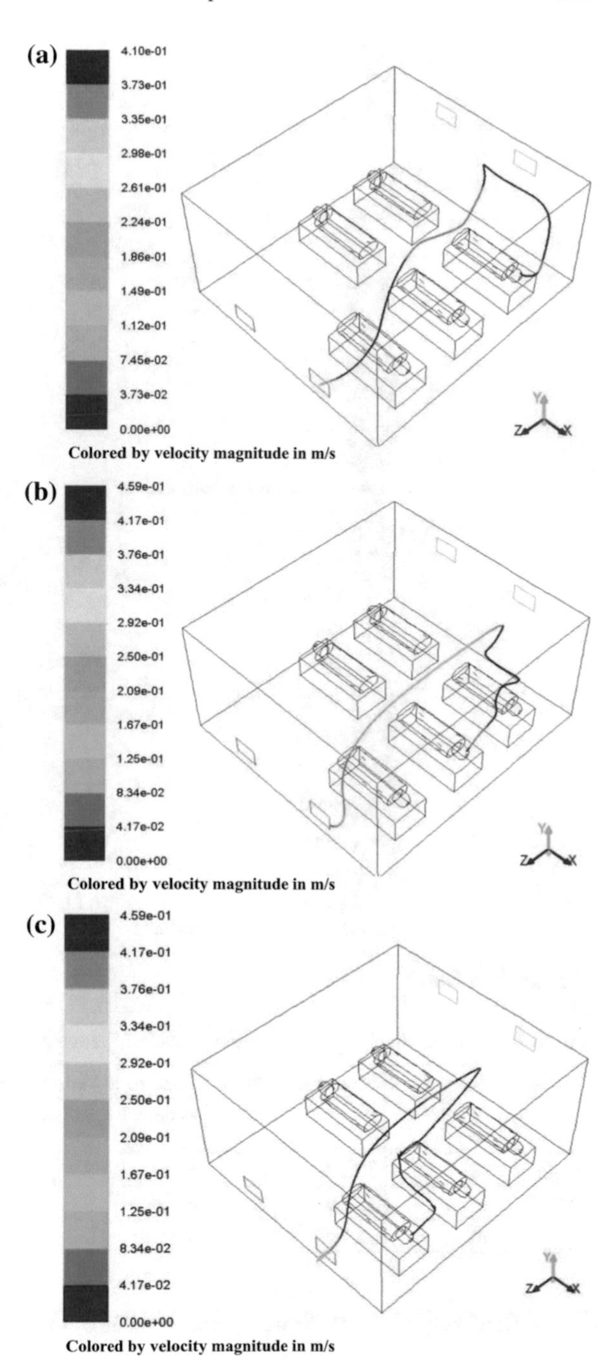

Fig. 10 **a** Movement of contaminant in case 1 through patient 1. **b** Movement of contaminant in case 1 through patient 2. **c** Movement of contaminant in case 1 through patient 3. **d** Movement of contaminant in case 1 through patient 4. **e** Movement of contaminant in case 1 through patient 5

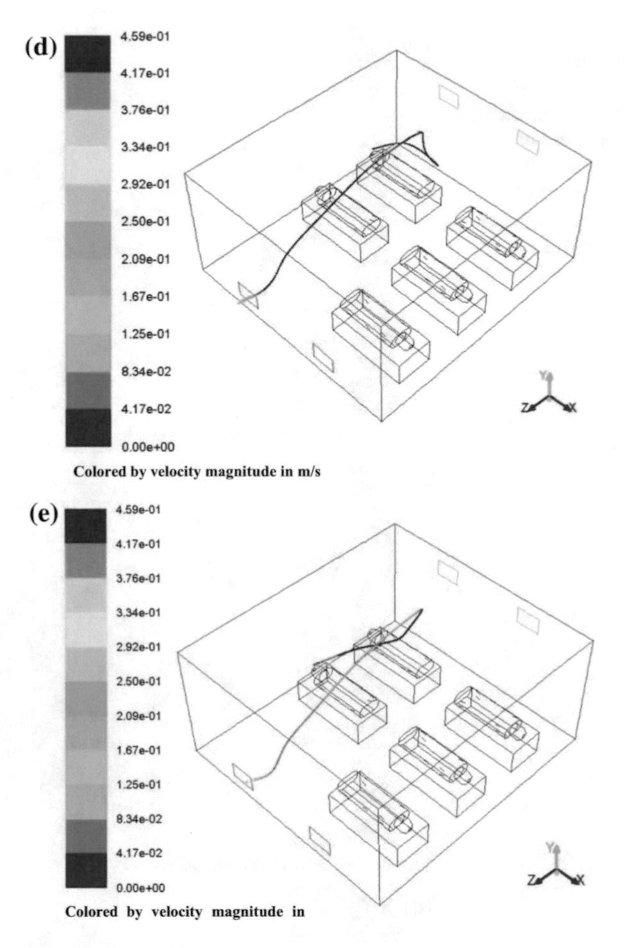

Fig. 10 (continued)

Table 2 Time required by contaminated particle to leave ICU for case (3)

Cases	Air velocity (W)	Reynolds No.	ACH	Time to leave the ICU (in minute)				
				Patient 1	Patient 2	Patient 3	Patient 4	Patient 5
Light source	28	12,289	2	6	8	9	7	8
	36	12,289	2	6	8	9	7	8
	40	12,289	2	5	8	9	7	8

8 Conclusion and Scope for Future Work

The studies have been carried out for inlet velocity 0.2–1 m/s in the hospital for two cases (case 1 and case 3) having a different position of the outlet and the heat source with different rating with air velocity 0.2 m/s for case 2. Tables 1 and 2 show the time taken by contaminant to leave the ICU room for different cases of ventilation. It is clear from the Table 1, the maximum time required by the contaminant to leave the ICU, when extra vent is provided and air velocity is 0.2 m/s. Proper ventilation depends upon the quality of the outdoor air provided to remove the contaminant.

In this context, the supply air should have sufficient proportion and flow rate of fresh air and its flow rate and distribution should enable it to dilute and remove contaminants from all parts of the hospital/isolation rooms. The air velocities should be within the acceptable limits of noise level and human comfort. As air flow rate increases, ventilation rate increases with specific speed depends upon case of hospital arrangement. By observing the movement of the contaminated particle, it is advisable to keep the nurse station and healthcare staff near the inlet to keep them healthy.

Numerical simulation of room airflow and heat transfer can be practiced in the air conditioning design of non-conventional buildings. CFD techniques make it potential to look into various aspects of the indoor air quality, comfort, and energy use and ventilation efficiency due to air conditioning. It can be utilized as an effective tool in diagnosing system problems, optimizing system designs, and improving system efficiency. In order to promote and widen the application and scope of study of room airflow, the following areas need further investigation:

- There are many aspects of air conditioning design which have never been investigated in details, e.g., interference by furniture, persons (stationary or moving), effects of beams, roughness, protrusions and many features of the building, combined supply air/return air ducts, effects of outlet diffusers.
- Numerical model can be improved by incorporating windows in the room for natural light. Natural illumination is always essential for the well-being of patients, staff, physician, etc.
- For each patient care, there should be a visual access to the outdoors. Workflow and clear-space requirements will drive design decisions about how best to meet family needs and integrate families into patient care.
- Multiple simultaneous chemical reactions can be modelled, with reactions occurring in the bulk phase (volumetric reactions) and/or on wall or particle surfaces, and in the porous region. Species transport modelling capabilities, both with and without reactions. The fluid phase is treated as a continuum by solving the Navier–Stokes equations, while the dispersed phase is solved by tracking a large number of particles, bubbles, or droplets through the calculated flow field. The dispersed phase can exchange momentum, mass, and energy with the fluid phase.
- In order to improve the experimental work, thermocouple, particle image velocimetry (for measurement of velocity), high camera resolution (for capturing better flow details), can be used for better and accurate results.
- The flow has been considered to be steady and isothermal boundary conditions have been applied to the walls. In actual practice due to the variation of incident solar radiation, nocturnal cooling and variation of outdoor air temperature and relative humidity, the wall temperatures are never constant.

References

1. Working Group 16 (2003) Ventilation, good indoor air quality and rational use of energy, European Commission Joint Research Centre, Institute For Health and Consumer Protection, Physical and Chemical Exposure Unit, Report No. 23, EUR 20741 EN
2. WHO (2010) WHO guidelines for indoor air quality: selected pollutants, Europe
3. Gupta JK, Lin CH, Chen Q (2010) Characterizing exhaled air flow from breathing and talking. Indoor Air 20:31–39
4. Shetabivash H (2015) Investigation of opening position and shape on the natural cross ventilation. Energy Build 93:1–15
5. Pulat E, Ersan HA (2015) Numerical simulation of turbulent airflow in a ventilated room: Inlet turbulence parameters and solution multiplicity. Energy Build 93:227–235
6. Prakash D, Ravi Kumar P (2015) Analysis of thermal comfort and indoor air flow characteristics for a residential building room under generalized window opening position at the adjacent walls. Int J Sustain Built Environ 4:42–57
7. Romano F, Marocco L, Gusten J, Joppolo CM (2015) Numerical and experimental analysis of airborne particles control in an operating theatre. Build Environ 89:369–379
8. Balocco C, Lio P (2011) Assessing ventilation system performance in isolation rooms. Energy Build 43:246–252
9. He Q, Niu J, Gao N, Zhu T, Wu J (2011) CFD study of exhaled droplet transmission between occupants under different ventilation strategies in a typical office room. Build Environ 46:397–408
10. Yau YH, Chandrasegaran D, Badarudin A (2011) The ventilation of multiple-bed hospital wards in the tropics: a review. Build Environ 46:1125–1132
11. Robinson M, Stilianakis NI, Drossinos Y (2012) Spatial dynamics of airborne infectious diseases. J Theor Biol 297:116–126
12. Bhamjee M, Nurick A, Madyira DM (2013) An experimentally validated mathematical and CFD model of a supply air window: forced and natural flow. Energy Build 57:289–301
13. Nielsen PV, Li Y, Buus M, Winther FV (2014) Risk of cross-infection in a hospital ward with downward ventilation. Build Environ 45:2008–2014
14. Sinha SL, Arora RC, Roy S (2000) Numerical simulation of two dimensional room air flow with and without buoyancy. Energy Build 32(1):121–129
15. Sinha SL (2001) Behavior of inclined jet on room cooling. Build Environ 36:569–578
16. Thool SB, Sinha SL (2014) Performance evaluation of conventional mixing ventilation systems for operating room in the view of infection control by numerical simulation. Int J Bio-sci Bio-technol 6(4):87–98
17. Verma TN, Sinha SL (2015) Trajectory of contaminated particle in intensive care unit of hospitals using numerical modelling. Int J Design Manuf Technol 9(1), January 2015
18. Verma TN, Sinha SL (2015) Numerical simulation of contaminant control in multi-patient intensive care unit of hospital using computational fluid dynamics. J Med Imag Health Inform 5:1–5
19. Awbi HB (1991) Ventilation of building. Chapman and Hall, London
20. Verma TN (2015) Numerical simulation of contaminant control in intensive care unit (ICU) of hospitals, National Institute of Technology. Ph.D. thesis
21. Patankar SV (1980) Numerical heat transfer and fluid flow. McGraw Hill, Washington
22. Versteeg H, Malalasekera W (1995) An introduction to computational fluid dynamics. Longman, London
23. Ghoshdastidar PS Computer simulation of flow and heat transfer. Tata McGraw-Hill Publishing Company Limited
24. HVAC Design manual for Hospital and Clinics, ANSI/ ASHRAE, Standard 55-2010 (2010) Thermal Environment Condition for Human Occupancy, American Society of Heating, Refrigerating and Air-conditioning Engineers, Inc.

Part V
Miscellaneous

A Review on Clean Combustion Within Porous Media

Niraj Kumar Mishra, P. Muthukumar and Snehasish Panigrahy

Abstract Technological growth of any nation demands more fossil fuels which cause two major threats. First one is the shortage of fossil fuel, and the second is environmental pollution. Recently, the age-old conventional combustion process is being substituted by an innovative combustion technology, called porous medium combustion. This surge of interest in porous radiant burner is driven by some of the important benefits such as high thermal efficiency, low emission characteristics, high power modulation range, extended flammability limits and high power density. In the last two decades, there has been a significant development in the research aiming at changing the operating parameters and the design configurations of the porous radiant burners to attain lower emissions and higher thermal performances. Various burners based on porous medium combustion have been developed for industrial and domestic applications and showed beneficial over their conventional burner counterparts. Porous radiant burners based on porous medium combustion technology showed good emission characteristics and offer higher thermal efficiencies. Although, durability of few burners is still a matter of concern which results in non-commercialization of these products. This chapter summarizes the development of various porous radiant burners used in both industrial and cooking applications.

Keywords Combustion · Porous radiant burner · Premixed combustion
CO emissions · NO_x emissions

N. K. Mishra (✉)
Department of Mechanical Engineering, National Institute of Technology Uttarakhand, Srinagar, India
e-mail: nkm@nituk.ac.in

P. Muthukumar · S. Panigrahy
Department of Mechanical Engineering, Indian Institute of Technology Guwahati, Guwahati, India

N. Sharma et al. (eds.), *Air Pollution and Control*, Energy, Environment, and Sustainability, https://doi.org/10.1007/978-981-10-7185-0_12

Abbreviations

PMC Porous medium combustion
PM Porous matrix
PMB Porous medium burner
CB Conventional burner
ppm Parts per million
LPG Liquefied petroleum gas
Al_2O_3 Alumina
ZrO_2 Zirconia
PSZ Partially stabilized zirconia
SiC Silicon carbide
PRB Porous radiant burner
PZ Preheating zone
CZ Combustion zone
CO Carbon monoxide

Nomenclature

d_m Equivalent pore diameter (mm)
c_p Specific heat (kJ/kg K)
ρ Density (kg/m^3)
k Thermal conductivity of the fuel-air mixture (W/m K)
S_L Laminar flame speed (m/s)

1 Introduction

In developing countries, majority of the energy requirements is fulfilled by the conventional combustion devices working on fossil fuels. The diminishing fossil fuel reserves and the rise in environmental imbalance due to pollutants originating from combustion processes have demanded the necessity to look for alternative sources of energy. At the same time, some design modifications are also required to make the existing combustion systems more efficient. The main focus of these design modifications is to improve the thermal efficiency to the maximum possible extent and to minimize the emissions. Global warming and accompanying climate change due to the pollutants formed from the combustion have become a serious alarm. Hence, a continuous effort on improving the performance of such devices has remained an utmost interest for the policy makers and researchers dealing with environmental pollution and energy conservation.

Pollutants produced from the conventional combustion devices affect the health of the users and the environment in several ways. The formation of pollutants occurs in two ways: one is primary air pollutants which are emitted directly from the source, and the other is secondary pollutants which are formed through the reactions involving primary pollutants in the atmosphere. The primary pollutants discussed in this section are carbon monoxide (CO), nitric oxide (NO) and nitrogen dioxide (NO_2) because they dominate during the gas combustion process.

CO is the most abundant pollutant in the lower atmosphere, and its effect varies from normal cough to death depending upon the exposure levels. Figure 1 [1] illustrates the different effects of CO exposure levels. CO generally is the major species formed during the rich combustion. For stoichiometric and somewhat lean mixtures, CO is found in considerable measures at typical combustion temperatures as a result of the dissociation of CO_2. CO concentrations rapidly fall with temperature. According to Basu et al. [2], 0.1 s of residence time is necessary for complete combustion. In furnaces, for example, where the residence time is measured in seconds, to reach conditions for complete combustion is not difficult. In IC engines, the temperature rapidly falls and the residence time is not sufficiently long to approach conditions for complete combustion. In order to decrease the formation of CO concentrations in flue gases, the residence time must be as long as possible. Higher combustion temperatures are one more advantageous for CO reduction, and lean air mixture (excess air) ratio is also preferred for CO reduction. At lower air–fuel (AF) ratio, the CO emission is higher and the CO equilibrium concentration is high. For high AF, the oxidation velocity decreases owing to the lower temperatures in the combustion zone. One should emphasize that the above considerations are valid only for premixed combustion processes.

Nitrogen oxides (NO_x) are also one of the important pollutants formed from the combustion of fossil fuels. Everyone is exposed to small extents of NO_x in ambient

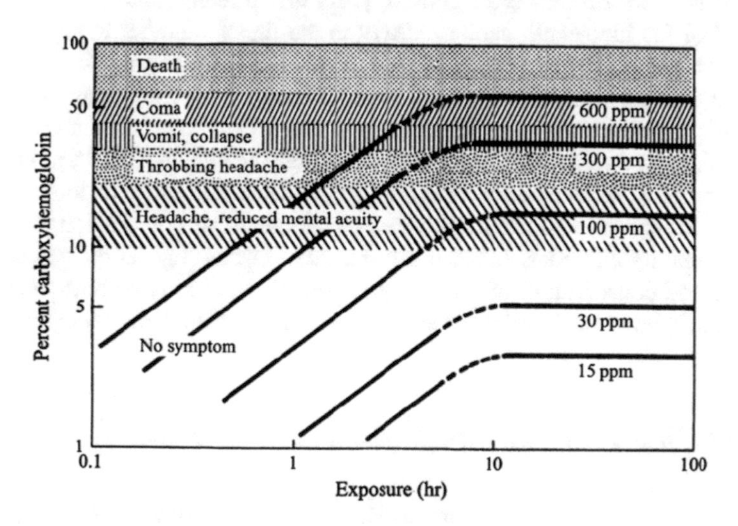

Fig. 1 Effects of different levels of CO exposure on humans [1]

air. Higher exposure may occur by burning kerosene, wood, near gas stoves or if one smokes. It is important to note that there is no antidote for NO poisoning. CO and NO_x formed during combustion affect the environment and health in numerous ways.

Over last two decades, many researchers have been explored porous medium combustion (PMC) technology for both industrial and domestic applications due to its multidimensional benefits, particularly lower emission of pollutants and higher thermal efficiency. In this chapter, a brief discussion about the working principle of PMC and followed by some of the important applications of the burners based on PMC are presented. Major breakthroughs achieved from PMC technology and future research focus are also highlighted at the end of this chapter.

2 Porous Medium Combustion (PMC)

In the late eighties and early nineties, researchers were more focused in understanding the heat transfer process and combustion mechanism within the porous matrix (PM) and at the beginning of the twenty-first century, the research focus was shifted towards the understating of combustion stabilization mechanism and various burner developments. In this section, the history, working principle, material used and advantages of the PMC are presented.

2.1 History of PMC

The PMC is a two-century-old technology. Its first research activity started in the beginning of the nineteenth century. Davy is the first researcher to report a work relevant to the PMC [3]. He made two important inventions. His first finding concluded that the combustion cannot occur in tubes below a certain radius. This minimum radius was named as "quenching radius". His second finding established the fact that even without a flame, a gas can be burnt below its ignition temperature. This was termed "flameless combustion". In the recent times, both these concepts have been widely used in the design and development of combustion devices for some practical applications. In recent years, due to strict legislation on ecological protection and conservation of fuel, major research attention has been focused on this technology and many of its practical applications have been recognized.

2.2 Working Principle and Classification of PMC

Recently, a novel concept of combustion called PMC has received significant attention as it offers multifaceted benefits such as flameless combustion, high

thermal efficiency, low emissions and uniform heat flux over the conventional combustion technology working on free-flame mode. The PMC operates a novel concept of using a 3-D PM in the combustion zone for improved heat transport from the burned to unburned portion of the air-fuel mixture. As the PM has high thermal conductivity and good radiative properties, the contributions of radiation and conduction in the PMC are very significant. Also, due to a large surface area of the PM and high heat transfer coefficient, the convective heat transfer is also better than the free-flame combustion. The better heat transport (through the combined modes of conduction, convection and radiation) results in a homogeneous temperature distribution in the combustion zone. Depending on the flow rate and thermo-physical properties of the porous material, the flame may become stable either inside or on the surface of the PM.

The PMC is characterized by a high burning velocity, reduced temperature drop across the reaction zone, high radiant output, high peak flame temperature and reduced enthalpy of flue gas. Unlike the conventional burners, in the burner based on PMC, a low calorific fuel can also be combusted. In general, based on the type of media used, the PMC is categorized into catalytic and inert. The former one utilizes a catalytic medium, which is coated on the solid matrix and participates in the combustion process; whereas the latter one remains inert but augments the heat transfer process. Further, depending upon whether the flame is stabilized above the surface or within the PM, the PMC can be classified as surface stabilized or matrix stabilized. In matrix stabilized combustion, the flame stabilizes close to the inlet and the combustion takes place completely inside the PM. The gas temperature reaches its maximum value in the reaction zone and decreases in the downstream due to cooling. Unlike this, in surface-stabilized combustion, the flame stabilizes on the downstream surface of the PM and the maximum volumetric heat release there.

The dependency of the flame location on porosity has directed to the idea of a double-layered porous radiant burner (PRB) having different porosities [4]. A double-layered PRB consists of a preheating zone (PZ) and a combustion zone (CZ) (Fig. 2). The PZ, which has a low porosity, prevents possibility of ignition and flame propagation, thereby occurrence of flashback. The air-fuel mixture is

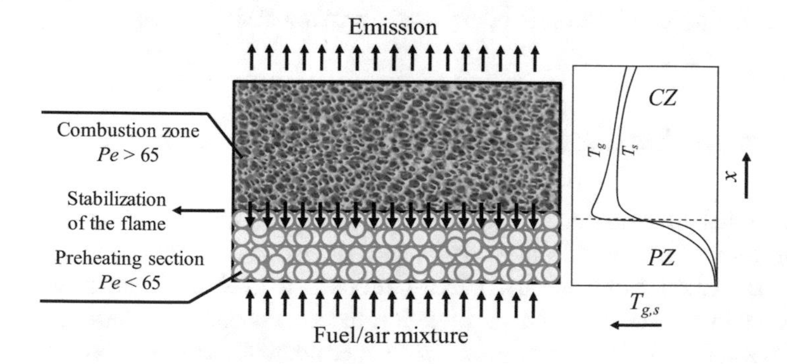

Fig. 2 Schematic of double-layered PRB

preheated in this zone, and it helps in improving the combustion efficiency and enhances the flammability limit. Further, preheating improves the stability of the combustion regime too. In the CZ, due to high porosity, combustion takes place and flame propagates. In this, the reaction zone is enlarged and because of homogenization of temperature, the formation of NO_x is reduced. The interface of the two zones serves as a flame holder.

To gain insight into the complex combustion mechanism within the PRB, a conceptual illustration of the PMC has been provided in Fig. 2. Fuel-air mixture enters the PRB through its bottom surface of the preheating section and in no time the combustion caused by a spark spreads all over the burner, which eventually stabilizes at the intersection of the PZ and the CZ. The heat generation due to the combustion process within the CZ gives in rise to the convective heat transfer from the gaseous-phase to the solid matrix, leading to augmentation of solid-phase temperature. Owing to the improved heat recirculation from the CZ to the PZ via conduction and radiation heat transfer, the solid-phase temperature (T_s) increases, which consequently elevates the unburnt gaseous fuel-air mixture temperature (T_g), as shown in Fig. 2.

Babkin et al. [5] provided the criterion for flame stabilization inside a PRB as a function of Peclet number, *Pe* (ratio of heat flow by convection transport to heat flow by conduction), based on mean pore diameter. They proposed the following limiting condition for the flame propagation

$$Pe \geq 65 \quad \text{where,} \quad Pe = \frac{S_L d_m c_p \rho}{k} = Re_L \, \mathrm{Pr} \tag{1}$$

where S_L is the laminar flame speed, d_m is the mean pore diameter, c_p is the specific heat, ρ is the density and k is the thermal conductivity of the gas. For a stable operation, a double-layered PRB should be designed in such a way that the Pe for the preheating zone should be less than 65, and the same for combustion zone should be greater than or equal to 65. The optimization of porosities as well as the thermal conductivities of the two zones gives rise to a better heat utilization and temperature homogenization. Under certain conditions, the heat recirculation gives rise to a kind of flame called excess enthalpy flame. Nowadays, the concept of excess enthalpy combustion is being used in many PMC devices.

2.3 Excess Enthalpy Combustion

Weinberg [6] first introduced the concept of excess enthalpy. He reported that minimization of heat loss from a combustion device to the surrounding through the heat recirculation from exhaust gases generates a kind of flame called "excess enthalpy flame". The combustion, in which the excess enthalpy flame generates, is called "excess enthalpy combustion" or "super adiabatic combustion" (SAC). The typical characteristic of an excess enthalpy flame is that its maximum temperature is higher than the corresponding adiabatic flame temperature.

Weinberg [6] proposed several recuperative schemes, in which heat from exhaust could be extracted and used for preheating the incoming reactants. Among all the schemes, the one with internal heat recirculation appeared to be very promising as it led to the generation of excess enthalpy flame. He listed some of the benefits of excess enthalpy flame such as high combustion and thermodynamic efficiencies, lower emission of pollutants. Hardesty and Weinberg [7] found that owing to a high heat feedback, for a given equivalence ratio, at high gas velocities, the peak temperature of the combustible mixture could be increased beyond the adiabatic flame temperature. At high gas velocities, the reaction zone is also found to widen.

2.4 *Advantages of PMC*

The major advantages of PMC are as follows:

- It has a good heat transfer properties and keeps the burner's surface temperature lower, and, as a result, reduces the formation of NO_x emissions and due to increase in residence time, the formation of CO emission is low [8].
- A portion of combustion heat is transported to the upstream section which preheats the incoming fresh mixture leading to greater flame speeds [9, 10].
- The burner can be operated on a wide range of power modulation with multifuel options [11].
- This method improves the heat transfer process due to good radiation and conduction properties of porous medium [9–11].

2.5 *Materials Used in PMC*

Durst and Trimis [10] investigated different porous structures viz. ceramic pebbles, ceramic foams, metal foils and wires for use in the PMC. They reported that the heat transport properties of the porous medium depend on the form of the structure and the foam material. Metallic materials are found less suitable for the PMC since they are thermally unstable.

Ceramics (shown in Fig. 3) are the most appropriate materials for PMC applications as they offer high operating temperatures, resistance to erosion and wear, chemical stability and thermal shock [11]. The most common ceramics used for PMC are alumina (Al_2O_3), silicon carbide (SiC) and zirconia (ZrO_2) and their respective thermo-physical properties are summarized in Table 1. Al_2O_3 (Alumina) is the utmost common porous media, employed both in a packed bed and as a lamella structure. It has a high capability to temperature resistance and has good

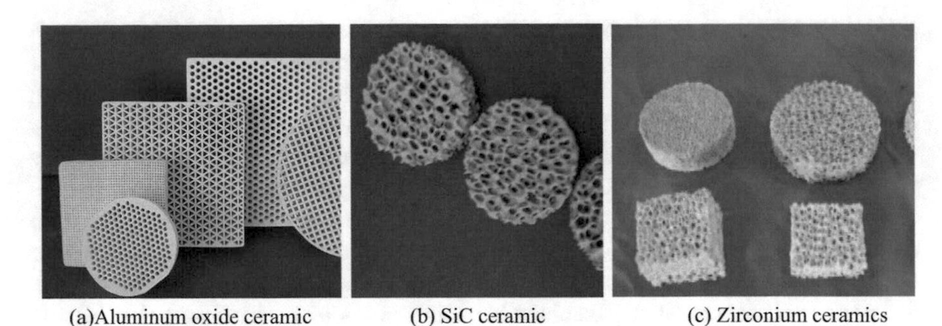

(a)Aluminum oxide ceramic (b) SiC ceramic (c) Zirconium ceramics

Fig. 3 Different types of ceramics

Table 1 Thermo-physical properties of most commonly used ceramics for PMC [4]

Parameter	Units	Al_2O_3	SiC	ZrO_2
Thermal expansion coefficient α (20–1000 °C)	10^{-6} 1/K	8	4–5	10–13
Thermal conductivity λ at 20 °C	W/m K	20–30	80–150	2–5
Thermal conductivity λ at 1000 °C	W/m K	5–6	20–50	2–4
Specific thermal capacity	J/g K	0.9–1	0.7–0.8	0.5–0.6
Thermal stress resistance parameter	10^{-3} W/m	3	23	1
Total emissivity at 2000 K	–	0.28	0.9	0.31

resistance to wear and erosion, along with cost-effective too. Ceramics having higher silica contents will have typically moderate operating temperatures and higher thermal conductivities. SiC, SiSiC (silicon infiltrated silicon carbide) and ceramics oxidize are commonly used for combustion zone in two-layer PRB. In comparison with alumina, they have high thermal conductivity, emissivity and lower coefficient of thermal expansion. Whereas, zirconia-based ceramics have a very high operating temperature but a low thermal conductivity, high coefficient of thermal expansion, and reasonable thermal shock resistance and emissivity.

3 Applications of PRB

Owning to numerous benefits, the applications of the PRBs are widespread. They are used both in industrial and domestic sectors. Recently, some researchers explored the use of PRB in domestic applications such as household water heating, cooking and lightning. Gas-fired PRBs are used in a number of industrial applications such as paper drying, paper finishing, powder and paint curing, baking, textile drying. They are also used in IC engines, gas turbine combustion chamber, steam generator and electricity generation. Some of the important applications of the PRB are discussed in the following sections.

3.1 Domestic Applications

The old conventional devices based on free-flame used globally by millions of people for different domestic purposes are causing severe indoor pollution resulting in many serious health problems. To overcome this problem, many researchers have explored the use of PRB based on PMC technology in cooking, household heating and water heating applications.

Jugjai and Rungsimuntuchart [12] applied the idea of the PMC in a LPG stove for improving the efficiency. In doing so, they proposed a new concept named as semi-confined porous radiant re-circulated burner (PRRB) where the primary air was preheated to higher temperatures than CB. With the former, they could achieve 12% higher thermal efficiency than the 30% for the CB, and with the latter, it was just double the CB. This led to an energy saving of around 50%. The NOx and CO emissions were found significantly low in the PRRB than CB. Qui and Hayden [13] designed and fabricated a PRB made of fiber felt along with a recuperator using natural gas. The main objective was to study the performance of the existing gas-fired lanterns equipped with porous medium.

Pantangi et al. [14, 15] developed a double-layered PRB which comprises of SiC foam as combustion zone and Al_2O_3 balls as preheating zone and tested the thermal efficiencies of the burners working at a different firing rate and equivalence ratios. They concluded that the efficiency was a function of equivalence ratio and the burner diameter. Muthukumar et al. [16] demonstrated that the efficiency of the stove could be further improved to 71% by using a ceramic block instead of Al_2O_3 balls in the PZ. In the same development [16], the effect of CZ porosity was investigated by Muthukumar and Shyamkumar [17], and they reported that at 90% porosity, the thermal efficiency of the burner was 75% and emissions were also lower.

Mishra et al. [18] developed medium-scale (5–10 kW) LPG cooking stove with two-layer PRB, and reported that the burner produces stable and flameless combustion for the equivalence ratio range of 0.54–0.72. The maximum improvement of thermal efficiency was found to be 28% higher than the available CB for the same input thermal load of 5 kW. In the kerosene stove, Sharma et al. [19] reported the optimal thickness of the combustion zone PM to be 20 mm to have better emissions and higher efficiencies than the usual thermal efficiencies of the stoves existing in the market. The PRBs developed by Pantagi et al. [14, 15], Muthukumar et al. [16, 17], Mishra et al. [18] and Sharma et al. [19] required external air supply for their operation.

Towards providing solutions to the limitations of the aforementioned PRB-integrated cooking burners, Mishra and co-researchers [20–22] developed a self-aspirated LPG and kerosene cooking stoves with PRB, which worked on natural draft eliminating any external appliances such as compressor or blower for air entrainment. With the use of self-aspirated PRB cooking burners instead of the conventional stoves available in the Indian market, the maximum increment of thermal efficiency was found up to 14% for thermal load input of 1–3 kW and up to

26% for 5–15 kW, respectively. While for kerosene-fired PRB stove, efficiency was found up to 8% higher than their conventional pressure stove counterparts, for power input in the range of 1.5–3 kW [22].

Mujeebu et al. [23–25] compared the thermal efficiencies of the burners operating with surface combustion and submerged combustion mode and stated that the surface combustion caused in better thermal performance. Yoksenakul and Jugjai [26] developed a self-aspirating, porous medium burner (SPMB) with the packed bed of alumina spheres. They used the same mixing tube and the same fuel nozzle as used in CB. Burner performance and combustion characteristics of SPMB were done for of CO and NO_x emissions at the burner exit and radiation efficiency. Avdic [3] used the concept of PMC in household heating and water heating purposes. Qiu and Hayden [13] developed a unique ceramic fibre felt for lighting application.

To summarize, it can be stated that burners with high efficiency have been developed over the last two decades. There has been considerable improvement in the design of the conventional burners with integration with the burner works on PMC technology which improves the efficiency and reduces the emissions significantly.

3.2 Industrial Applications

Many industrial burners based on old combustion devices are causing severe outdoor pollution affecting the environment in numerous ways like depletion in ozone layer, global warming. It is also affecting the human health causing several respiratory and chronic diseases. Impressed from the aids of PMC particularly for low emission, many investigators have explored the various burner's designs for different applications.

Tanaka et al. [27] explored the possible application of the PRB in the combustor of the second stage of the chemical gas turbines developed by Arai et al. [28]. The aim of their investigation was to utilize the unused chemical energy of the exhaust in the second-stage combustor equipped with porous medium, so that the flammability limits would widen and thermal efficiency be improved. From their results, they concluded that use of the porous medium in the chemical gas turbine would be a good choice. Delalic et al. [29] successfully used PM heat exchangers as an integral component in the low temperature (condensate) boilers. The results showed a better heat exchange efficiency and combustion stability. The thermal efficiency was found to be higher than 95% for both 5 and 9 kW burners at different excess air ratios. NO_x and CO_2 emissions were low due to improved energy efficiency. The system was found to be applicable for central heating system also.

Mjaanes et al. [30] experimentally investigated the possible use of the PMC as a reformer in the fuel cell. They used PRBs of two types: double-layered alumina foam and double-layered alumina beads to convert methanol, methane, octane and automotive grade petrol for the generation of hydrogen which is used in the fuel cell. The experiments showed that the alumina beads had long lifetime than foams and the conversion efficiency of the burner was also high. Raviraj and Janrt [31] did an experimental study with an aim of achieving higher conversion efficiency along with the parameters affecting the conversion efficiency. They could achieve 65% net conversion efficiency for a particular equivalence ratio and inlet velocity of the methane. For better conversion efficiency, they suggested low specific heat and low thermal conductivity porous media.

Liu and Hsieh [32] conducted experiments with PRBs using LPG under steady-state and transient conditions. In their experiments, they introduced cooling tubes in the post-flame zone to recover the maximum heat used to preheat the incoming air-fuel mixture unlike conventional PMC. The heating of working fluid passing through the cooling tubes has found many applications such as in water and air heaters, boilers and chemical processes. Durst and Weclas [33] first applied PRB technology to the diesel engines. Weclas [34] explored the application of the PMC in IC engines, and he found improvements in many respect. The main benefit was the reduction of NO_x and elimination of soot. The combustion was found homogeneous and flameless. The investigations showed encouraging results.

The idea of combined heat and power with the PRB was put forward by Echigo et al. [35], and a system was developed by Hunt et al. [36] for hybrid electrical vehicles. Recently, few investigators have used micro- and meso-scale applications in PMC technology [37, 38]. Dobrego et al. [39–41] made a remarkable contribution by removing highly polluting volatile organic compounds such as formaldehyde, benzole, phenol, acetone through oxidation in porous radiant burner. Ismail et al. [42] investigated the application of PRB with micro-cogeneration system. They showed that the PRB can be implemented for heating and cell phone charging. Welch [43] developed a PRB for the improved combustion of gas and other inflammable vapours. Mitchell [44] patented the burner for the combustion of mineral oils in furnaces. Ruby [45] developed a PRB for the application of steam generators, and Hays [46] developed a PRB for surface combustor applications.

Combustion in porous media has also been investigated for oil and gas industry. Sanmiguel et al. [47] investigated experimentally on the various methods and relevant controlling mechanisms related with gas phase combustion in porous media, particularly at elevated pressures. The feasibility of burning solid fine powder inside porous media has been numerically investigated by Kayal and Chakravarty [48]. They investigated the effects of absorption coefficient, emissivity of medium, flame position and reaction enthalpy flux on radiative energy output efficiency. Wawrezenik et al. [49] used the PMC for the HCl synthesis from H_2 and Cl_2. In a concise way, it can be stated that various burners with high efficiency have been developed over last two decades. In the near future, the PRB will be the most potential ingredient for combined heat and power generation applications.

4 Environmental Aspects of PMC as Clean Combustion

Global warming and associated climate change due to the pollutants generated from combustion have become a serious concern [50]. PRBs based on PMC technology show good emission characteristics and offer higher thermal efficiencies. Owing to the preheating effect, the heat release in a PRB per unit area is increased and this leads to lower emissions of unburnt hydrocarbon (UHC). Further, because of wider reaction zone, lower surface temperature of the burner and homogenization of temperature within the PRB, the NO_x formation is reduced. Moreover, the longer residence time of the hot flue gases inside the PM leads to reduction in CO emissions.

Temperature and residence time are greatly dependent on the operating parameters such as, equivalence ratio, firing rate and also on the physical properties of the PM, which ultimately affect the CO and NO_x formation [51]. There have been several studies where the variations of NO_x or CO emissions as a function of firing rate, equivalence ratio, residence time etc. have been reported [22, 50]. Recently, Panigrahy et al. [52, 53] conducted both experimental and numerical investigation on LPG-fired PRB-integrated cooking stove for different equivalence ratios and thermal load inputs. They observed that to meet the accepted indoor air quality regulation prescribed by World Health Organisation, the cooking stove had to be operated below equivalence ratio of 0.4.

In order to improve the emission features of the existing cooking stoves, Mishra et al. [20, 21] used PRB in the domestic LPG burners (1–3 kW) and medium-scale cooking burners (5–15 kW). Due to improved combustion efficiency and homogeneous burner temperature, CO emissions of the PRB for domestic scale were found to be in the range of 30–140 ppm and NO_x emissions were measured in the range of 0.2–3.5 ppm, which were much lower than the emissions generated from the conventional domestic cooking burners. The CO emissions of the PRB for medium-scale cooking stove were found in the range of 60–190 ppm and NO_x emissions were in the range of 2–10 ppm.

Xiong et al. [54] used 60 kW bench-scales PM combustor heaters with two rows of water-cooled tube coils and reported ultra-low emissions. NO_x and CO were found less than 15 ppm, and total hydrocarbon (THC) was less than 3 ppm. Mital et al. [55] studied the emission characteristics of CO, HC and NO_x for PMC. The CO and HC pollution indices were observed to be relatively low, CO: 0.1–3.6 g/kg and HC: 0.1–1.2 g/kg. With decrease in equivalence ratio and increase in the firing rate, the CO and HC emissions were reduced. Whereas, NO_x emission was found to decrease with decrease in equivalence ratio, but increased with firing rate. The decrease in emission was due to higher heat release rate per unit area and an increase in preheating temperature of unburned fuel-air mixture at the higher firing rates. Scribano et al. [56] studied the performance of self-recuperative radiant tube burner with natural gas operated in non-premixed condition. The author observed that the burner was able to reduce the NO_x emission by 50% for a wide range of power 12–18 kW and equivalence ratio of 0.5–0.8. In addition, CO emission was also found very low (<50 mg/Nm^3 3% O_2). It is concluded from this section that lower emissions have been reported by many researchers by operating the burner with excess air and optimizing the thickness and porosity of PRB.

5 Conclusions

In order to overcome the undesirable features such as low power density, low power modulation range, low combustion efficiency and high level of pollutant emissions of conventional burners based on free-flame combustion, the need for PMC which addresses the above problems has been discussed. Towards the understanding of basic principle of the PMC, a brief history of research and developments of porous radiant burners in various fields of applications are presented. The PMC has been proved as one of the realistic options to overcome the environmental pollution problem to a great extent. The major conclusions and future research directions are summarized below:

- Lower emissions can be easily achieved by operating the burner with excess air and optimizing the thickness and porosity of PRB.
- SiC, ZrO_2 and Al_2O_3 are suitable for the fabrication of PRB. However, the lifespan of these materials is yet to be tested.
- Investigations on PRB are yet to be performed with non-premixed combustion, and their outcome with respect to stability and efficiency is to be also explored.
- Critical reviews on various applications of PRB have been carried out, and it has been found that among the gaseous fuels only LPG has received great attention while the development of biogas or natural gas stoves with PRB is limited. A very few study has been done with liquid fuel used in PRB, so there is a need to focus on vapourization technique of the liquid fuel.
- More research focus on micro- and meso-scale applications on PMC technology should be given. As, this will enable in commercialization of combustion systems used in micro-electro-mechanical applications.
- Applications of PMC technology with their low emissions characteristics will help in resolving present and future concerns of environmental pollutant emission regulations.

References

1. Turns SR (2000) An introduction to combustion: concepts and applications, 2nd edn. McGraw-Hill, New York
2. Basu P, Kefa C, Jestin L (2000) Boilers and burners: design and theory. Springer, New York
3. Avdic F (2004) Application of the porous medium gas combustion technique to household heating systems with additional energy sources. Ph.D. thesis, University of Erlangen-Nuremberg

4. Pickenacker O, Pickenacker K, Wawrzinek K, Trimis D, Pritzkow WEC, Muller C (1999) Innovative ceramic materials for porous-medium burners. Interceram, Germany 48:424–433
5. Babkin VS, Korzhavin AA, Bunaev VA (1991) Propagation of premixed gaseous explosion flames in porous media. Combust Flame 87:182–190
6. Weinberg FJ (1971) Combustion temperatures: the future. Nature 233:239–241
7. Hardesty DR, Weinberg FJ (1974) Burners producing large excess enthalpies. Combust Sci Technol 8:201–214
8. Keramiotis C, Stelzner B, Trimis D, Founti M (2012) Porous burners for low emission combustion: an experimental investigation. Energy 45:213–219
9. Sathe SB, Peck RE, Tong TW (1990) Flame stabilization and multimode heat transfer in inert porous media: a numerical study. Combust Sci Technol 70:93–109
10. Durst F, Trimis D (2002) Combustion by free flames versus combustion reactors. In: 16th International conference on clean air and environ (Clean Air 2002), vol 3. Christchurch, New Zealand, 19–22 Aug 2002, pp 1–20
11. Wood S, Harris AT (2008) Porous burners for lean-burn applications. Prog Energy Combust Sci 34:667–684
12. Jugjai S, Rungsimuntuchart N (2002) High efficiency heat-recirculating domestic gas burners. Exp Therm Fluid Sci 26:581–592
13. Qui K, Hayden ACS (2006) Premixed gas combustion stabilized in fiber felt and its application to a novel radiant burner. Fuel 85:1094–1100
14. Pantangi VK, Mishra SC, Muthukumar P, Reddy R (2011) Studies on porous radiant burners for LPG cooking applications. Energy 36:6074–6080
15. Mishra SC, Muthukumar P, Pantangi VK (2013) Porous Radiant burner for domestic LPG cooking device with improved thermal efficiency and reduced emissions of CO and NO_x. Patent Application No: 73/KOL/2013
16. Muthukumar P, Anand P, Sachdeva P (2011) Performance analysis of porous radiant burners used in LPG cooking stove. Int J Energy Environ 2:367–374
17. Muthukumar P, Shyamkumar P (2013) Development of novel porous radiant burners for LPG cooking applications. Fuel 112:562–566
18. Mishra NK, Mishra SC, Muthukumar P (2015) Performance characterization of a medium-scale liquefied petroleum gas cooking stove with a two-layer porous radiant burner. Appl Therm Eng 89:44–50
19. Sharma M, Mahanta P, Mishra SC (2016) Usability of porous burner in kerosene pressure stove: an experimental investigation aided by energy and exergy analyses. Energy 103:251–260
20. Mishra SC, Muthukumar P, Mishra NK (2015) Self-aspirated LPG domestic cooking stove with a two-layer porous radiant burner. Patent No: 543/KOL/2015
21. Mishra SC, Muthukumar P, Mishra NK, Panigrahi S (2015) Medium-scale self-aspirated improved air entrainment LPG Cooking stove with a two-layer porous radiant burner. Patent Application No: 201631037245
22. Mishra SC, Muthukumar P, Sinha GS, Sharma M, Mishra NK (2016) Self aspirated pressurized kerosene cooking stove with a Porous Radiant Burner. Patent Application No: 201631037245
23. Mujeebu MA, Abdullah M, Mohamad A (2011) Development of energy efficient porous medium burners on surface and submerged combustion modes. Energy 36:5132–5139
24. Mujeebu MA, Abdullah MZ, Bakar MZA, Mohamad AA (2011) A mesoscale premixed LPG burner with surface combustion in porous ceramic foam. Energy Sources Part A 34:9–18
25. Mujeebu MA, Abdullah MZ, Zuber M (2013) Experiment and simulation to develop clean porous medium surface combustor using LPG. J Ther Sci Technol 33:55–61
26. Yoksenakul W, Jugjai S (2011) Design and development of a SPMB (self-aspirating, porous medium burner) with a submerged flame. Energy 36:3092–4000
27. Tanaka R, Shinoda M, Arai N (2001) Combustion characteristics of a heat recirculating ceramic burner using low-calorific fuel. Energy Convers Manage 42:1897–1907

28. Arai N, Shinoda M, Churchill SW (1999) The characteristics of heat recirculating burner. Trans CSME 23:147–158
29. Delalic N, Mulahasanovic DZ, Ganic EN (2004) Porous media compact heat exchanger unit —experiment and analysis. Exp Therm Fluid Sci 28:185–192
30. Mjaanes HP, Chan L, Mastorakos E (2005) Hydrogen production from rich combustion in porous media. Int J Hydrogen Energy 30:579–592
31. Raviraj SD, Janrt LE (2006) Numerical and experimental study of the conversion of methane to hydrogen in a porous medium reactor, combust. Flame 144:698–709
32. Liu JF, Hsieh WH (2004) Experimental investigation of combustion in porous heating burners. Combust Flame 138:295–303
33. Durst F, Weclas M (2001) A new type of internal combustion engine based on the porous-medium combustion technique. J Automobile Eng IMechE Part D 215:63–81
34. Weclas M (2005) Porous media in internal combustion engines, Cellular ceramics-structure, manufacturing, properties and applications. Wiley-VCH-Publication
35. Echigo R, Yoshida H, Tawata, H, Tada S (1993) In: 12th international conference on thermoelectrics, Yokohama, Japan, 9–11 Nov
36. Hunt TK, F, Sievers RK (1994) AMTEC auxiliary power unit for hybrid electric vehicles. In: Proceedings 29th intersociety energy conversion engineering conference, Monterey, U.S.A, 7–11 Aug 1994
37. Marbach TL, Agrawal AK (2006) Heat-recirculating combustor using porous inert media for meso-scale applications. J Propul Power 22:145–150
38. Sadasivuni V, Agrawal AK (2009) A novel meso-scale combustion system for operation with liquid fuels. Proc Combust Inst 32:3155–3162
39. Dobrego KV, Gnezdilov NN, Kozlov IM, Bubnovich VI, Gonzalez HA (2005) Numerical investigation of the new regenerator–recuperator scheme of VOC oxidizer. Int J Heat Mass Transf 48:4695–4703
40. Dobrego KV, Gnezdilov NN, Kozlov IM, Shmelev ES (2006) Numerical study and optimization of the porous media VOC oxidizer with electric heating elements. Int J Heat Mass Transf 49:1–10
41. Dobrego KV, Gnezdilov NN, Kozlov IM (2007) Parametric study of recuperative VOC oxidation reactor with porous media. Int J Heat Mass Transf 50:2787–2794
42. Ismail KA, Abdullah MZ, Zubair M, Ahmad ZA, Jamaludin AR, Mustafa KF, Abdullah MN (2013) Application of porous medium burner with micro cogeneration system. Energy 50:131–142
43. Welch W (1890) Improvements in burners for the use of gas and other inflammable vapors. British Patent No. 5293
44. Mitchell A (1898) Improvements in furnaces or grates for the consumption of mineral oils. British Patent No. 7078
45. Ruby CF (1902) Steam-generator. US Patent No. 737279
46. Hays JW (1933) Surface combustion process. US Patent No. 2095065
47. Sanmiguel JE, Mehta SA, Moore RG (2003) An experimental study of controlled gas phase combustion in porous media for enhanced recovery of oil and gas. ASME Trans 125
48. Kayal TK, Chakravarty M (2007) Combustion of suspended fine solid fuel in air inside inert porous medium: a heat transfer analysis. Int J Heat Mass Transf 50:3359–3365
49. Wawrzenik K, Kesting A, Kunzel J, Pickenäcker K, Pickenäcker O, Trimis D (2001) Experimental and numerical study of applicability of porous combustors for HCl synthesis. Catal Today 69:393–397
50. Hansen J, Sato M (2016) Regional climate change and national responsibilities. Environ Res Lett 11:1–9
51. Trimis D, Durst F (1996) Combustion in a porous medium-advances and applications. Combust Sci Technol 121:153–168
52. Panigrahy S, Mishra NK, Mishra SC, Muthukumar P (2016) Numerical and experimental analyses of LPG (liquefied petroleum gas) combustion in a domestic cooking stove with a porous radiant burner. Energy 95:404–414

53. Panigrahy S, Mishra SC (2016) Analysis of combustion of liquefied petroleum gas in a porous radiant burner. Int J Heat Mass Transf 95:488–498
54. Xiong TY, Mark JK, FF Fish (1995) Experimental study of a high-efficiency, low emission porous matrix combustor-heater. Fuel 74:1641–1647
55. Mital R, Gore JP, Viskanta R (1997) A study of the structure of submerged reaction in porous ceramic radiant burners. Combust Flame 11:175–184
56. Scribano G, Solero G, Coghe A (2006) Pollutant emissions reduction and performance optimization of an industrial radiant tube burner. Exp Therm Fluid Sci 30:605–612

An Overview of Current Knowledge Concerning the Environmental Consequences of the Nuclear Pollution: Sources, Effects and Control

S. K. Verma, S. L. Sinha and D. K. Chandraker

Abstract Nuclear power raises a number of fundamental environmental issues. The main problem is how to deal with the quantities of highly radioactive wastes which are produced from nuclear power plants. Discharges from nuclear power plant can cause substantial climatic contamination danger and hazard for individuals' lives and well-being. In this chapter, different techniques for modelling and control of hazards have been presented. The modelling is in view of recreation and perception of spreading of air pollutants, estimation of the source term for atomic and compound fiascos, and the hazard appraisal of unsafe substances. This chapter will include the principle of modelling the nuclear and chemical disasters, optimal control of theoretical frame with example, various modelling techniques, challenges associated with measurement of pollutants, etc. Finally, solution and recommendation of good model will be presented. The inclusion of related references provides a starting point for the interested reader/researchers/industrialists.

Keywords Atmospheric pollution hazard · Modelling · Concentration
Nuclear · Chemical disaster

1 Introduction

Atomic power plants (APPs) under common working condition liberate little quantities of radioactive effluents during their lifetime [1]. The significance of crisis vigilance has been accepted in the last two decades, mainly after the Chernobyl accident. In the improbable incident of an accident like loss of coolant accident

S. K. Verma (✉) · S. L. Sinha
Mechanical Engineering Department, National Institute of Technology,
Raipur 492010, Chhattisgarh, India
e-mail: mailme.shashi09@gmail.com

D. K. Chandraker
Reactor Design and Development Group, Bhabha Atomic Research Centre,
Hall-7, Trombay, Mumbai 400085, Maharashtra, India

N. Sharma et al. (eds.), *Air Pollution and Control*, Energy, Environment,
and Sustainability, https://doi.org/10.1007/978-981-10-7185-0_13

(LOCA) or the successive failure of physical barriers, can consequence in the release of radioactive material. During the nuclear emergency, the accurate evaluation of the pollutant dispersion and the dose received by the public are vital to measure the potential health risk. For the accurate prediction of nuclear pollutants, CFD is becoming increasingly popular. It offers a huge potential as compared to existing models. For calculating the concentrations at different locations, dispersion modelling uses mathematical equations. This equation relates the environment dispersion and chemical and physical processes within the plume. The pollutant dispersion can be studied with the help of full-scale field measurements, wind tunnel experiments and computational fluid dynamics (CFD). For field measurement situations, it is tough to concurrently control operative and intertwined parameters' effects such as atmospheric conditions (wind speed, wind direction), topography and geography (underlying surface roughness, mountain height, and width, shape). The need to utilize hazardous radioactive sources and also the trouble in creating suitable boundary conditions' similarity may limit the efficiency of wind tunnel experiments. The laboratory experiments need time and acquire costs for predicting plume dispersion. On the other hand, with the fast expansion of computer technology and technological advancements, the numerical simulation technique is regarded as an efficient tool for accurately predicting the plume dispersion behaviour. However, CFD works fine for these circumstances. CFD has been proved to be an extremely powerful and efficient tool for the studies of radionuclide's dispersion with the factors considered individually or in combination with the wind field effect [2]. Several previous studies have involved simulations of the atmospheric dispersion of nuclear power plant (NPP) emissions from the Chernobyl and Fukushima accidents. Sources of nuclear pollution include both natural and man-made as shown in Fig. 1.

Causes of radioactive pollution are shown in Fig. 2.

The electricity generation based on nuclear fuel is one of the world's environmentally damaging activities. Emissions to the atmosphere have been the major centre of energy impact studies other considerable impacts such as land disorder and population dislocation jointly with their economic and social implications are less emphasized. Impacts of radioactive pollution are shown in Fig. 3.

Preventative and control measures to take for radioactive pollution are shown in Fig. 4.

The present chapter deals with the study of various modelling techniques and numerical model used by different researchers. Finally, some of the good numerical approach and their solutions will be presented in depth.

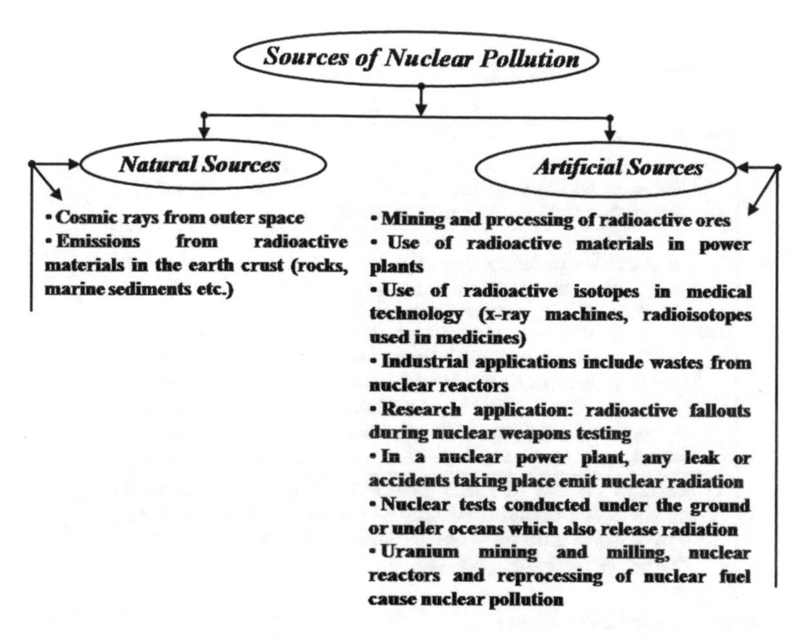

Fig. 1 Sources of nuclear pollution

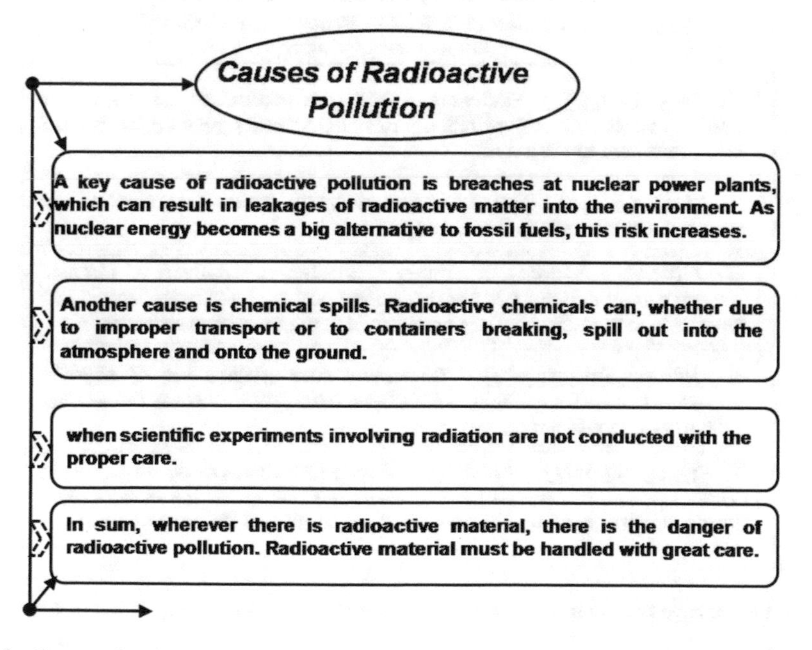

Fig. 2 Causes of radioactive pollution

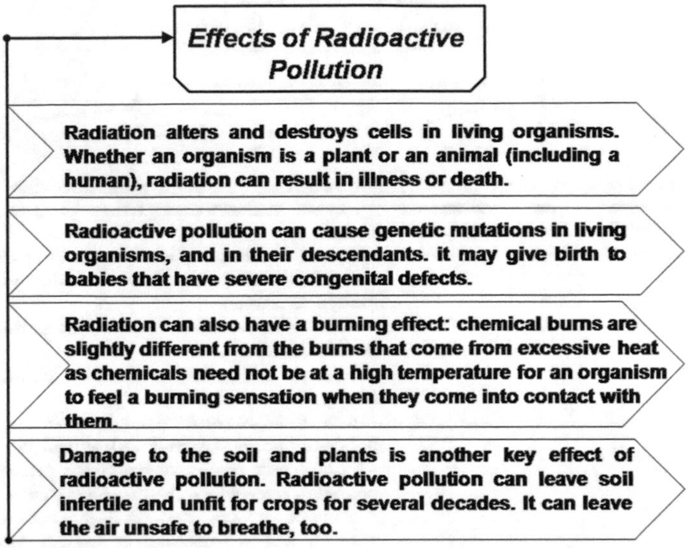

Effects of Radioactive Pollution

Radiation alters and destroys cells in living organisms. Whether an organism is a plant or an animal (including a human), radiation can result in illness or death.

Radioactive pollution can cause genetic mutations in living organisms, and in their descendants. it may give birth to babies that have severe congenital defects.

Radiation can also have a burning effect: chemical burns are slightly different from the burns that come from excessive heat as chemicals need not be at a high temperature for an organism to feel a burning sensation when they come into contact with them.

Damage to the soil and plants is another key effect of radioactive pollution. Radioactive pollution can leave soil infertile and unfit for crops for several decades. It can leave the air unsafe to breathe, too.

Fig. 3 Effects of radioactive pollution

Preventative and Control Measures to take for Radioactive Pollution

1. ***Safe Storage*** : Radioactive pollution needs to be stored in specialized containers, which do not break easily and which do not allow radiation to seep out.
2. ***Clear Warnings*** : Clear warnings about the presence of radioactive material will prevent this from happening.
3. ***Alternative Sources of Energy*** : Seeking alternatives to nuclear energy will reduce the number of nuclear power plants in the world and thus reduce the risks that are associated with radioactive material.
4. ***Nuclear Disarmament*** : Disarming and getting rid of nuclear weapons is a very significant way of eliminating the hazards associated with radioactive material.
5. ***Proper Disposal of Radioactive Waste*** : Radioactive waste needs to be disposed of in a particular way. It cannot just be buried in the soil or thrown into the garbage bin as it would contaminate the environment.

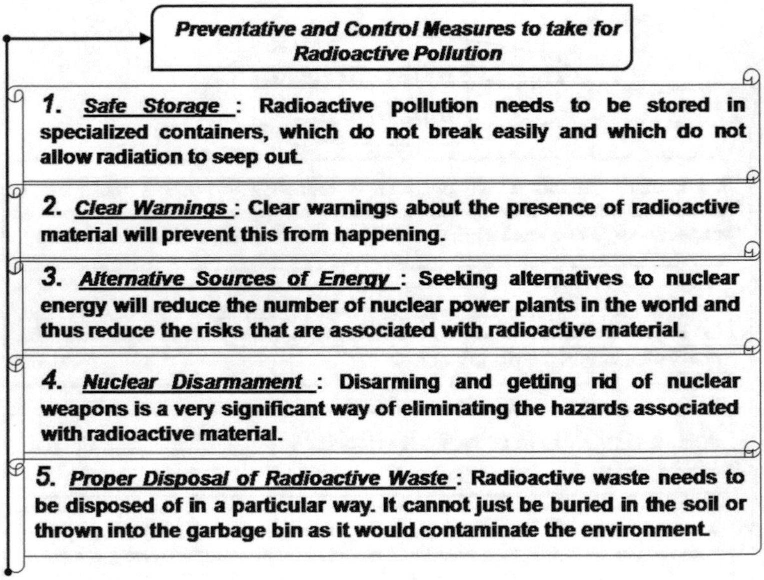

Fig. 4 Preventative and control measures for radioactive pollution

2 Air Pollution Models

The vast quantity of harmful substances may be released into the atmosphere due to nuclear disasters. These hazardous substances are transported and diffuse with the motion of air stream into the atmosphere. Liu and Huang [3] concluded that in order to accurately calculate the hazard of the disasters, a number of kinds of models like source term model, the meteorological field forecast model, the atmospheric dispersion model and the hazard evaluation model need to be investigated. A broad review of atmospheric dispersion methodology is specified by Holmes and Morawska [4].

2.1 Box Models

Box model uses the conservation laws for evaluating the mass balance of a given system. It is a simplest model in view of the information of protection of mass and preservation of energy. The model is capable to cover the complicated chemistry, and the treatment of transport is simplified. For the analysis inside a domain, it is assumed that the pollutant is homogeneous and air mass is well mixed. Figure 5 shows the evaluation of box model on the basis of different parameters. Boundaries of the domains are boxes, where particles of pollutant are transferred from one area

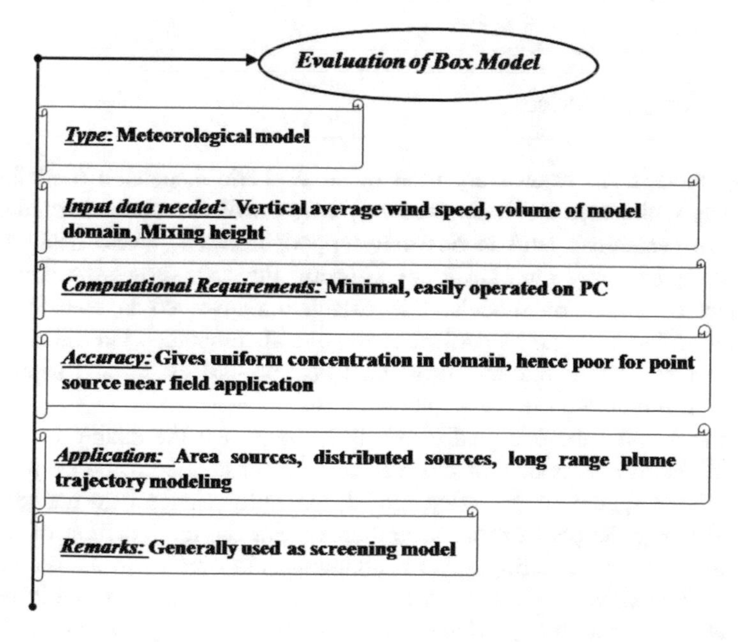

Fig. 5 Evaluation of box model on the basis of different parameters

of the environment to another. For each pollutant, the mass balance equation can be written as

Input rate of mass = Output rate of mass + Transformation rate of mass + Accumulation rate of mass

2.2 Lagrangian and Eulerian Models

The Lagrangian and Eulerian models explain the similar advection–diffusion equation. The dissimilarity among Lagrangian and Eulerian approaches to modelling consists in the dissimilar behaviour of the frame of reference. The Lagrangian approach depends on concentration of the property of a specific liquid by taking after its direction. Lagrangian models are like the crate models, where the district of air containing an underlying convergence of toxins is considered as a case [5]. The container is thought to be advected with the stream, and the model takes after the direction of the crate. Lagrangian models use wind field, buoyancy and turbulence effects for calculating the trajectories of air pollutants. The ordinary differential equations (ODEs) are used for calculating the trajectories of air pollutants as a substitute of using partial differential equations (PDEs) in the unique dispersion problem, which is computationally a simple task and avoids spatial truncation errors and numerical diffusion. Molecule scattering model is a case of reasonable execution of a Lagrangian display [6].

2.3 Gaussian Models

Gaussian models are extensively used in atmospheric dispersion modelling and have a very rapid response time, since they only calculate a single formula. These models are generally useful in decision support software, where healthy model set-up and quick reply time are a key concern. Figure 6 shows the comparative assessment of dispersion models. The calculation involved in these models is almost immediate and can be performed in general computers. On the other hand, some parameterizations like meteorological data pre-processing and sophisticated turbulence can enhance the computational cost.

Keeping in mind the end goal to precisely conjecture the danger of the catastrophes, different assortments of models must be constrained to be examined, the source term show, the air scattering model, the earth science field gauge demonstrate and thusly the peril examination display. For the reproduction of the close range atmospherical scattering, CFD is changing into even extra across the board [2, 7, 8]. Punitha et al. [8] solve the conservation of mass of a single pollutant species of concentration (C), which is given by the following equation.

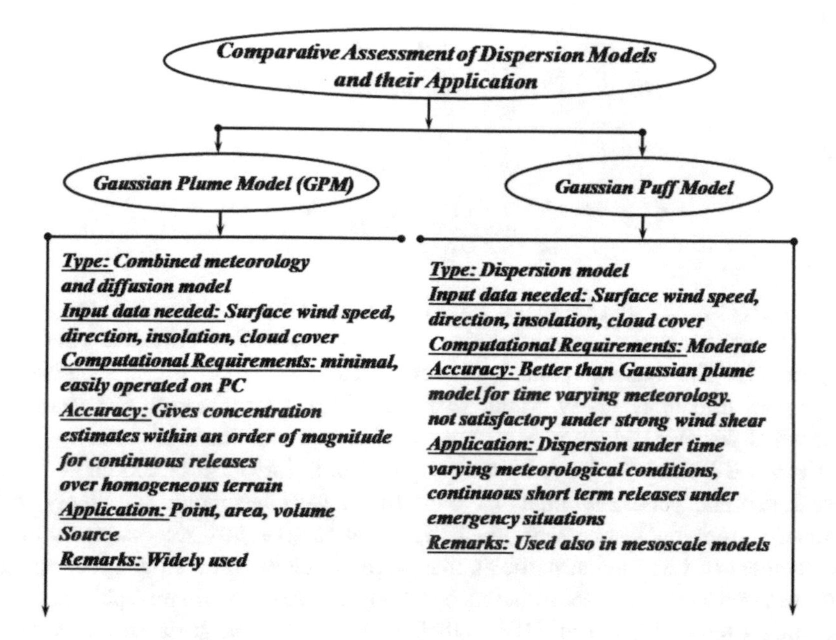

Fig. 6 Comparative assessment of dispersion model

$$\rho \sum_i \frac{\partial (u_i C)}{\partial x_i} = \rho \sum_i \frac{\partial}{\partial x_i}\left(D \frac{\partial C}{\partial x_i} \right)$$

$$D = D_{\mathrm{L}} + D_{\mathrm{T}}$$

where D_{L} and D_{T} denote laminar and turbulent diffusivity coefficients, respectively. The isotropic eddy viscosity parameter which is present in both the momentum and pollution transport equations was solved by standard k–ε turbulence model. In dispersion studies, they show strong correlations between ground-level concentration of sodium aerosols at various source elevations (5, 10, 20 and 30 m). Finally, concluded that the maximum ground-level concentration is impropriation to the source height.

Nakayama and Nagai [7] concluded that the numerical simulation model requires considerable CPU time statistics of concentrations. They used LES model as a practical matter for emergency responses. In numerical model, the equations for simulating air flow and material dispersion in the atmosphere are the continuity equation, the Navier–Stokes equation and the transport equation of concentration expressed, respectively, as

$$\frac{\partial u_i}{\partial x_i} = 0,$$

$$\frac{\partial u_i}{\partial t} + u_j \frac{\partial u_i}{\partial x_j} = -\frac{1}{\rho}\frac{\partial P}{\partial x_i} + \frac{\partial}{\partial x_j}\upsilon\left(\frac{\partial u_i}{\partial x_j} + \frac{\partial u_j}{\partial x_i}\right)$$

and

$$\frac{\partial c}{\partial t} + u_j \frac{\partial c}{\partial x_j} = \frac{\partial}{\partial x_j} D\left(\frac{\partial c}{\partial x_j}\right)$$

where u_i, t, P, ρ, υ, c and D are the wind velocity, the time, the pressure, the density, the kinematic viscosity, concentration and the molecular diffusion coefficient, respectively. The subscript i stands for coordinates (1: streamwise, 2: spanwise and 3: vertical direction). They found that the mean concentration decreases with downwind distance, which is similar to wind tunnel experimental data.

Vach and Duong [9] played out a progression of CFD reproductions to list the base focus and statement fields of uninvolved particles which are likely to be produced from nuclear power plant. The radioactive tuft was depicted as an arrangement of Lagrangian particles that were circulated on a slow speed foundation acquired abuse RANS turbulence displaying. An indistinguishable approach was trailed by Gallego et al. [10] in their endeavour to break down the voyaging separation of radioactive particles, radiated by a nuclear power plant. Regardless of the undeniable reality, no gamma measurements appraisal was performed in these investigations. Raza and Avila [11] outlined that the ensuing measurement from these particles will be processed by regarding each as some degree supply and by including the commitment from every molecule inside the area. Xie et al. [12] took after partner Eulerian approach together with a RANS reproduction of the twist field to mimic the scattering of ^{222}Rn released from a metal mine ventilation shaft. They consider the average wind convection, turbulent diffusion and radionuclide decay in their numerical model. The Eulerian equation representing the conservation of mass as applied to ^{222}Rn concentration was given by

$$\frac{\partial C}{\partial t} + u\frac{\partial C}{\partial x} = D\frac{\partial^2 C}{\partial x^2} - \lambda C + Q$$

where u denotes the velocity vector of the ^{222}Rn, C is the ^{222}Rn average concentration in the air, D is the effective diffusion coefficient of ^{222}Rn in air, Q is the source term, and the radon decay constant is expressed by λ. They got the correlation between radon concentration at different wind speed and as a function of distance downstream from the uranium mine shaft outlet.

After, the powerful rate to the overall population was approximate by increasing the following fixation with measurements change variables [13]. Duarte et al. [14] found the concentration of the radioactive plume by using the ANSYS CFX 14.0 software, which uses Reynolds-averaged Navier–Stokes (RANS) equations. They used the following Gaussian model for calculating the concentration of the radioactive plume in kg/m^3.

$$\langle C\rangle(x,y,z) = \frac{Q_{\mathrm{m}}}{2\pi\sigma_y\sigma_z u} \times \exp\left[-\frac{1}{2}\left(\frac{y}{\sigma_y}\right)^2\right] \times \left\{\exp\left[-\frac{1}{2}\left(\frac{z-H_{\mathrm{r}}}{\sigma_z}\right)^2\right] + \exp\left[-\frac{1}{2}\left(\frac{z+H_{\mathrm{r}}}{\sigma_z}\right)^2\right]\right\}$$

where Q_{m} is stationary source term (kg/s) at a height H_{r} above ground level, in *m*, with wind direction x and constant wind speed u in m/s, σ_y and σ_z are the Pasquill–Gifford dispersion coefficients for open field, which depend on the dispersion direction x. These coefficients are usually valid for distances in the range 10^2–10^4 m from the source.

Instead of RANS, de Sampaio et al. [3] utilized LES demonstrating and analysed the time development of the scattering of radionuclide's inside the area of nuclear vitality plants, outlining the significance of the local wonders on the scattering downside.

In physical model, de Sampaio et al. [3] defined that ϕ be the concentration of the first radionuclide and φ be the concentration of the second one. In that case, the transport equations for ϕ and φ can be written as

$$\frac{\partial\phi}{\partial t} + u_b\frac{\partial\phi}{\partial x_b} + \frac{\partial\xi_b}{\partial x_b} + \lambda\phi = 0$$

$$\frac{\partial\varphi}{\partial t} + u_b\frac{\partial\varphi}{\partial x_b} + \frac{\partial\eta_b}{\partial x_b} - \lambda\phi = 0$$

where λ indicates the rate of decay of the first radionuclide, u is the velocity. By the Fick's law, diffusive fluxes of ϕ and φ can be written as

$$\xi_b = -\psi\left(\frac{\partial\phi}{\partial x_b}\right) \quad \text{and} \quad \eta_b = -\zeta\left(\frac{\partial\varphi}{\partial x_b}\right)$$

where ψ and ζ are the corresponding molecular diffusivities. They found a correlation among the concentration of radionuclide and non-dimensional time for different chimney heights and distance down flow from the chimney. Fuka and Brechler [15] found a strong correlation between observed and calculated results. They finally developed a correlation between dispersion of pollutant from a point source in various positions with respect to the obstacle. Table 1 shows the significant footprints of the different modelling techniques and their findings.

Furthermore, Fuka and Brechler [15] and Nakayama et al. [16] used LES to mimic the scattering of hot matter. The most essential gifts of LES over RANS are its expanded precision and furthermore the undeniable truth that scattering as a result of turbulent swirls is best caught. Not one or the other, nonetheless, coupled the LES model to a gamma rate model to ascertain the radiological dosage from cloud sparkle.

Table 1 Important footprints of the different modelling techniques and their findings

S. No.	Published year	Researcher	Modelling techniques	Concluding remarks
1	1981	Kalmaz and Barbieri	Chemical species dispersion model	They discussed the key toxic radioactive chemicals concerned in the uranium recovery operations and presented computer simulation results of their hydrodynamic dispersion, migration and prediction of future material movements
2	1992	Kelly et al.	Gaussian plume dispersion model	Moss samples showed a region of better deposition of radioactive aerosol adjoining the factory, with a diameter of about 3 km The anticipated zone of improved ^{238}U fixations in the environmental airborne was around oval fit as a fiddle, ex-keeping an eye on a separation of 1.5–3 km from the production line
3	1996	J. Brandt et al.	Lagrangian meso-scale model and a long-range transport Eulerian model	The air contamination caused by a solitary, however, solid discharge source is examined. The consistency of the aggregate model is tried both by utilizing the outstanding turn test and by applying it to learn both the Chernobyl mischance and the two ETEX-discharges
4	2005	Talerko	Model of atmospheric transport LEDI (Lagrangian–Eulerian diffusion model)	The real elements of spatial and worldly varieties of radioactive sullying fields over the domain of Ukraine on the provincial scale, together with the development of the real substantial scale spots of radioactive pollution caused by dry and wet affidavit

(continued)

Table 1 (continued)

S. No.	Published year	Researcher	Modelling techniques	Concluding remarks
5	2007	Canepa et al.	SAFE_AIR (simulation of air pollution from emissions_above inhomogeneous regions)	Safe air view shows difficulties in simulating the plume rise in low wind conditions
6	2008	Sampaio et al.	Large eddy simulation	In order to elucidate the local effects, the physical model used includes the equations that govern the flow
7	2009	G. de	$k–\varepsilon$ turbulence model	Concentration levels for particles less than 1 μm are unaltered by the particle diameter
8	2013	Melo et al.	Pasquill–Gifford Gaussian model and CFD model	The application of computational fluid dynamics for atmospheric dispersion calculations enables a more practical simulation of ground conditions and velocity profile as compared to the Gaussian model
9	2014	Mészáros et al.	Gaussian, Lagrangian, Eulerian and CFD models	The additional development of dispersion models will shift towards multi-scale modelling approaches that can contain complicated parameterization of cloud physical processes and biogeochemical relations
10	2015	Ricciardi et al.	RANS simulations using SST $k–\omega$ turbulence model	This investigation demonstrates the capacity of the displaying approach that has been received (as far as computational space, lattice discretization, limit conditions and turbulence show) to agreeably anticipate wind-actuated weight impacts on an assortment of geometries of atomic structures and fireplace debilitates

(continued)

Table 1 (continued)

S. No.	Published year	Researcher	Modelling techniques	Concluding remarks
11	2015	A. Sadiq Aliyu et al.	AERMOD dispersion model which is based on the Gaussian plume model (GPM)	Under normal operation, the NPP does not pose any important public health and environmental impacts. However, accidental conditions considered by precipitation will lead to discernible radiological risks within the NPP sites emergency planning zone
12	2015	Vervecken et al.	Large eddy simulation (LES)	The turbulent inconsistency within the wind field will result in dose estimates that square measure underestimated by up to an element of 4 once standard semipermanent measurements are used to estimate the dose from short exposures

2.4 *Source Term Models*

Nuclear and chemical disasters were usually companied with explosion, air mass jet, liquid evaporation and such phenomena, and also the initial physical properties have composite options [17]. The physics and mechanical effects will influence the parameters of the harmful sources, like unharness rate, initial rise height, initial volume. These supply term parameters can have an effect on the dispersion of risky substances within the atmosphere [18]. By applying principles of explosion mechanics, fluid mechanics and physics, the supply term models for nuclear and chemical disasters were created to see supply term parameters (Fig. 7).

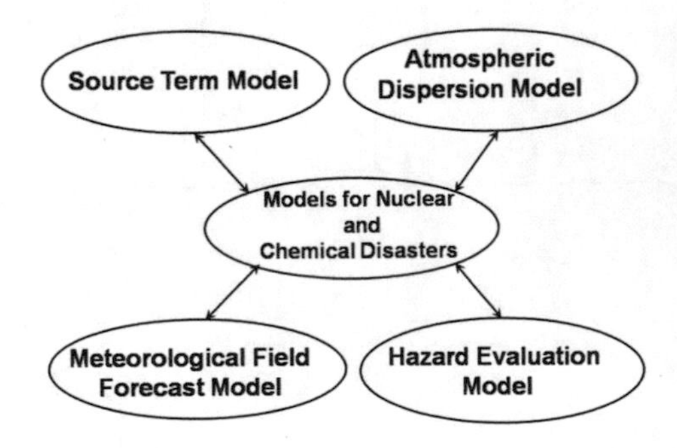

Fig. 7 Types of model for predicting the hazard of the disaster

2.5 Dispersion of Effluents in the Atmosphere

The effluents' discharge from the stack or at the ground level might be in the form of continuous or instantaneous plume/puff. The effluents are transported by wind and diffused by turbulence close to the atmosphere. The dispersion may be defined as a collective effect of transport and diffusion mechanism. The effluents may be in the form of noble gases or reactive vapours or particulates in case of nuclear facilities. Radioactive decay is a further form of decrease of radioactive effluent concentration. The revision of the discharge of radioactive effluents may be separated into four dissimilar areas as shown in Fig. 8.

Figure 9 shows the basic governing factors of dispersion.

2.6 Atmospheric Dispersion Modelling

Atmospheric dispersion modelling gives the information about dispersion of air pollutants in the ambient atmosphere with the help of numerical simulation. For solving equations, it uses the computer codes and then telling the circulation of pollutants. The preliminary situation may be the meteorological situation such as wind speed and direction, precipitation. The process circumstances (heat capacity of the plume, terrain roughness, etc.) are active in the atmospheric boundary layer [19]. In the case of radioactive pollutants, the output is recognized in expressions of

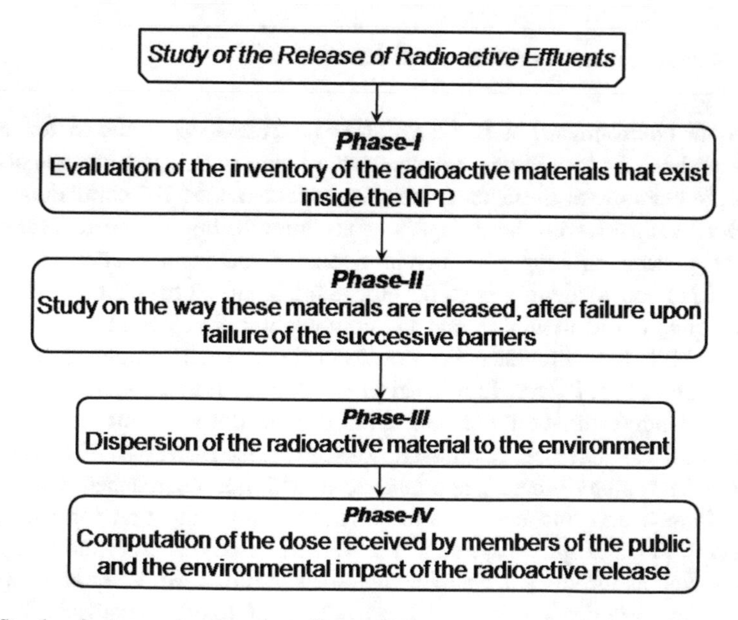

Fig. 8 Steady of release of radioactive effluents [19]

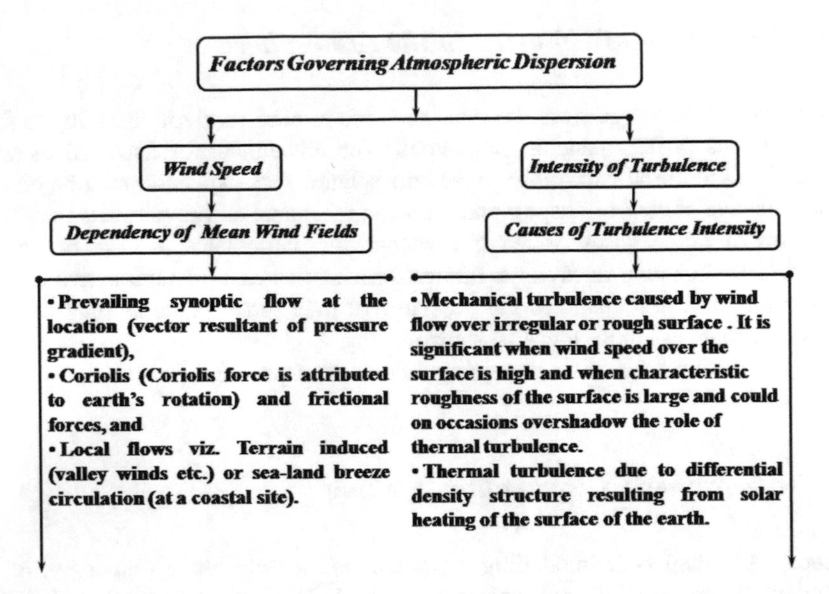

Fig. 9 Basic governing factors of dispersion [19]

activity concentration in air. From the source, the concentrations of pollutants are predicted in the downwind directions. Figure 10 shows the brief summary of atmospheric dispersion model [19].

2.7 *Issues, Controversies, Problems*

To save our environment, human lives and to decrease economic losses, computational models of air pollutant dispersion are being developed to appreciate and forecast the outcome of these phenomena and accidents. Model simulations should be quick and must have a high degree of accuracy to be used in real-time applications (e.g. decision support). Therefore, one of the main challenges of atmospheric dispersion modelling is to develop models and software that can provide numerical predictions in an accurate and computationally efficient way. Disaster at Chernobyl NPP has stimulated the development of such accidental release and decision support software like Real-time Online Decision Support System (RODOS). Underestimating the concentrations or doses of air pollutants (e.g. radionuclides) can have serious health consequences. However, in case of overestimation in regions where important dose will not be reached would waste valuable human and financial resources. The transport and transformation of air pollutants in the atmosphere are mainly governed by advection (wind field). Other processes such as turbulence, chemical reactions, radioactive decay and deposition can also play significant roles in the dispersion of toxic substances. Therefore,

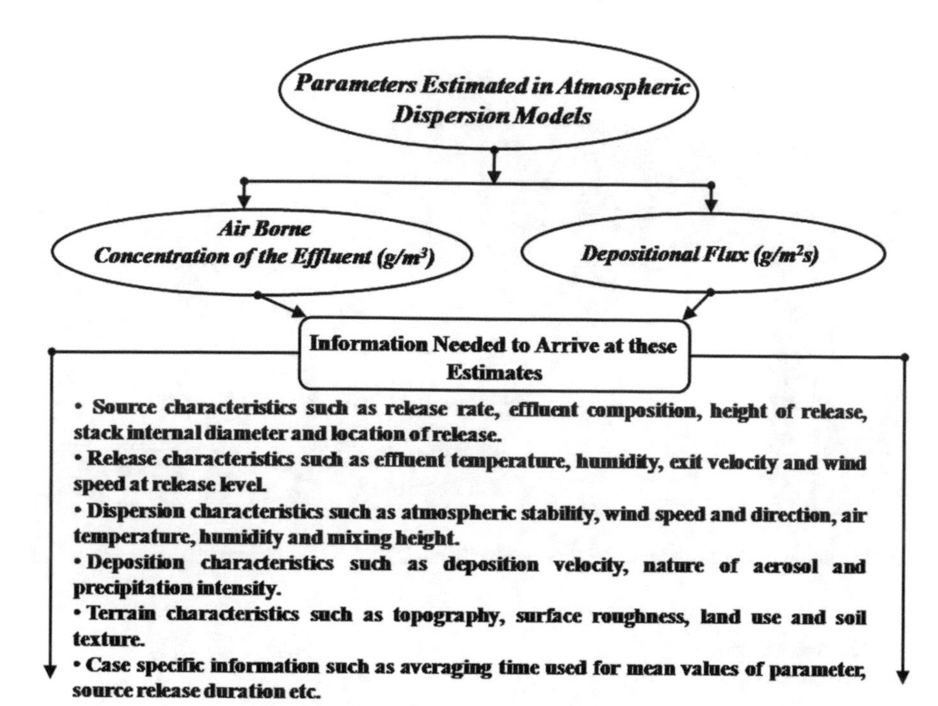

Fig. 10 Brief summary of atmospheric dispersion model [19]

strategies for model development require an interaction between researchers from different fields (e.g. meteorology, geophysics, chemistry, IT).

One more significant issue in modelling the spreading of air pollutants is the analysis of the lifetime of the chemical species and the feature distance that species can be transported, which is correlated with the lifetime. If the emitted species have short lifetime (minutes–hours) in the atmosphere (usually reactive species or aerosols), it cannot be transported in long range and their effects will be concentrated on local scale (e.g. effect of the particulate matter). The behaviour of effluents (from NFs in gaseous or particulate form) released to the atmosphere is shown in Fig. 11. However, chemical species such as NO_x, SO_2, which have longer lifetimes (hours–days), can have wider impact zones. Therefore, in such cases, regional or continental modelling approaches are essential. Some pollutants (e.g. CO, NH_3) have long lifetimes (days–weeks) resulting in long-range transport, so for these pollutants, continental and global transport models need to be employed. For simulating the dispersion of air pollutants, ranges of modelling approaches have been developed. The main aim of this chapter is to offer a brief review of air pollution modelling tools and their application.

Vervecken et al. [20] used LES to observe the time evolution of the turbulent dispersion of hot gases within the part physical phenomenon, and it's coupled to a gamma rate model that's supported the point-kernel technique with buildup factors.

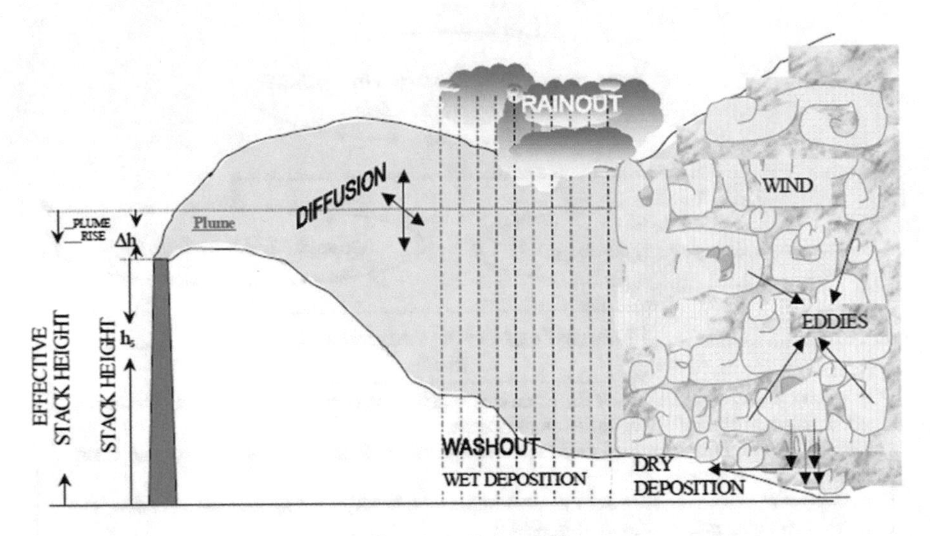

Fig. 11 Study of the liberation of radioactive effluent [22]

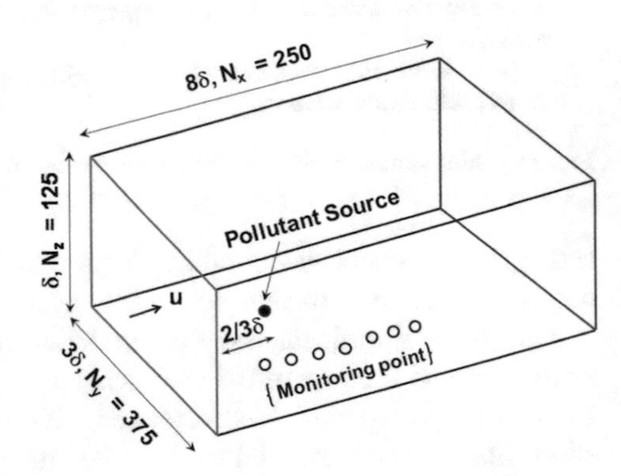

Fig. 12 Sketch of the simulated domain [20]

The pure mathematics of the cases thought of is incredibly simple as shown in Fig. 12, i.e. associate open field with none obstructions from buildings or vegetation. The simulation domain preferred is of size $8\delta \times 3\delta \times \delta$ where δ is the boundary layer thickness. The total number of grid cells was 11.7 million. Two noble gases are particular as the subjects of this learn, i.e. ^{41}Ar and ^{133}Xe. These gases are inert and do not set down. Both are regularly emitted in regular process of a variety of nuclear amenities and mainly ^{133}Xe might be emitted in huge quantities in nuclear accidents. By taking L as the pollutant liberate elevation, U as the average airstream speed at elevation L and R as the pollutant liberation rate, the non-dimensional time, distance and concentration can be distinct as $t^* = tU/L$, $x^* = x/L$ and $c^* = cUL^2/R$, correspondingly. An Eulerian approach is employed to formulate the dispersion of the radionuclide as a transient three-dimensional advection–diffusion downside with disintegration.

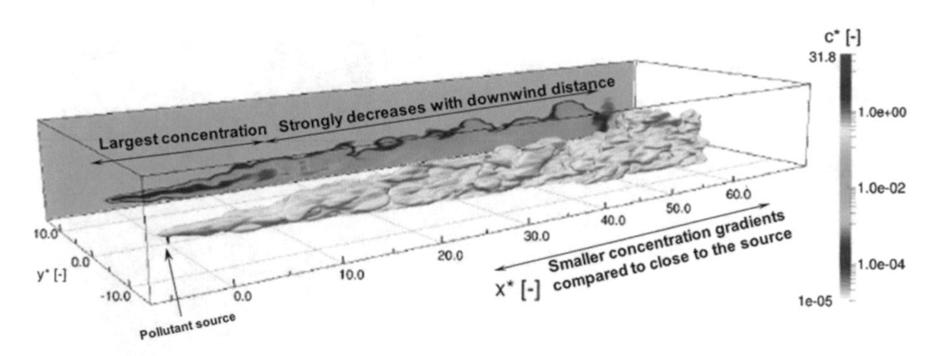

Fig. 13 Instantaneous [41]Ar concentration c^* (= cUL^2/R), released from 0.1δ altitude [20]

Fluence rates were monitored at ground level (2×10^{-03} δ or 1.5 m for $\delta = 750$ m) at multiple distances from the purpose of unleash on streamwise direction (u). The dose assessment downside was simulated mistreatment the OpenFOAM finite-volume ASCII text file simulation platform. The pollutant supply is modelled as some extent supply by simply adding a source term to the corresponding cell within the domain. Periodic boundary conditions area unit set within the horizontal directions for all variables, so simulating a site of an infinite depth and breadth. For the concentration, Neumann boundary conditions area unit applied at the opposite boundaries. An instant, three-dimensional iso-surface of the concentration $c^* = 10^{-5}$ is shown in Fig. 13.

Sampaio et al. [3] numerically simulated the two cases (for controlled emission and harmful event). Figure 14 illustrates domain decomposition for the case of four processors. The quantitative relation "number of shared nodes/number of interior nodes" tends to reduce as a given discretization is developed, which means that we will expect higher parallel presentation because the variety of degrees of freedom is magnified. However just in case of a harmful event, wherever they postulated the failure of all barriers, together with the break of the reactor construction walls. Within the examples analysed in Fig. 15, the radionuclides square measure discharged at a really extreme temperature. Their simulated results show that the plume behaviour was sturdily obsessed with the Richardson variety, which shows the comparative power among natural and forced convection. Among the massive variety of radionuclides generated in an exceedingly NPP, they set to follow during this example the concentration of Iodine 136 (I-136). This part could be a typical fission product with a half-life of eighty three.4 s, decaying through beta emission into chemical element 136 (Xe-136).

Case A was the only model. The investigation was constrained to a site wherever there are not any obstacles to the flow past the chimney. The stuff was discharged at identical temperature of the encircling 25 °C (298 K) that means that buoyancy forces weren't gift just in case A.

Case B considers obstacles within the flow, about to symbolize associate uneven topography and also the presence of little buildings. As just in case A, the fabric

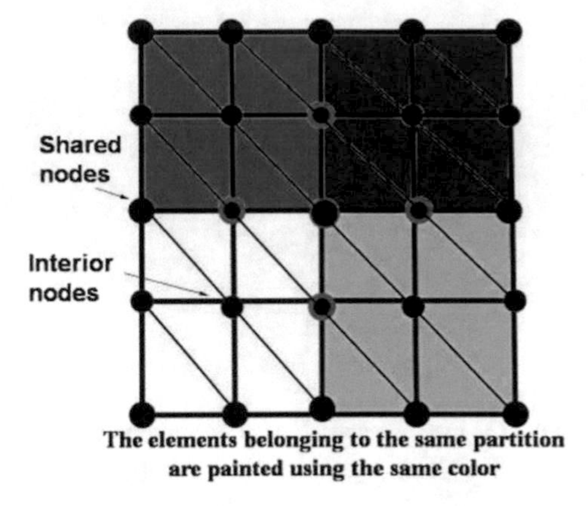

Fig. 14 Domain decomposition using four processors [3]

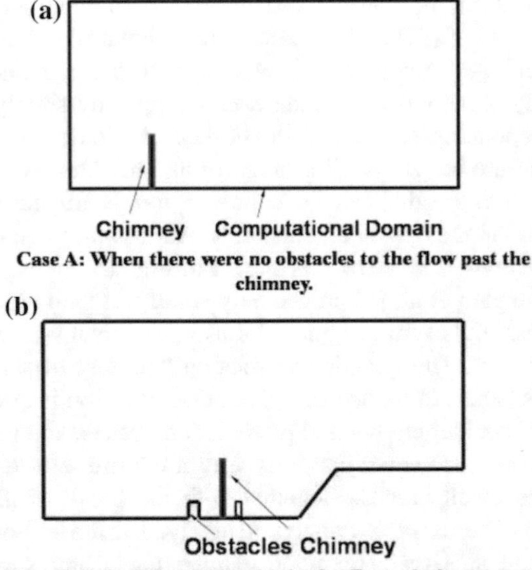

Fig. 15 Domains used for the analyses of controlled release through the chimney [3]. **a** Case A, **b** Case B and C

was discharged at identical temperature of the surroundings. Thus, buoyancy was not active just in case B. This case most closely resembles the discharge of radio nuclides in traditional operation of a NPP, since there have been continually obstacles close to the ability plant and also the temperature distinction between the fabric discharged and also the near atmosphere was negligible.

Case C presents the mixture of obstacles within the flow and also the release of hot stuff through the chimney. Identical irregular topography utilized in case B has been utilized in case C; however, they assume that the temperature of the fabric discharged was 100 K beyond the part temperature. During this example, they need the interference of the flow with the irregular piece of ground shared through extreme buoyancy forces that thrust the discharged material upwards. Case C depicts a rigorous accident wherever the plant operators are forced to unleash some hot stuff to the atmosphere so as to regulate the interior pressure within the reactor building. This was the foremost attention grabbing of the three examples analysed, because it involves unequal topography and buoyancy effects. Thus, they set to gift the results of case C in bigger detail. The ϕ-concentration on the monitored points is bestowed for the three cases simulated so as to create a comparative assessment of the consequences of piece of ground irregularities and buoyancy forces on the part dispersion. We tend to recall that there are not any buoyancy forces in cases A and B.

The concentrations ϕ expected at the three observed points for cases A (when there were no obstacles to the flow past the chimney), B (when obstacles in the flow, aiming to represent an irregular topography and the presence of small buildings) and C (grouping of obstacles in the flow and the discharge of warm radioactive material through the chimney) were compared in Fig. 16. This shows

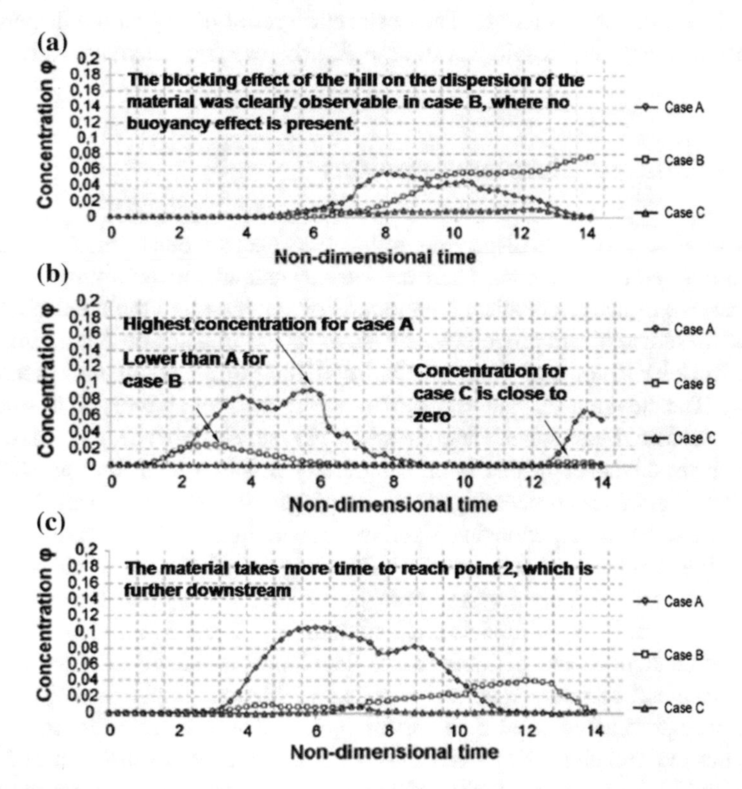

Fig. 16 Concentration ϕ at the monitored points as a purpose of the non-dimensional time ($t^* = tu_0/L$) [3]. **a** Point 1 (189 m to the right of the chimney and 67.5 m high), **b** Point 2 (339 m to the right of the chimney and 67.5 m high), **c** Point 3 (489 m to the right of the chimney and 67.5 m high)

the radionuclide concentrations on selected points for cases A, B and C. Within the subsequent, they plan to justify a number of the differences determined within the lightweight of the flow fields obtained within the CFD simulation.

The concentrations revealed in Fig. 16a match up to purpose 1 (189 m to the right of the chimney and 67.5 m high. In fact, it absolutely was affordable to expect terribly low concentration values for case C at purpose one, because the buoyancy forces drive the fabric upwards. On the opposite hand, there are not any buoyancy possessions in cases A and B. The completely different the various topography in cases A and B ends up in different flow patterns that designate the terribly different concentrations determined in Fig. 16a.

Figure 16b shows the concentrations ϕ comparable to purpose two (67.5 m high and 339 m to the right of the chimney). A similar argument bestowed for purpose one, concerning the concentration levels discovered for cases A, B and C, conjointly pertain for purpose two. As usual, though, the fabric takes longer to succeed in purpose two, that is any downstream. Purpose three is found 489 m from the chimney, however conjointly at the peak of 67.5 m. The concentrations ϕ obtained for purpose three are shown in Fig. 16c. This time is nearer to hill enclosed within the simulation of cases B and C. The obstruction result of hill on the dispersion of the fabric is clearly noticeable just in case B, wherever no buoyancy result is gift.

3 Solutions and Recommendations

Modelling of air pollutants dispersion in the atmosphere is one of the important and demanding scientific problems. There are various natural and anthropogenic events, where passive or active chemical compounds are emitted into the atmosphere. The effect of these chemical compounds can have severe impacts on environment and human. In order to forecast these effects, modelling of air pollutants dispersion is required. The advantages and drawbacks of some model tools and strategies, namely Gaussian, Lagrangian, Eulerian and CFD models, have been discussed. The several recent advances in this multidisciplinary research field, like parallel computing using graphical processing units, or adaptive mesh refinement, have been focussed here. However, choosing a suitable model in a particular case is a key to handling the problem and successfully estimating the dispersion of air pollutants.

Mészáros et al. [21] summarized that the validation is an essential step in progress to test capabilities of models and use models in decision-making. There are a number of challenges and possible future developments for air quality modelling. As we discussed former, numerical weather prediction model (NWP) and CFD models provide reliable wind data for all scales of atmospheric dispersion simulations, but their relationship proved to be difficult. Recent developments of NWPs achieve more thorough resolution. CFD models with enhanced computational capacity, parallel computing and LES simulation for anisotropic turbulence are becoming even improved for planetary boundary layer simulations.

4 Conclusion

The aim of finding out the radionuclide's dispersion from an APP is to determine knowledge for safety analysis. So underneath traditional operational conditions, the radiological dose received by the receiver is safe to the extent recognized by existing regulation. The extension of CFD software for atmospheric studies hold assure, and these are current topics of research in both the meteorological and environmental engineering field. Such a modelling approach can reduce the uncertainty in the dispersion pattern of air pollutants. Finally, one should emphasize that the further growth of dispersion models will shift towards multi-scale modelling approaches that can contain complicated parameterization of cloud physical processes and biogeochemical connections. In order to develop the procedure for validation of the code to be worn for quantitative predictions in atomic reactor protection, still effort is required. The validated and tested workstation code can be useful for the diffusion of radio nuclides in other radioactive installations and laboratories, e.g. nuclear waste deposits, research reactors, nuclear fuel reprocessing plants, cyclotrons, food irradiation industries.

The application of computational fluid dynamics tool for atmospheric dispersion calculations enables a more realistic simulation of ground conditions and velocity profile as compared to the Gaussian model. However, the Gaussian model is extremely traditional and leads to excessive and exaggerated preventive measures. CFD, on the other hand, allows for a wide variety of considerations and modelling capabilities and thus has been extensively used for pollutant dispersion evaluations.

References

1. Lamarsh J (1975) Introduction to nuclear engineering. Addison-Wesley, NY
2. de Sampaio AB, Junior MA, Lapa CM (2008) A CFD approach to the atmospheric dispersion of radionuclides in the vicinity of NPPs. Nucl Eng Des 238:250–273
3. Liu F, Huang SX (2011) Optimization theories and applications for atmospheric environmental risk control. Beijing: China Meteorological Press.
4. Holmes N, Morawska L (2006) A review of dispersion modelling and its application to the dispersion of particles: an overview of different dispersion models available. Atmos Environ 40(30):5902–5928
5. Gurjar B (2008) Air pollution: health and environmental impacts. CRC
6. Zheng D, Leung J, Lee B, Lam H (2007) Data assimilation in the atmospheric dispersion model for nuclear accident assessments. Atmos Environ 41(11):2438–2446
7. Nakayama H, Nagai H (2009) Development of local-scale high-resolution atmospheric dispersion model using large-eddy simulation part 1: turbulent flow and plume dispersion over a flat terrain. J Nucl Sci Technol 46:1170–1177
8. Punitha G, Sudha AJ, Kasinathan N, Rajan M (2008) Atmospheric dispersion of sodium aerosol due to a sodium leak in a fast breeder reactor complex. J Power Energy Sys 2:889–898
9. Vach M, Duong VM (2011) Numerical modeling of flow fields and dispersion of passive pollutants in the vicinity of the temelin nuclear power plant. Environ Model Assess 16:135–143

10. Gallego E, Barbero R, Cuadra D, Domingo J, Iranzo A (2010) Modelling with a CFD code the nearrange dispersion of particles unexpectedly released from a nuclear power plant. In: proceedings 3rd European IRPA Congress (Helsinki), pp 14–18
11. Raza S, Avila R (2001) A 3D lagrangian particle model for direct plume gamma dose rate calculations. Radiol Prot 21:145–154
12. Xie D, Wang H, Kearfott KJ (2012) Modeling and experimental validation of the dispersion of ^{222}Rn released from a uranium mine ventilation shaft. J Atmos Environ 60:453–459
13. Xie D, Wang H, Kearfott KJ, Liu Z, Mo S (2014) Radon dispersion modeling and dose assessment for uranium mine ventilation shaft exhausts under neutral atmospheric stability. J Environ Radioact 129:57–62
14. Duarte JP, Frutuoso e Melo PFF, Alves ASM, dos Passos EM (2013) Atmospheric dispersion and dose evaluation due to the fall of a radioactive package at a LILW facility. Int J Energy Eng, 3(3):119–126 doi:10.5923/j.ijee.20130303.01
15. Fuka V, Brechler J (2012) Large eddy simulation modelling of the dispersion of radioactive particulate matter. Int J Environ Pollut 48:156–163
16. Nakayama H, Jurcakova K, Nagai H (2013) Development of local-scale high-resolution atmospheric dispersion model using large-eddy simulation. Part 3: turbulent flow and plume dispersion in building arrays. J Nucl Sci Technol 50:503–519
17. Shunxiang H, Feng L, Qingcun Z, Fei H, Jiang Z, Zifa W (2015) Modeling and optimal control of atmospheric pollution hazard in nuclear and chemical disasters. Procedia IUTAM 17:79–90
18. Raskob W, Ehrhardt J (2007) Status of the RODOS system for off-site emergency management after nuclear and radiological accidents. In: The first international conference on risk analysis and crisis response
19. Atomic Energy Regulatory Board (AERB) (2008) Atmospheric dispersion and modelling, Guide No. AERB/NF/SG/S-1
20. Vervecken L, Camps J, Meyers J (2015) Dynamic dose assessment by large eddy simulation of the near-range atmospheric dispersion. Published in J Radiol Prot 35:165–178. doi:10.1088/0952-4746/35/1/165
21. Leelossy Á, Molnár F, Izsák F, Havasi Á, Lagzi I, Mészáros R (2014) Dispersion modeling of air pollutants in the atmosphere: a review. Central Eur J Geosci 6(3):257–278. doi:10.2478/s13533-012-0188-6
22. Lewis EE (1977) Nuclear power reactor safety. Wiley, NY

Scope of Pyrolysis Process as a Sustainable Method to Dispose Waste Tires: A Review

Raghuram Kommineni, Hemanth Boddapu and Shijo Thomas

Abstract Vehicles on roads are increasing tremendously and so are the waste and pollution. Every year billions of waste tires are being produced which is either left as such as solid waste or used for land fillings or burnt. Open tire fires are polluting air by emitting carbon monoxide, sulfur dioxide, oxides of nitrogen, organic pollutants, and poly-aromatic hydrocarbons (PAH). These gases can have chronic health affects like skin rashes, irritation of eyes, and respiratory problems. There is a growing concern for development of sustainable waste tire disposable methods. Tire being a polymer can be subjected to pyrolysis process to derive feedstock materials. It was found by many researchers that the products from pyrolysis can be used in different applications. There are three main products of pyrolysis of tires, namely gaseous products, liquid products, and the solid product called as char. These products when collected properly could be used for various other purposes. The gaseous products mainly consist of hydrogen, lighter hydrocarbons (HC), and carbon monoxide. These gases have the potential to generate energy. The liquid products when distilled produce gasoline like fuel and diesel like fuel. Char being porous can act as a gasification catalyst and when upgraded can be used to generate carbon black (CB) which is reutilized in manufacturing of tires. Also, for tire pyrolysis to be a sustainable process for disposing tires, hybrid technologies at industrial scale have to be explored. This paper provides an insight on tire pyrolysis processes and a detailed overview on latest research being done on the products of pyrolysis stating the use of these products to make the pyrolysis process economically viable.

Keywords Tire pyrolysis · Waste tires · Hybrid technologies · Char
Carbon black

R. Kommineni · H. Boddapu · S. Thomas (✉)
School of Nano Science and Technology, NIT Calicut, Calicut 673601, Kerala, India
e-mail: shijo@nitc.ac.in

N. Sharma et al. (eds.), *Air Pollution and Control*, Energy, Environment, and Sustainability, https://doi.org/10.1007/978-981-10-7185-0_14

List of Abbreviations

BSFC	Brake specific fuel consumption
CB	Carbon black
HC	Hydrocarbons
MCM	Mobile composition of matter
PAH	Poly-aromatic hydrocarbons
TG	Thermo gravimetric
TPO	Tire pyrolysis oil

1 Introduction

With the rapid growth in automobile industry, scrap tire generation has been increasing tremendously every year. Around 1.5 billion tires are being produced annually across the globe, most of these eventually end up as waste [1]. Scrap tire disposal has been a major issue as, tires being durable and almost immune to biological degradation, they cannot be disposed without taking proper measurements. Scrap tire has been used in land filling which lead to landslides. Tires have high heat value and can be used energetically but the gasses emitted on combustion of scrap tire are harmful to the environment and is not eco-friendly. As an alternative approach pyrolysis of scrap tire is effective and can be economical as it converts tire into useful products like lighter hydro carbon gases, oils, and char. Studies showed different pyrolysis methods like non-catalytic pyrolysis, catalytic pyrolysis [2], hydrous pyrolysis, plasma pyrolysis [3] with different reactors like fluidized bed, fixed bed, and rotary kilns. Many researches are done to find the potential uses of the products of tire pyrolysis. Studies showed that pyrolysis oil is a potential source of chemical feedstock for industrial products like benzene, xylene, toluene, limonene.

Tire pyrolysis oil (TPO) can be used as a fuel directly or as a blended mixture with gasoline or diesel in boilers and internal combustion engine and many studies are being done in this area. Lot of research work is being done in enhancing TPO properties by using different catalysts. Pyrolysis char mainly comprises of CB filler and inorganic compounds which are found initially in tire and also carbonaceous compounds that are formed after pyrolysis. Studies done by Choi et al. [4] and Hadi et al. [5] investigated the production and utilization of tire char. Some upgraded tire char to activated carbon through activation process, some used tire char as catalyst in gasification and pyrolysis processes, and some studied the ability of tire char in adsorbing the pollutants.

Though the tire pyrolysis products have the capacity to be reutilized, there are some limitations in doing so. Sulfur content presents in oil and char is one of the major limitations. TPO could not be used directly because of the strict emission control. Instead, it is being used in blended forms in lower proportions. The ash content in tire char is responsible for its limited usage as CB. In order to make these products useful, these products are to be additionally processed to remove the undesirable compounds.

The idea behind this review paper is to combine the work done by various authors in this field of research and to identify the usefulness of products obtained from tire pyrolysis. This paper also compares the work done by different researchers to support the facts and finally justifying the pyrolysis of scrap tire.

2 Pyrolysis of Tire

Pyrolysis is a process of heating feedstock materials like biomass, plastics, tires to obtain high quality fuels. It is generally done under inert atmosphere in a specially designed reactor. This process is environment friendly compared with incineration or combustion. The products of pyrolysis were found to have economic value as fuels, catalysts, and raw materials for CB. Thermo-gravimetric (TG) analysis reported by Juma et al. [6] signifies that typically tire pyrolysis occurs in the range of 250–550 °C, while Lopez et al. [3] have stated that 475 °C is the required temperature to completely de-volatize tire rubber. Nitrogen gas is commonly used to maintain an inert atmosphere during the pyrolysis.

2.1 *Types of Pyrolysis*

The pyrolysis can be classified as thermal pyrolysis, catalytic pyrolysis, co-pyrolysis, flash pyrolysis, and microwave pyrolysis. Thermal pyrolysis is the conventional type of pyrolysis where feedstock is subjected to high heating under inert atmosphere without any catalyst to alter the rate of reactions. This can be referred some times as thermolysis or non-catalytic pyrolysis. In catalytic pyrolysis feedstock is valorized in presence of a catalyst, usually zeolites with optimum pore sizes and high surface area, to alter the rate of reactions like hydrocracking, aromatization, monomerization. Co-pyrolysis is the process in which feedstock is composed of two or more types of raw materials and is performed to improve product quality. Extensive research is being done on co-pyrolysis with tire as one of the feedstock raw material. In flash pyrolysis feedstock is subjected to sudden heating, by continuously feeding the tire into high-temperature reactor. Some authors have stated that flash pyrolysis is the best method to produce high economic value products. Microwave pyrolysis is done by microwave heating of the feedstock at about 2.45 GHz. In this, heating occurs by continuous rotation of dipoles and hence few authors have reported that microwave pyrolysis provides easy control over reactions because of uniform heating over the feedstock.

2.2 *Pyrolysis Reactors*

Some of the most common types of reactors used for tire pyrolysis are discussed here.

2.2.1 Fixed Bed Reactor

In fixed bed reactors, the catalyst is either mixed with the reactant material or it is placed above the reactant in the form of a bed. The reactants when passed through the catalyst get converted into products. In this type of reactors, the catalyst bed does not have any movement. There are various configurations in which the catalysts can be placed inside the reactor depending on the conditions to be maintained. This reactor is the simplest of all types of reactors, in construction and is the most commonly used reactor in industries for production of various chemicals.

2.2.2 Fluidized Bed Reactor

Fluidized bed reactors also fall under heterogeneous catalytic reactors. In this reactor the catalyst is fluidized, i.e., a pressurized gas is passed through the catalyst bed which makes the catalyst suspended and cause it to behave like a fluid. This allows extensive mixing in all directions which results in excellent temperature stability, increased mass transfer, and reaction rates. These reactors are capable of handling large amounts of feed and catalyst. These are commonly used in catalytic cracking processes. The main disadvantages of this is high construction and maintenance cost.

2.2.3 Continuous Stirred Tank Reactor

In continuous stirred tank reactor there is a continuous flow of reactor and products. These reactors range from 50 ml to 100 l in capacity. In this reactor an agitator is present which is hydraulic controlled or motor controlled. This agitator continuously mixes the content in the reactor, thus maintaining a homogeneous condition throughout. These are often used in water treatment plants.

2.2.4 Rotary Kiln Reactor

Rotary kiln reactor consists of a cylinder which is slightly inclined with respect to horizontal plane. It is fed from the upper end of the cylinder. The material while being processed reaches the bottom end of the cylinder and undergo mixing and stirring. Usually a flame or an external heating furnace is employed. For pyrolysis reaction external heaters are required.

2.2.5 Conical Spouted Bed Reactor

This reactor helps in fast pyrolysis and is an apt replacement for fluidized bed reactor. The vigorous cyclic particle movement ensures that there are higher rates of heat and mass transfer between phases. The advantage that conical spouted bed reactor have over fluidized bed is that the particle size can vary much and can be of bigger dimensions. This reactor can handle particles with irregular texture and having agglomeration problems.

Each reactor has its own unique features to transfer heat to the feedstock and few additional advantages. Some are costly due to their complicated designs. The products of pyrolysis reportedly depend on the reactor design.

3 Products of Pyrolysis

When tire material is valorized at temperatures close to 450 °C, many gases are released and a solid residue is left over in the reactor which is called char. Among the hot mixture of gasses released, a fraction of them is condensable into liquid at room temperatures. Hence, we get a liquid fraction from tire pyrolysis. The non-condensable gasses are treated as gas fraction from tire pyrolysis.

3.1 Liquid Fraction of Tire Pyrolysis

TPO contains a wide variety of compounds ranging from molecules with 5 carbon atoms to more than 20 carbon atoms. It can be identified that, aromatics are one of the major fraction in oil. Since oil is a highly complex mixture of olefins, mono-aromatics, poly-aromatic compounds, and tar compounds, it cannot be directly used as a fuel. Many researches have reported the various factors governing the pyrolysis reactions to optimize the yields specific to the purpose.

The percentage of oil obtained from tire pyrolysis can be varied significantly depending on the reactor used, catalyst, heating rate, etc. Arabiourrutia et al. [7] have reported oil yield to be about 60% when conical spouted bed reactor was used. Shah et al. [8] reported oil yields to be lesser than 30% by weight when a batch reactor is employed. Considering the quality of oil produced, catalytic pyrolysis can reduce the tar content and PAH. But the liquid yield reduces because of increased cracking to give more gas yields as reported by Williams et al. [9] with ZSM-5 catalyst at 400–500 °C. Arabiourrutia et al. [7] reported that the usage of catalysts like HY and HZSM-5 in a conical spouted bed reactor resulted in liquid yield above

40% with reasonably good quality. Many authors have reported that liquid yields decrease with increase in temperature and get converted into gas yields. To obtain light olefins and decrease PAH content Witpathomwong et al. [10] used Ru/ MCM-48 as a catalyst. Trongyong and Jitkarnka [11] could increase naphtha and kerosene fractions by improving selectivity toward mono-aromatic compounds using Al-MCM-48 as a catalyst.

Martínez et al. [12] stated that low heating rate leads to formation of secondary products due to high residence time of vapors. Experiments performed with high heating rate showed that lighter oils could be produced but at the expense of percentage yield. However, compared with conventional pyrolysis Lopez et al. [3] reported that flash pyrolysis gave a 58% liquid yield at 475 °C. Ahoor and Zandi-Atashbar [2] have reported that $Mgcl_2$ as catalyst can be used to produce oil comparable with Euro-III diesel standards. Overwhelmingly, researchers have pointed that TPO can be an alternative to diesel fuel. Also, to improve the PAH and olefin content in bio fuels, Wang et al. [13] have experimented co-pyrolysis of tire and bamboo residue using a dual catalytic stage of CaO and HZSM-5.

3.1.1 TPO as Fuel in Diesel Engines

Investigations on direct injection diesel engine with blends of TPO and diesel were performed by several researchers. Frigo et al. [14] have reported that blends of TPO with diesel in 20:80 ratio by volume resulted in no significant variations with respect to pure diesel engine performance. Also, it was reported that 40:60 blend of tire oil and diesel showed a worse engine combustion quality. However, no significant carbon deposits were observed. To investigate performance of TPO in turbo-charged engine, [15] used the pyrolysis liquid fraction in 190–350 °C range. It was reported that pure tire oil can be used in turbo-charged engine without intercooling at high loads, but sulfur dioxide emissions were extremely higher than vehicle specifications. With intercooler, turbo-charged engine reportedly showed poor performance, hence its usage is limited to power generation turbo-charged systems. Murugan et al. [16] performed post pyrolysis refining of TPO to improve the quality of oil produced. Desulfurization and vacuum distillation were performed and about 80% of oil was diesel like fuel. The engine was able to run using a diesel blend of up to 90% TPO, and usage of 100% TPO in the engine was a failure. TPO is also being used with biodiesel and diesel blends to reduce emissions by Koc and Abdullah [17]. The brake specific fuel consumption (BSFC) was reported to be significantly lower when engine was run with 10% tire oil—10% biodiesel—80% diesel blend.

Frigo et al. [14] and Murugan et al. [16] have reported that blends of tire oil have longer ignition delays compared to pure diesel engine performance. This is because of low cetane number of TPO. Hence researchers have experimented engine performance with cetane improvers like di-ethyl ether. Hariharan et al. [18] administered di-ethyl ether into the intake manifold when 100% TPO was used as a fuel, which showed similar operational characteristics as diesel. Sharma and Murugan

[19] and Tudu et al. [20] have experimented with Jatropha methyl ester and di-ethyl ether, respectively, as cetane improvers added to tire oil blends. Both have reported positive results with better efficiencies as the cetane number of TPO with cetane improvers and that of diesel are closer and hence engine can be run more efficiently with same injection delay.

3.2 Gas Yields from Tire Pyrolysis

Analysis of gas fractions released during tire pyrolysis reveals that, it mainly constitutes of H_2, CO, and light HCs (like methane, ethane, ethane, propane, and butane). Also trace amounts of H_2S were reported by few authors. The percentage of gas yields can vary largely depending on the process parameters as shown in Table 1. Arabiourrutia et al. [7] reported gas yields to be lower than 16%, while Boxiong et al. [21] have stated that with HY zeolites on a dual stage (pyrolyser + gasifier) fixed bed reactor, gas yields have significantly increased with respect to non-catalytic pyrolysis. Few authors have reported that mobile composition matter (MCM) catalysts have increased gas yields significantly. Dũng et al. [22] reported that Ru/MCM-41 increased gas yields in comparison with MCM-41. Witpathomwong et al. [10] experimented with Ru/MCM-48 and reported similar results.

Song et al. [23] worked on microwave pyrolysis and stated that H_2 production was 54% higher than conventional pyrolysis. It was also mentioned that 90% of the gas yields by volume constituted H_2, CO, CH_4. While Laresgoiti et al. [24] reported that HCs were a major fraction of gas yields. Leung et al. [25] reported that CH_4 was 44.5% and H_2 was 20.7% at 800 °C. Elbaba and Williams [26] reported that an increase in temperature from 600 to 900 °C in a two stage pyrolysis—catalytic steam gasification using Ni/Al_2O_3 catalyst, H_2 and CO production increased with decrease in hydrogen percentage. It was also reported that optimum steam input can increase H_2 production. Mozafari et al. [27] has performed modeling of tire thermolysis and air gasification and reported that cold gas efficiency of air gasification was about 73% while that of thermolysis was only 33%. It was stated that two stage thermolysis and air gasification can give better results. Kaewluan and Pipatmanomai [28] stated that possibility of power production from tire pyrolysis gas in a gas turbine. Few authors have reported that pyro-gas can be used to provide a fraction of heat required during the pyrolysis (Fig. 1).

3.3 Tire Pyrolysis Char

The solid residue-char is either collected in the reactor after the pyrolysis or collected from the container placed beneath the reactor. Carbon is the main constituent of char which is around 80% as stated by Hadi et al. [5]. They performed elemental

Table 1 Percentage yields of tire pyrolysis

Author	Reactor type	Temperature	Catalyst	Char (%)	Oil (%)	Gas (%)	Remarks
[35]	Fixed bed reactor	430–600 °C	HZSM-5	42	43	15	Less liquid yield, more char
			HY	45	39	16	Less liquid yield, more char
[7]	Conical spouted bed reactor		HZSM-5	34.1	58.3	7.6	More liquid yield
			HY	34.5	62.2	3.3	More liquid yield
[21]			HY	37	32	31	More gaseous products
[22]			MCM-41	47	38	15	Less liquid yield, more char
[8]			Al_2O_3	55.7	23.2	21.1	More char yield
			SiO_2	56.3	28.2	15.5	More char and gaseous products
[11]			Al-MCM-48	42	43	9	More char
[3]	Conical spouted bed reactor	425 °C	Non-catalytic	38	58	4	Flash pyrolysis gave good liquid yield
		475 °C		36	57	7	
		575 °C		35	55	10	
[36]	Double fixed bed reactor	500 °C	Non-catalytic	39	36	25	OS-oyster shells heating rate: 1–5 °C powdered catalysts catalyst: tire = 1:30, less liquid yield
			$CaCO_3$/OS	45	25	30	
			ZSM-5/OS	44	34	22	
			Al_2O_3/OS	38	34	28	
			MgO/OS	38	33	29	
[14]	Pilot scale pyrolyser	300–500 °C	Non-catalytic	50	45	5	More char

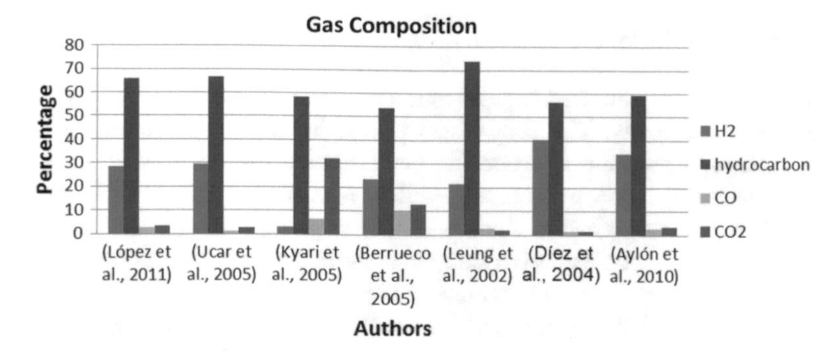

Fig. 1 Gas composition obtained by various authors [25, 37–42]

analysis on char and has depicted the results which showed higher % of carbon and traces of compounds like sulfur, metals like zinc and titanium. The presence of metals was also observed by Seng-eiad and Jitkarnka [29] and Al-Rahbi and Williams [30]. A lot of research work is being done in this area to use char in the most economical and viable way. The properties of char like high carbon content, porosity, and large surface area make them potential for various uses.

3.3.1 As Carbon Black

CB is an amorphous carbon material of quasi-graphitic structure. It finds its main market in rubber and tire industry as filler and pigment in rubber products. Hence, this CB can be separated and can be used in tire manufacturing by enhancing its properties. Due to high sulfur and ash content, utilization of tire pyrolysis char as CB is limited. Char finds usage as a low grade carbon black. CB produced from char is having high ash content compared to commercially available CB. Martínez et al. [31] worked on this and stated that by doing acidification/basification treatment on char ash content can be reduced considerably and increases carbon content which are in comparable to that of commercially available carbon blacks. Martínez et al. [31] also found that the size of CB particles were bigger because of ash and carbonaceous deposits on the particles. A milling process can also be used to reduce the size of the particles.

Choi et al. [4] did work on optimizing the conditions in preparing CB from char. The two parameters that he took into consideration are the pyrolysis temperature and duration of pyrolysis. It is stated that less sulfur content is present in char when pyrolysis took place at higher temperature and thus gave a better CB. Also, though the temperature of pyrolysis has no effect on surface area, it showed an increase in pore diameter.

3.3.2 Activated Carbon

Activated carbon is often used in purification processes attributed to its high surface area and micro porosity. Tire char which is rich in carbon source and also being porous have high potential in producing porous adsorbent material by using some activating agents to enhance its porosity and surface areas. Hadi et al. [5] and Lopez et al. [32] used steam whereas Choi et al. [4] used CO_2, as the activating agent. Hadi et al. [5] performed extensive study and characterized activated carbons using N_2 adsorption–desorption experiments at −196 °C. The elemental analysis showed presence of around 80% carbon and also sulfur in char. Lopez et al. [32] in his paper mentioned that the sulfur content decreased after activation process. Although, Fouladi Tajar et al. [33] in his paper reported that sulfur functional groups increases absorbability of the material. This, in addition, favors the adsorbing ability of the material. A maximum surface area of 733 m^2/g was obtained at an activation temperature of 975 °C and for Lopez et al. [32] surface area obtained was over 500 m^2/g. Both Hadi et al. [5] and Lopez et al. [32] plotted the variation of surface area with respect to the parameters they have considered. Most of the researches used steam as activating agent, but Choi et al. [4] used CO_2 to utilize greenhouse gas. Choi et al. [4] in their paper compared the properties of acid treated and activated char and obtained a maximum surface area of 437 m^2/g at an activation temperature of 950 °C. All the authors stated that activation process increased surface area but using steam the surface area obtained was on higher side.

3.3.3 As Catalyst

Pertaining to the high porosity and surface area, researchers have been working on tire char's catalytic properties in gasification and pyrolysis processes. Seng-eiad and Jitkarnka [29] used tire char as catalyst for tire pyrolysis and Al-Rahbi and Williams [30] studied the effect of tire ash minerals on syngas production from biomass gasification. Seng-eiad and Jitkarnka [29] tested tire char's catalytic properties by treating it with HNO_3. They observed increased production of gasoline and diesel like fuels, although gas yields are on higher side with treated char as the acidity, pore size, and surface area got enhanced, resulting in greater cracking. Author also observed that char effectively removed sulfur from the oil.

Al-Rahbi and Williams [30] worked on the effect of various parameters on the production of syngas including tire ash minerals. He also studied the effect of both treated and untreated chars and the results are similar to that of Seng-eiad and Jitkarnka [29]. The results showed that the catalytic action of treated char has been reduced in tar cracking in biomass gasification and they stated that removal of metals after the treatment of char with acid is the reason. The metals became the active spots of the catalyst, but since the metals are removed, treated char showed less catalytic effect in tar cracking.

Results from both the papers showed that tire char improved the results and can act as a good catalyst. Treated char though did not show any better results compared to untreated char can still be used based on the product expected.

4 Economics of Pyrolysis

Tough harnessing fuels from scrap tires is a novel idea to harness energy from waste; pyrolysis is an energy intensive process. The process is economical only when energy harnessed is sufficiently greater than energy consumed for the total process. Conventional automotive fuels are extracted and refined with minimal amounts of energy inputs. There is no economic edge for TPOs over conventional petroleum-based fuels until and unless low cost methods to provide heating are implemented into pyrolysis process. Major running costs of tire pyrolysis involve costs associated with tires, catalysts, inert gas, and electricity. Electricity takes up major share in variable costs. If catalysts used are high grade quality, catalyst associated costs are to be considered significant. Since process parameters also play a vital role in yield quantities, there is a probability for optimizing pyrolysis without catalysts. Electricity costs have to be reduced drastically to operate at industrial scales. Few authors have mentioned that pyro-gas can be used to provide a fraction of required heat. Also, if heat can be supplied by high temperature exhausts of industrial processes, overall process can be tuned to be very economical. Luo and Feng [34] investigated the oil and gas production by waste heat of blast furnace slag. It is a very economical process, as no electrical energy for heating is used; moreover, blast furnace slag constitutes calcium-magnesium complexes which act as catalysts required for pyrolysis process.

Among the fixed costs associated with the process, type of reactor and scale of operation are significant decisions to be made. Depending on the scale of operation and type of product required, reactor design should be chosen. Conical spouted bed reactor and fluidized bed reactors can support continuous feeding which is crucial at industrial scale plants. At laboratory scale, fixed bed and fluidized bed reactors are economical as conical spouted bed reactor, rotary kiln, screw type reactors need high operational and maintenance skills. Revenue form tire pyrolysis products have to be improved by identifying market requirements of activated carbon products, alternative automotive fuels, and syngas applications. More research is required to understand the process in more detail, to innovate on refining of pyrolysis products, make fuel harnessing from tire pyrolysis economical and set pathway to dispose tires in an environment friendly manner.

5 Conclusion

With growing environmental concern for tire disposal methods, pyrolysis provides an eco-friendly method to utilize waste tires with valuable products. Pyrolysis of tire involves complex reactions occurring simultaneously, which is why process parameters can alter the yields significantly. Depending on the requirement, suitable operation of reactors has to be made to extract the optimum yield. Catalytic pyrolysis has a huge potential to improve the product quality and quantity. Overwhelming research supports the possibility of diesel blended with TPO as a good fuel for CI engines. Dual stage pyrolysis and steam gasification can be used to produce high calorific value gases also to produce hydrogen gas. Char from tire pyrolysis can be used as a gasification catalyst, and can be sold as an alternative to commercially available CB after refining or can be reused in manufacturing tires. More research has to be done to create an economical pathway for industrial scale tire pyrolysis. Based on market requirements and supply chain models of all tire pyrolysis products, one has to take strategic decisions for process to be sustainable and a successful method to dispose waste tires.

References

1. Pilusa J, Shukla M, Muzenda E (2014) Economic assessment of waste tyres pyrolysis technology: a case study for Gauteng Province, South Africa. Int J Res Chem Metall Civ Eng 1(1): 1442–1450
2. Ahoor AH, Zandi-Atashbar N (2014) Fuel production based on catalytic pyrolysis of waste tires as an optimized model. Energy Convers Manag 653–669
3. Lopez G et al (2017) Waste truck-tyre processing by flash pyrolysis in a conical spouted bed reactor. Energy Convers Manag 523–532
4. Choi G-G, Jung S-H, Oh S-J, Kim J-S (2014) Total utilization of waste tire rubber through pyrolysis to obtain oils. Fuel Process Technol 57–64
5. Hadi P et al (2016) Sustainable development of tyre char-based activated carbons with. J Environ Manag 1–7
6. Juma M et al (2006) Pyrolysis and combustion of scrap tire. Pet Coal 48(1):15–26
7. Arabiourrutia M et al (2008) HZSM-5 and HY zeolite catalyst performance in the pyrolysis of tires in a conical spouted bed reactor. Ind Eng Chem Res 7600–7609
8. Shah J, Jan MR, Mabood F (2009) Recovery of value-added products from the catalytic pyrolysis of waste tyre. Energy Convers Manag 991–994
9. Williams PT, Besler S, Taylor DT (1990) The pyrolysis of scrap automotive tyres: the influence of temperature and heating rate on product composition. Fuel 1474–1482
10. Witpathomwong C, Longloilert R, Wongkasemjit S, Jitkarnka S (2011) Improving light olefins and light oil production using Ru/MCM-48 in catalytic pyrolysis of waste tire. Energy Procedia 245–251
11. Trongyong S, Jitkarnka S (2016) Enhanced sulphur removal from tyre-derived oil using aluminosilicate MCM-48 with pyrolysis of waste tyres. Renew Sustain Energy Rev 684
12. Martínez JD et al (2014) Performance and emissions of an automotive diesel engine using a tire pyrolysis liquid blend. Fuel 490–499
13. Wang J et al (2017) Co-pyrolysis of bamboo residual with waste tire over dual catalytic stage of CaO and co-modified HZSM-5. Energy 90–98

14. Frigo S, Seggiani M, Puccini M, Vitolo S (2014) Liquid fuel production from waste tyre pyrolysis and its utilisation in a Diesel engine. Fuel 399–408
15. Vihar R, Seljak T, Opresnik SR, Katrasnik T (2015) Combustion characteristics of tire pyrolysis oil in turbo charged compression ignition engine. Fuel 226–235
16. Murugan S, Ramaswamy MC, Nagarajan G (2008) A comparative study on the performance, emission and combustion studies of a DI diesel engine using distilled tyre pyrolysis oil–diesel blends. Fuel 2111–2121
17. Koc AB, Abdullah M (2014) Performance of a 4-cylinder diesel engine running on tire oil–biodiesel–diesel blend. Fuel Process Technol 264–269
18. Hariharan S, Murugan S, Nagarajan G (2013) Effect of diethyl ether on tyre pyrolysis oil fueled diesel engine. Fuel 109–115
19. Sharma A, Murugan S (2013) Investigation on the behaviour of a DI diesel engine fueled with jatropha methyl ester (JME) and tyre pyrolysis oil (TPO) blends. Fuel 699–708
20. Tudu K, Murugan S, Patel SK (2016) Effect of diethyl ether in a DI diesel engine run on a tyre derived fuel-diesel blend. J Energy Inst 525–35
21. Boxiong S et al (2007) Pyrolysis of waste tyres with zeolite USY and ZSM-5 catalysts. Appl Catal B Environ 150–157
22. Dũng NA, Klaewkla R, Wongkasemjit S, Jitkarnka S (2009) Light olefins and light oil production from catalytic pyrolysis of waste tire. J Anal Appl Pyrol 281–286
23. Song Z et al (2017) Gaseous products evolution during microwave pyrolysis of tire powders. Int J Hydrogen Energy (in press)
24. Laresgoiti MF et al (2000) Chromatographic analysis of the gases obtained in tyre pyrolysis. J Anal Appl Pyrol 43–54
25. Leung DYC et al (2002) Pyrolysis of tire powder: influence of operation variables on the composition and yields of gaseous product. Fuel Process Technol 141–155
26. Elbaba IF, Williams PT (2012) Two stage pyrolysis-catalytic gasification of waste tyres: influence of process parameters. Appl Catal B Environ 136–143
27. Mozafari A, Tabrizi F, Farsi M, Mousavi SAHS (2017) Thermodynamic modeling and optimization of thermolysis and air gasification of waste tire. J Anal Appl Pyrol (in press)
28. Kaewluan S, Pipatmanomai S (2011) Gasification of high moisture rubber woodchip with rubber waste in a bubbling fluidized bed. Fuel Process Technol 92(3):671–677
29. Seng-eiad S, Jitkarnka S (2016) Untreated and HNO_3-treated pyrolysis char as catalysts for pyrolysisof waste tire: in-depth analysis of tire-derived products and charcharacterization. J Anal Appl Pyrol
30. Al-Rahbi AS, Williams PT (2017) Hydrogen-rich syngas production and tar removal from biomass. Appl Energy 501–509
31. Martínez JD, Murillo R, Garcia T (2013) Production of carbon black from the waste tires. Research Gate
32. Lopez G et al (2009) Steam activation of pyrolytic tyre char at different temperatures. J Anal Appl Pyrol 539–543
33. Fouladi Tajar A, Kaghazchi T, Soleimani M (2009) Adsorption of cadmium from aqueous solutions on sulfurized activated carbon prepared from nut shells. Hazard Matter 165:1159–1164
34. Luo S, Feng Y (2017) The production of fuel oil and combustible gas by catalytic pyrolysis of waste tire using waste heat of blast-furnace slag. Energy Convers Manag 136:27–35
35. Williams PT, Brindle AJ (2002) Catalytic pyrolysis of tyres: influence of catalyst temperature. Fuel 2425–2434
36. Kordoghli S et al (2017) Impact of different catalysis supported by oyster shells on the pyrolysis of tyre wastes in a single and a double fixed bed reactor. Waste Manag (in press)
37. Aylón E et al (2010) Valorisation of waste tyre by pyrolysis in a moving bed reactor. Waste Manag 30(7):1220–1224
38. Berrueco C et al (2005) Pyrolysis of waste tyres in an atmospheric static-bed batch reactor: analysis of the gases obtained. J Anal Appl Pyrol 74(1–2):245–253

39. Díez C, Martínez O, Calvo LF, Morán A (2004) Pyrolysis of tyres. Influence of the final temperature of the process on emissions and the calorific value of the products recovered. Waste Manag 24(5): 463–69
40. Kyari M, Cunliffe A, Williams PT (2005) Characterization of oils, gases, and char in relation to the pyrolysis of different brands of scrap automotive tires. Energy Fuels 19:1165–1173
41. López FA, Centeno TA, Alguacil FJ, Lobato B (2011) Distillation of granulated scrap tires in a pilot plant. J Hazard Mater 190(1–3):285–292
42. Ucar S, Karagoz S, Ozkan AR, Yanik J (2005) Evaluation of two different scrap tires as hydrocarbon source by pyrolysis. Fuel 84(14–15):1884–1892

Printed in the United States
By Bookmasters